Completeness of root functions
of regular differential operators

Pitman Monographs and
Surveys in Pure and Applied Mathematics 71

Completeness of root functions of regular differential operators

Sasun Yakubov

*Afula Research Institute,
University of Haifa*

CRC Press
Taylor & Francis Group
Boca Raton London New York

CRC Press is an imprint of the
Taylor & Francis Group, an **informa** business
A CHAPMAN & HALL BOOK

**Copublished in the United States with
John Wiley & Sons, Inc.**

First published 1994 by Longman Group UK Limited

Published 2019 by CRC Press
Taylor & Francis Group
6000 Broken Sound Parkway NW, Suite 300
Boca Raton, FL 33487-2742

© 1994 by Taylor & Francis Group, LLC
CRC Press is an imprint of Taylor & Francis Group, an Informa business

No claim to original U.S. Government works

ISBN 13: 978-0-582-23692-9 (hbk)

Visit the Taylor & Francis Web site at
http://www.taylorandfrancis.com

and the CRC Press Web site at
http://www.crcpress.com

Copublished in the United States with John Wiley & Sons Inc.

AMS Subject Classifications: 34, 35, 46, 47

ISSN 0269-3666

British Library Cataloguing in Publication Data

A catalogue record for this book is
available from the British Library

Library of Congress Cataloging-in-Publication Data

Yakubov, S. (Sasun)
  Completeness of root functions of regular differential operators /
S. Yakubov.
    p. cm. -- (Pitman monographs and surveys in pure and applied
mathematics, ISSN 0269-3666 ;   )
  Includes bibliographical references and index.
  1. Differential equations--Numerical solutions. 2. Differential
equations, Partial--Numerical solutions. 3. Polynomial operator
pencils. I. Title. II. Series.
QA372.Y35 1993
515'.35--dc20   93-35715
                                                       CIP

CONTENTS

PREFACE

This book gives the first systematic treatment of the questions connected with completeness of elementary solutions of mathematical physics problems.

In particular, the completeness problem for root vectors of unbounded polynomial operator pencils, and the coercivity and completeness of root functions of boundary value problems for differential equations, both ordinary and partial, are investigated.

Conditions sufficient to guarantee multiple completeness, in the sense of M. V. Keldysh for root vectors of a system of unbounded operator pencils, are also established.

The obtained abstract results enable us to study principally boundary value problems for both ordinary and partial differential equations with a polynomial spectral parameter. Corresponding conditions guaranteeing coercivity of the considered problems and completeness of elementary solutions are also derived. These results, in particular, improve and generalize S. Agmon's results on completeness of root functions of elliptic boundary value problems.

We also consider elliptic boundary value problems in tube domains, i.e., in nonsmooth domains. Our results are new relative to those of P. Grisvard's book, which is devoted to this question.

This book is intended for scientists and graduate students in Functional Analysis, Differential Equations, Equations of Mathematical Physics and related topics. It would undoubtedly be very useful for mechanics and theoretical physicists.

Recommendations and some words of advice by Professor Shmuel Agmon helped me to overcome the difficulties in connection with the publication of this book. I am greatly indebted to Professor Shmuel Agmon for his critical remarks and suggestions for improvement of this book. I wish to thank Professors Selim Krein, Israel Gohberg, Jonathan Arazy, Mikhail Agranovich and Alexander Markus for encouragement, helpful comments and references for papers concerning the manuscript. I am very grateful to my son Dr. Yakov Yakubov for pointing out some mistakes in the original proofs of some of the theorems and rendering invaluable technical assistance.

This book was written in the Azerbaijan Institute of Mathematics and Mechanics (Baku, Azerbaijan) and in the Afula Research Institute of Haifa University (Afula, Israel). I would like to thank my colleagues at these Institutes, especially my doctoral students, for many years of discussion of some parts of the book.

Supported in part by the Ministry of Absorption (Israel), Ministry of Science and Technology (Israel) and the Rashi Foundation (Israel-France).

INTRODUCTION

The ideas and methods of functional analysis, penetrating into different subjects of mathematics, give rise to new fields of mathematics. In just such a way the theory of systems of unbounded polynomial operator pencils and the theory of differential-operator equations arose as a result of penetration of functional analysis into the theory of differential equations. The interpolation theory of spaces and operators also derived from the theory of differential equations. A short review of this interpolation theory and some closely connected problems of analysis that have been well-studied in monographs, is given in chapter 1.

For a long time two theories, regarded as basic results in the theory of liner operators with a discrete spectrum, were the Jordan theory, which states that a system of root vectors of linear operators acting in finite dimensional spaces is a basis, and the Hilbert's result, which states that a system of eigenvectors of a compact selfadjoint operator acting in a Hilbert space is an orthogonal basis. Only in 1951 did M. V. Keldysh [30] prove the completeness of root vectors for one class of non-selfadjoint operators. At the same time M. V. Keldysh introduced the important notion of n-fold completeness of root vectors and proved the fundamental theorem on n-fold completeness for a polynomial operator pencil, the principal part of which is generated by one selfadjoint operator. This work gave rise to a series of works of other authors. However, these abstract results cannot be considered to be effectively applicable to both ordinary differential equations and partial differential equations.

When initial boundary value problems for non-stationary partial differential equations are studied with the help of the Laplace transform, the corresponding boundary value problem for stationary partial differential equations with a polynomial spectral parameter in both an equation and boundary value conditions should be studied first. That is why, starting from the first works of G. D. Birkhoff [7] and J. D. Tamarkin [69], the theory of boundary value problems for ordinary differential equations with a polynomial spectral parameter has been intensively developed in the last 100 years.

We would like to emphasize that in the works [7], [69] and some others the statement of the expansion problem in root functions was not complete. Only one function was expanded with the help of a "very non-minimal" system of root functions. Theory of boundary value problems for elliptic partial differential equations with a polynomial spectral parameter remained almost constant for 1960 years. Coercivity in both a spectral parameter and space variables in Sobolev spaces was proved for the regular elliptic boundary value problems with a parameter, by S. Agmon and L. Nirenberg [2], and M. S. Agranovich and M. I. Vishik [3]. Traditional problems such as multiple completeness of root functions, expansion in the root functions, asymptotics of eigenvalues, etc., have been studied intensively, not only for regular ordinary differential equations in the sense of Birkhoff-Tamarkin, but also for some irregular boundary value problems for ordinary differential equations. None of them have been studied for elliptic boundary value problems with a parameter or studied within the framework of the Keldysh theory [29], [30]. Such a disproportion is due to the fact that when studying boundary value problems for ordinary differential equations one can use a rich theory of analytic functions, and due to the simplicity of the problem resolvent structure. A well developed analytical scheme for the one-dimensional case, namely the resolvent estimation on expanding contours, does not work for boundary value problems for partial differential equations because of the complicated structure of the resolvent. M. V. Keldysh's abstract results [29], [30] and his numerous modifications [5], [16], [56] cover mainly those problems, equations of which have, in their principal parts, the form $A(x, D) + \lambda^n$, where $A(x, D)$ is an elliptic operator and an operator generated by $A(x, D)$ and boundary value conditions, is selfadjoint. In the second chapter of the book this gap is eliminated, and the theory of multiple completeness of root vectors of a system of unbounded polynomial operator pencils is established. The theory establishes the multiple completeness of root functions not only of regular elliptic boundary value problems with a polynomial spectral parameter but also of regular boundary value problems for ordinary differential equations. So, a universal approach was created to establish the multiple completeness of root functions of boundary value problems for both ordinary and partial differential equations. This abstract theory can be successfully applied to establish the multiple completeness of root functions of irregular boundary value problems [93] and of boundary value problems for equations without a weight [80], [92].

In chapter 2 the results about completeness of linear operator root vectors are given. Unlike the books by I. C. Gohberg and M. G. Krein [19], and N. M. Dunford and Ja. T. Schwartz [13], the auxiliary material has been minimized. The results on completeness of root vectors of unbounded polynomial operator pencils system are also stated in chapter 2. These results, if even one operator pencil is available, strengthen the analogous results given in the books by I. C. Gohberg and M. G. Krein [19], and A. S. Markus [46] and in the paper by G. V. Radzievskii [56]. We do not assume that the operator pencil

$$L(\lambda) = \lambda^n I + \lambda^{n-1} A_1 + \cdots + A_n$$

has a weight, that is the estimate

$$\|A_k u\| \leq C\|A^k u\|, \qquad u \in D(A^k),$$

where $A \geq \gamma^2 I$ in H, is not assumed.

M. V. Keldysh [30] and F. E. Browder [9] proved completeness of root functions of an elliptic boundary value problem if the principal part is selfadjoint. S. Agmon [1] proved completeness of root functions of an elliptic boundary value problem if the principal part is non-selfadjoint (see also G. Geymonat and P. Grisvard [17]).

In this book completeness of root functions is proved if the principal part of problems is an elliptic boundary value problem with a polynomial spectral parameter. In this case some difficulties appear, one of which is the following: given, in the principal part, differential operators

$$L_p u = \sum_{|\alpha|=m_p} b_{p\alpha}(x')D^\alpha u(x') + T_p u|_{x'}, \qquad x' \in \Gamma, \quad p = 1,\ldots,m ,$$

where Γ is the $(r-1)$-dimensional smooth boundary of a bounded domain $G \subset R^r$ and the operators T_p from $W_q^{m_p}(G)$ into $L_q(\Gamma)$ are bounded. It is necessary to prove that the set

$$H_0 = \{u | u \in W_q^{2m}(G), \quad L_p u = 0, \quad x' \in \Gamma, \quad p = 1,\ldots,m\}, \qquad m_p \leq 2m - 1,$$

is dense in $L_q(G)$. If the T_p are differential operators, i.e.,

$$T_p u = \sum_{|\alpha| \leq m_p - 1} b_{p\alpha}(x')D^\alpha u(x')$$

then it is known, since $C_0^\infty(G) \subset H_0 \subset L_q(G)$ and $\overline{C_0^\infty(G)} = L_q(G)$. The general case is not so simple.

In chapter 3 a principally boundary value problem in a segment for an ordinary differential equation with a polynomial spectral parameter, in both an equation and boundary conditions, is studied. In addition to the main question – establishing descreteness of the spectrum and multiple completeness of root functions of the problem – the theorems on an isomorphism are proved and regions of coercivity in a spectral parameter are found. It is shown that regularity of the problem provides both coercivity in a spectral parameter (maximal decrease of the "resolvent" in a spectral parameter at infinity) and multiple completeness of root functions of the problem.

It should be noted that the class of regular problems in our sense is wider than those in the books by J. D. Tamarkin [69], M. L. Rasulov [57] and others. It should also be noted that our technique is distinguished from the traditional technique of the books [13, 53, 57, 69], which are based on the Green function construction. It is a small modification of the well-known technique for regular elliptic boundary value problems, stated, for instance, in the articles by S. Agmon and L. Nirenberg [2], and M. S. Agranovich and M. I. Vishik [3].

In chapter 4 multiple completeness of root functions of regular elliptic boundary value problems with a parameter in a bounded domain is studied. As is known, the problem of multiple completeness of root functions system arises when the problems of mathematical physics are being solved by the Fourier method. This was first encountered by M. V. Keldysh [30].

In spite of the fact that the theory of solvability of elliptic boundary value problems with a polynomial spectral parameter both in an equation and in boundary value conditions was completed in the works of S. Agmon and L. Nirenberg [2], and M. S. Agranovich and M. I. Vishik [3] in the 1960s, multiple completeness of root functions of such problems was not proved yet. In 1962 S. Agmon [1] proved completeness of root functions of elliptic boundary value problems, when boundary conditions do not depend on a spectral parameter and the spectral parameter is entered in the equation linearly. In the case when the spectral parameter is entered in the equation polynomially, multiple completeness of root functions was proved by S. Ya. Yakubov in 1986 [88]. For an elliptic pseudo-differential system (in terms of Douglis-Nirenberg) on compact manifolds without boundary, completeness of root functions was proved by A. N. Kozhevnikov in 1973 [36].

In chapter 4 multiple completeness of root functions of elliptic boundary value problems, with a polynomial spectral parameter in both an equation and the boundary conditions, was also proved. Thus, the problem was completely solved. It will undoubtedly be of great importance in solving problems of mathematical physics and differential equations.

In chapters 3 and 4 the theorems of density of some sets in the product of Sobolev spaces are also proved, and can be useful for various applications. In the basic results – theorems on multiple completeness – it is not required that orders of boundary value conditions should be less than the equation order. This fact enabled us to strengthen a classical result of S. Agmon's [1], where a spectral parameter enters linearly in an equation. A. N. Kozhevnikov [36, 37], and L. A. Kotko and S. G. Krein [35] proved the completeness of root functions of an elliptic boundary value problem, when a spectral parameter enters linearly both in an equation and in boundary value conditions.

Further, it is shown that the irregularity of a problem can influence the character of the problem. For instance, a spectrum of such problems in contrast to the regular problems can have a finite point of accumulation. As it is known, such a situation does not arise in the theory of ordinary differential equations, since a degree of the principal term of eigenvalue asymptotics of boundary value problems, for ordinary differential equations in a finite segment, does not depend on the boundary value conditions but depends only on the order of the equation.

There are many works devoted to parabolic and hyperbolic differential-operator equations. A few examples are the monographs by S. G. Krein [39], H. Tanabe [70], A. Pazy [54], and H. O. Fattorini [15]. There are fewer publications concerning the elliptic differential-operator equations. For example, the results in books by S. G. Krein [39], S. Ya. Yakubov [85] and others do not cover boundary value problems for elliptic differential-operator equations of the 4-th order when the principal part of the equation has the form $u''''(t) + Au''(t) + Bu(t)$, where $AB^{-\frac{1}{2}}$ is a bounded operator and is not compact. Moreover, we prove not only correct solvability, but coercive solvability of considered problems.

In chapters 5 and 6 we find algebraic conditions on boundary value problems for differential-operator equations of the 4-th order on an interval, and for boundary value problems for partial elliptic equations of the 4-th order in cylindrical domains, to be fredholm. In particular, a boundary value problem for the biharmonic equation, which is important for mechanics, is considered. Note that, in contrast to the general theory of elliptic boundary value problems, our boundary conditions can be partially non-local.

In chapter 5 a solution of the Cauchy problem for differential-operator equations of parabolic type is approximated by the linear combinations of elementary solutions. This problem is solved in a case when application of the Fourier method becomes very difficult. This is the case when the corresponding spectral problem is principally non-selfadjoint.

A lot of monographs and articles have been devoted to the solvability questions of regular elliptic boundary value problems. However, the results on solvability of boundary value problems in the cylindrical domains for elliptic equations of the 4-th order, which are presented in chapter 6, are new. They are new even if the problem is differential, since boundary conditions are non-local. Algebraic conditions of the solvability are found. We have considered elliptic boundary value problems in the tube domains, hence in non-smooth domains. Our results are new in comparison with the results in P. Grisvard's book [21], which is devoted to this question. Note that in the book by P. Grisvard [21] and papers by V. A. Kondratiev [32], V. A. Kondratiev and O. A. Oleinik [33], and A. A. Shkalikov [64] a complementary condition arisis. More precisely, determinants of auxiliary "Kondratiev problems" do not have zeros on the line $\text{Re}\lambda = 3$. Equations and problems that are considered in chapter 6 belong to known types only in their principal parts. Algebraic conditions of the solvability or completeness of elementary solutions are found. Theorems of completeness of elementary solutions are new even for simple boundary conditions. Similar theorems were proved in papers by D. D. Joseph [25], Y. A. Ustinov and V. I. Yudovich [74], G. Geymonat and P. Grisvard [18], and A. A. Shkalikov [64].

The numeration of the statements and the formulas is given by two figures in each chapter. The first figure denotes the number of paragraph and the second one denotes the number of statement (formula) in the paragraph. When formulas (theorems) taken from other chapters are referred to, triple numeration is used. For instance, the reference to formula (2.4) of the second chapter is written as follows: (2.2.4). The completion of the proof is denoted by ■.

Chapter 1

Auxiliary Results

Below a number of notions, terms and facts are given, which will be used throughout the book.

1.1. General notions from functional analysis

1.1.1. Linear spaces. A *linear space* E is a set of elements $u, v, \ldots$ for which linear operations are defined (summation $u + v$ of the two elements u, v and multiplication λu of the element u by the complex number λ), and the operations are subject to the general rules. A zero element is denoted by 0, as a zero scalar.

Elements $u_1, \ldots, u_n$ are called *linearly independent*, if their *linear combination* $\lambda_1 u_1 + \cdots + \lambda_n u_n$ is equal to zero if and only if $\lambda_1 = \cdots = \lambda_n = 0$, otherwise these elements are *linearly dependent*. Dimension of the space E, denoted by $\dim E$, is the maximum number of linearly independent elements in E. If there is no such finite number, then we suppose that $\dim E = \infty$.

A set E_0 from E is called a *linear manifold*, if E_0 is itself a linear space with respect to the induced linear operations in E. For any set M from E a set of all possible finite linear combinations of elements from M is a linear manifold. This is called a *linear span* of M.

For any sets M_1 and M_2 in E $M_1 + M_2$ denotes the *arithmetic sum* of M_1 and M_2, i.e., the set of all elements of the form $u_1 + u_2$, where $u_1 \in M_1$, $u_2 \in M_2$. One should note the difference between $M_1 + M_2$ and a union of M_1 and M_2, denoted by $M_1 \cup M_2$.

A linear manifold E_0 is said to be a *direct sum* of linear manifolds $E_1, \ldots, E_n$, if $E_0 = E_1 + \cdots + E_n$ and $\sum_{k=1}^{n} u_k = 0$, where $u_k \in E_k$, if and only if $u_k = 0$, $k = 1, \ldots, n$. The direct sum is denoted by $E_0 = E_1 \dot{+} \cdots \dot{+} E_n$.

1.1.2. Banach spaces. A linear space E is called a *linear normed space*, if each element $u \in E$ is connected with a real number $\|u\| \geq 0$, which is called the *norm* of the element u and has the following properties:

(1) $\|u\| = 0$ if and only if $u = 0$;

(2) $\|\lambda u\| = |\lambda| \|u\|$;

(3) $\|u + v\| \leq \|u\| + \|v\|$.

A sequence $u_n \in E$ is said to be convergent to an element $u \in E$ if $\lim\limits_{n \to \infty} \|u_n - u\| = 0$. It is written in the form $u_n \overset{E}{\to} u$. Definitions of closed, open, bounded, compact and precompact sets in E are introduced respectively.

The closure of a set M of E by the norm E is denoted by $\overline{M}|_E$, and sometimes by $\overline{M}$. The set M is called *dense* in E, if $\overline{M} = E$.

A sequence $u_n \in E$ is called *fundamental*, if $\lim\limits_{n,m \to \infty} \|u_n - u_m\| = 0$. A converging sequence is fundamental. If any fundamental sequence in the space E converges, then the space is called *complete*. A complete linear normed space is called *Banach*.

A *series* $\sum\limits_{n=1}^{\infty} u_n$ is said to be convergent to an element $u \in E$, if $s_k = \sum\limits_{n=1}^{k} u_n \overset{E}{\to} u$.

A sequence $u_n \in E$ is said to be a *basis* in E, if each element $u \in E$ is uniquely expanded in the converging series

$$u = \sum_{n=1}^{\infty} \lambda_n u_n.$$

A closed linear manifold E_0 of the Banach space E is called a *subspace of E*.

Let E and F be Banach spaces. The set $E \dotplus F$ of all two-dimensional vectors of the form (u, v), where $u \in E$, and $v \in F$, with usual coordinatewise linear operations and the norm

$$\|(u, v)\|_{E \dotplus F} = (\|u\|_E^2 + \|v\|_F^2)^{1/2}$$

is a Banach space and called a *direct sum* of E and F.

1.1.3. Operators in Banach spaces. Let E and F be Banach spaces. By an *operator*[1] A from E into F we mean a linear mapping from E into F, i.e., a mapping that associates each element u of some linear manifold $D \subset E$ with a certain element v of the space F and satisfies the condition

$$A(\lambda_1 u_1 + \lambda_2 u_2) = \lambda_1 A u_1 + \lambda_2 A u_2$$

---

[1] Since this book treats only linear operators, the word "linear" is dropped.

for all $u_1, u_2 \in D$ and all numbers $\lambda_1, \lambda_2 \in \mathbb{C}$. Set D is called a *domain of definition* of the operator A and is denoted by $D(A)$. The *range of values* $R(A)$ of the operator A is defined as the set of all elements from F of the form Au, $u \in D(A)$.

The *inverse* operator A^{-1} for the operator A from E into F is determined only when the mapping is one-to-one or, in other words, if from the equality $Au = 0$ it follows that $u = 0$. By definition, A^{-1} is an operator from F into E, mapping Au into u. Thus,

$$D(A^{-1}) = R(A), \quad R(A^{-1}) = D(A),$$

$$A^{-1}(Au) = u, \ u \in D(A), \quad A(A^{-1}v) = v, \ v \in R(A).$$

An operator A from E into F is *continuous* at a point $u_0 \in E$, if $D(A) = E$ and from $\|u_n - u_0\|_E \to 0$, $u_n \in E$ it follows that $\|Au_n - Au_0\|_F \to 0$. The operator A is continuous everywhere in E, if it is continuous in zero.

An operator A from E into F is *bounded*, if $D(A) = E$ and for some $C > 0$ $\|Au\|_F \leq C\|u\|_E$, $u \in E$. The operator A from E into F is continuous if and only if it is bounded. Let us denote by $B(E, F)$ the set of all bounded operators from E into F. With the natural definition of summation, multiplication by a scalar and the norm

$$\|A\|_{B(E,F)} = \sup_{\substack{u \in E \\ u \neq 0}} \frac{\|Au\|_F}{\|u\|_E}. \tag{1.1}$$

$B(E, F)$ becomes a Banach space.

An operator A from E into F is *compact*, if $D(A) = E$ and it maps every bounded set in E into a precompact set in F.

An operator A from E into F is *invertible*, if A^{-1} from F into E is bounded.

An operator A from E onto F is *isomorphic*, if the operators A from E into F and A^{-1} from F into E are bounded.

An operator A from $D(A)$ onto $R(A)$ is *isometric*, if

$$\|Au\|_F = \|u\|_E, \quad u \in D(A).$$

By the operator A in E we mean the operator A from E into E.

An operator A in E is *closed*, if from $\|u_n - u\| \to 0$, $u_n \in D(A)$, and $\|Au_n - v\| \to 0$ follows $u \in D(A)$ and $Au = v$.

For the operator A closed in E the domain of definition $D(A^n)$ of the operator A^n is turned into a Banach space $E(A^n)$ with respect to the norm

$$\|u\|_{E(A^n)} = \left(\sum_{k=0}^{n} \|A^k u\|^2 \right)^{1/2}.$$

The operator A^n from $E(A^n)$ into E is bounded. A bounded operator is closed.

If an operator A in E is not closed, then it has a closed extension (admits closure) if and only if from $u_n \in D(A)$, $u_n \xrightarrow{E} 0$, and $Au_n \xrightarrow{E} v$ it follows that $v = 0$. The smallest closed extension of the operator A is called its *closure* $\overline{A}$. If the operator A admits closure, then from $u_n \in D(A)$, $u_n \xrightarrow{E} u$, and $Au_n \xrightarrow{E} v$ it follows that $v = \overline{A}u$. Otherwise, it can be written as follows

$$\overline{A} \lim_{n\to\infty} u_n = \lim_{n\to\infty} Au_n,$$

if both limits exist. If A in E is closed, then

$$A \lim_{n\to\infty} u_n = \lim_{n\to\infty} Au_n,$$

if both limits exist.

A closed operator, defined on the whole space, is bounded. If A in E is closed and has an inverse operator, then A^{-1} in E is closed. If A in E is invertible, then A in E is closed.

1.1.4.   Dual spaces. A set $\mathbb{C}^n$ of all ordered collections $u = (\lambda_1, \ldots, \lambda_n)$ of complex numbers is an n-dimensional Banach space (a complex n-dimensional Euclidean space) with general coordinatewise linear operations and the norm

$$\|u\| = (|\lambda_1|^2 + \cdots + |\lambda_n|^2)^{1/2}.$$

Let E be a Banach space. Banach space $B(E, \mathbb{C})$ is called *dual* to E and denoted by E'. So, by definition $E' = B(E, \mathbb{C})$.

The elements of the space E' are called *continuous functionals* in E and denoted by u'. By (1.1) the norm of the functional u' is defined by the formula

$$\|u'\|_{E'} = \sup_{\substack{u \in E \\ u \neq 0}} \frac{|<u, u'>|}{\|u\|_E},$$

where $<u, u'>$ denotes the value of the functional $u' \in E'$ on the element $u \in E$. Therefore, for any $u \in E$, and $u' \in E'$, the following inequality holds:

$$|<u, u'>| \leq \|u'\|_{E'} \|u\|_E.$$

A space, dual to the space E', is called the *second dual* to E and denoted by E''. Each element $u \in E$ generates a continuous functional u'' in the space E'' by the formula $<u', u''> = <u, u'>$, moreover, $\|u''\|_{E''} = \|u\|_E$. So, the space E is isometrically and linearly mapped onto the space E''. If, in this case, the image of E coincides with the whole space E'', then the space E is called *reflexive*.

1.1.5. Adjoint operators. Let E and F be Banach spaces. The operator A from E into F is bounded. The bounded operator A^* from F' in E' is called *adjoint* to A, if for any $u \in E$, and $v' \in F'$

$$< Au, v' > \; = \; < u, A^*v' > .$$

There exists a unique operator conjugate to the operator A, bounded from E into F.

1.1.6. Hilbert spaces. A linear space E is called *pre-Hilbert*, if each pair of its elements u and v is associated with a complex number (u, v), called a *scalar product* of u and v, with the following properties:

(1) $(u, u) \geq 0$; $(u, u) = 0$ if and only if $u = 0$;

(2) $(u, v) = \overline{(v, u)}$;

(3) $(\lambda_1 u_1 + \lambda_2 u_2, v) = \lambda_1(u_1, v) + \lambda_2(u_2, v)$.

By the formula

$$\|u\| = (u, u)^{1/2}$$

the norm in a pre-Hilbert space is defined. A complete pre-Hilbert space is called *Hilbert* and usually denoted by H.

If $(u, v) = 0$, then the elements u and v are called *orthogonal* and written as $u \perp v$. Two sets M and N from the Hilbert space H are called *orthogonal*, if any element of M is orthogonal to any element of N. The orthogonality of M and N is denoted by $M \perp N$. H is said to be an *orthogonal sum* of the Hilbert spaces H_k, $k = 1, \ldots, n$, and written as $H = H_1 \oplus H_2 \oplus \cdots \oplus H_n$, if H is a direct sum of H_k and the scalar product in H is given by the formula

$$(u, v)_H = \sum_{k=1}^{n}(u_k, v_k)_{H_k},$$

where $u = (u_1, \ldots, u_n)$, $v = (v_1, \ldots, v_n)$.

A system of elements $\{u_k\}_1^\infty$ from H is called *orthonormal*, if

$$(u_k, u_m) = \delta_{km},$$

where $\delta_{km} = 1$ for $k = m$ and $\delta_{km} = 0$ for $k \neq m$. For the orthonormal basis the Parseval equality holds:

$$\|u\|^2 = \sum_{k=1}^{\infty} |(u, u_k)|^2.$$

1.1.7. Continuity of a vector-valued function. Let E be a Banach space. A mapping $x \to u(x)$: $[0,1] \to E$ is called *continuous* at a point $x_0 \in [0,1]$, if $\lim_{x \to x_0} \|u(x) - u(x_0)\| = 0$. The mapping $x \to u(x)$: $[0,1] \to E$ is called *weakly continuous* at the point $x_0 \in [0,1]$, if for any $u' \in E'$ $\lim_{x \to x_0} < u(x), u' > = < u(x_0), u' >$. The inequality

$$| < u(x), u' > - < u(x_0), u' > | \leq \|u'\| \|u(x) - u(x_0)\|$$

implies that a weak continuity follows from a strong continuity of the function. Let us give an example, illustrating that the inverse assertion does not hold. Let u_n be an orthonormal basis in H. Consider the function $u(x) = u_{[1/x]}$, $x \in (0,1]$, $u(0) = 0$, where $[y]$ denotes the integral part of the number $y \in \mathbb{R}$ ($y = [y] + \{y\}$, where $0 \leq \{y\} < 1$). Since for any $v \in H$ $\lim_{n \to \infty} (u_n, v) = 0$, then the function $u(x)$ is weakly continuous at the point $x_0 = 0$. On the other hand $\|u_n - u_m\| = 2^{1/2}$, from which follows that the function $u(x)$ is discontinuous at the point $x_0 = 0$.

Let us denote $B(E) = B(E, E)$. A mapping $x \to A(x)$: $[0,1] \to B(E)$ is called *strongly continuous* at a point $x_0 \in [0,1]$ if for any $u \in E$ the mapping $x \to A(x)u$: $[0,1] \to E$ is continuous at x_0.

1.1.8. Differentiable and analytic vector-valued functions. Let E be a Banach space. An element $v \in E$ is called the derivative of a function $u(x)$ with values from E at a point $x_0 \in [0,1]$, if

$$\lim_{h \to 0} \left\| \frac{u(x_0 + h) - u(x_0)}{h} - v \right\| = 0.$$

In this case we write $u'(x_0) = v$. If the function $u(x_0)$ at every point of the segment $[0,1]$ has a derivative, then the derivative $u'(x)$ is also a function with values from the Banach space E. So, the notions of *n-times differentiable* and *infinitely differentiable functions* are intrinsically introduced.

An element $v \in E$ is called the derivative of a function $u(\lambda)$ with values from E at a point $\lambda_0 \in G$, where G is a domain in the complex plane, if

$$\lim_{h \to 0} \left\| \frac{u(\lambda_0 + h) - u(\lambda_0)}{h} - v \right\| = 0.$$

The function $u(\lambda)$ is called *analytic* in the domain G if it has a derivative at each point of this domain. If $u(\lambda)$ is analytic in G, then for any $u' \in E'$ the scalar function $< u(\lambda), u' >$ is analytic in G. Unlike functions that are differentiable on a segment, the inverse assertion is also valid [23, Ch.III, §2.3.10]. This allows properties of the analytic functions with values in E to be obtained from the properties of the scalar analytic functions.

If A in E is closed, then

$$Au'(x) = (Au(x))'$$

if both derivatives exist.

A proper Riemann integral of a continuous vector function on the segment and an improper Riemann integral are introduced quite similarly to the scalar case, the only difference being that the limits arising in this case are understood in terms of convergence by the norm of the space E.

The numerous theorems of analysis of the properties of continuous and differentiable scalar functions, of the integrals depending on the parameter, and of Cauchy integrals, hold for the vector-valued functions too [39, p.3-15].

1.1.9. Measurability of a vector-valued function. A sequence $u_n(x)$ is said to be convergent to $u(x)$ *almost everywhere* on $[0, 1]$ if there exists a set Ω_0 of the measure zero, such that

$$\lim_{n \to \infty} \|u_n(x) - u(x)\| = 0$$

for any $x \in [0, 1] \setminus \Omega_0$.

The function $u(x)$ given on the segment $[0, 1]$, is called *countable-valued* if the values that it takes are not more than the countable number, moreover, each of its values other than zero $u(x)$ takes on some measurable set.

The function $u(x)$ given on $[0, 1]$ is called *measurable*, if there exists a sequence of countable-valued functions, converging to $u(x)$ almost everywhere on $[0, 1]$.

1.1.10. Bochner integral. A countable-valued function $u(x)$ given on $[0, 1]$ is called *integrable* (in the sense of Bochner), if the function $\|u(x)\|$ is integrable in the sense of Lebesque, and, by definition

$$\int_0^1 u(x) \, dx = \sum_{k=1}^{\infty} u_k m \Omega_k,$$

where $u(x) = u_k$ on Ω_k, $k = 1, \ldots, \infty$.

The function $u(x)$ given on $[0, 1]$ is called *integrable*, if there exists a sequence of countable-valued integrable functions $u_n(x)$, converging to $u(x)$ almost everywhere on $[0, 1]$ such that

$$\lim_{n \to \infty} \int_0^1 \|u_n(x) - u(x)\| \, dx = 0.$$

In this case, by definition

$$\int_0^1 u(x) \, dx = \lim_{n \to \infty} \int_0^1 u_n(x) \, dx.$$

It is known [23, Ch.III, §1, Th.3.7.4] that for the function $u(x)$ to be integrable it is necessary and sufficient that $u(x)$ be measurable and $\|u(x)\|$ be integrable.

if A in E is closed, then

$$A \int_0^1 u(x) \, dx = \int_0^1 Au(x) \, dx,$$

if both integrals exist. For more details see [23, Ch.III, §1].

1.1.11. Generalized derivative of a vector-valued function. A locally integrable function $v(x)$ with values from E is called a *generalized derivative* of the n-th order on $(0,1)$ of the locally integrable function $u(x)$ with values from E, if

$$\int_0^1 u(x)\varphi^{(n)}(x) \, dx = (-1)^n \int_0^1 v(x)\varphi(x) \, dx, \quad \varphi \in C_0^\infty(0,1),$$

where $C_0^\infty(0,1)$ denotes the set of infinitely differentiable finite scalar functions given on $(0,1)$.

1.1.12. Functional spaces of vector-valued functions. Let E be a Banach space.

$C^n([0,1], E)$ is a Banach space of n-times continuously differentiable functions $u(x)$ with values from E with the norm

$$\|u\|_{C^n([0,1],E)} = \max_{x \in [0,1]} \sum_{k=0}^n \|u^{(k)}(x)\|.$$

$W_p^n((0,1), E)$, $1 \le p < \infty$, is a Banach space of functions $u(x)$ with values from E which have generalized derivatives up to the n-th order inclusive on $(0,1)$ and the norm

$$\|u\|_{W_p^n((0,1),E)} = \sum_{k=0}^n \left(\int_0^1 \|u^{(k)}(x)\|^p \, dx \right)^{1/p}$$

is finite.

$W_{p,lok}^n((0,1), E)$ consists of functions with values from E, belonging to $W_p^n((a,b), E)$ for any $[a, b] \subset (0,1)$.

By virtue of [72, p.42] from $u \in W_p^n((0,1), E)$ it follows that the function $u^{(j)}(x)$, $j = 0, \ldots, n-1$, with values from E is absolutely continuous on $[0,1]$. Then by virtue of [23, Ch.III, §1, Th.3.8.6] the Newton-Leibniz formula holds:

$$u^{(j)}(x) - u^{(j)}(y) = \int_y^x u^{(j+1)}(\xi) \, d\xi, \qquad j = 0, \ldots, n-1.$$

1.1.13. Fourier multipliers. Consider the Fourier transform

$$(Fu)(\sigma) = (2\pi)^{-1/2} \int_{-\infty}^{\infty} e^{-i\sigma x} u(x) \, dx$$

and the inverse Fourier transform

$$(F^{-1}v)(x) = (2\pi)^{-1/2} \int_{-\infty}^{\infty} e^{i\sigma x} v(\sigma) \, d\sigma,$$

for functions $u, v \in L_1(\mathbb{R}, E)$.

The Fourier transform has a number of remarkable properties. One of them is the following

$$(Fu^{(k)})(\sigma) = (i\sigma)^k (Fu)(\sigma).$$

A mapping $\sigma \to T(\sigma) \; : \; \mathbb{R} \to B(E)$ is called a *Fourier multiplier* of the type (p, q) if

$$\|F^{-1}TFu\|_{L_q(\mathbb{R}, E)} \le C \|u\|_{L_p(\mathbb{R}, E)}, \qquad u \in L_p(\mathbb{R}, E).$$

It is known [13, p.1181] that in the case of the Hilbert space H, if the mapping $\sigma \to T(\sigma) \; : \; \mathbb{R} \to B(H)$ is continuously differentiable and

$$\|T(\sigma)\| \le C, \quad \|T'(\sigma)\| \le C|\sigma|^{-1}, \qquad \sigma \in \mathbb{R},$$

then the function $T(\sigma)$ is the Fourier multiplier of type (p, p) (the Mikhlin-Schwartz theorem).

1.2. Interpolation of spaces and operators

In the theory of differential equations there arise various spaces of differentiable functions and operators acting in these spaces. Over the recent years the abstract theory of interpolation has allowed the wide range of problems for such spaces to be studied from a unique point of view. Here we expound on some aspects of this theory, and for a more profound acquaintance we refer the reader to the books by H. Triebel [72] and S. G. Krein, Yu. I. Petunin and Y. M. Semenov [40].

1.2.1. Embedding of spaces. Let E_1 and E_2 be Banach spaces, in which a set-theoretical inclusion $E_1 \subset E_2$ holds, and the space E_2 induces on E_1 the linear space structure coinciding with the structure of the linear space E_1. Let us denote by J the operator that associates each element $u \in E_1$ with the same element of the space E_2.

If the operator J from E_1 into E_2 is continuous, we say that the *embedding* $E_1 \subset E_2$ *is continuous*.

If the operator J from E_1 into E_2 is compact, we say that the embedding $E_1 \subset E_2$ is compact.

Continuity of the embedding $E_1 \subset E_2$ is equivalent to the inequality

$$\|u\|_{E_2} \leq C\|u\|_{E_1}, \quad u \in E_1.$$

If the set E_1 is dense in E_2, we say that the embedding $E_1 \subset E_2$ is dense.

Lemma 2.1. *Let the following conditions be satisfied:*

(1) E_1, E_2, E_3 *are Banach spaces;*

(2) *the embedding $E_1 \subset E_2$ is compact;*

(3) *the embedding $E_2 \subset E_3$ is continuous.*

Then for any $\varepsilon > 0$

$$\|u\|_{E_2} \leq \varepsilon\|u\|_{E_1} + C(\varepsilon)\|u\|_{E_3}, \qquad u \in E_1. \tag{2.1}$$

Proof. Suppose, that (2.1) is not true. Then for some $\varepsilon_0 > 0$ there exist $u_n \in E_1$ and $C_n \to \infty$ such that

$$\|u_n\|_{E_2} > \varepsilon_0\|u_n\|_{E_1} + C_n\|u_n\|_{E_3}.$$

Putting $v_n = u_n/\|u_n\|_{E_1}$, we obtain

$$\|v_n\|_{E_2} > \varepsilon_0 + C_n\|v_n\|_{E_3}. \tag{2.2}$$

By virtue of condition 2 we have $\|v_n\|_{E_2} \leq C\|v_n\|_{E_1} \leq C$. Then from (2.2) follows

$$\|v_n\|_{E_3} \to 0. \tag{2.3}$$

Since $\|v_n\|_{E_1} = 1$, then by virtue of condition 2, there exists a subsequence v_{n_k}, converging in E_2. By virtue of condition 3 and (2.3) $\|v_{n_k}\|_{E_2} \to 0$. This contradicts (2.2). ∎

1.2.2. Embedding of dual spaces. If the embedding $E_1 \subset E_2$ is continuous, then the restriction on E_1 of any functional $u_2' \in E_2'$ naturally generates a linear functional on E_1. This functional is continuous in E_1. Indeed, for $u_1 \in E_1 \subset E_2$ we have

$$|<u_1, u_2'>| \leq \|u_2'\|_{E_2'}\|u_1\|_{E_2} \leq C\|u_2'\|_{E_2'}\|u_1\|_{E_1}. \tag{2.4}$$

Thus we obtain a linear mapping of the space E_2' into E_1'. If E_1 is not dense in E_2, then there exists a non-zero functional from E_2' that vanishes on E_1 and, therefore, is mapped into zero. In this case the mapping will not be injective. If E_1 is dense in E_2, then the mapping is injective, and the space E_2' can be regarded as embedded into the space E_1'.

Lemma 2.2. *Let the following conditions be satisfied:*

(1) *the space E_1 is reflexive;*

(2) *the embedding $E_1 \subset E_2$ is continuous and dense.*

Then the embedding $E_2' \subset E_1'$ is continuous and dense.

Proof. From (2.4) for $u_2' \in E_2'$ we have the estimate

$$\|u_2'\|_{E_1'} \le C\|u_2'\|_{E_2'},$$

i.e., the embedding $E_2' \subset E_1'$ is continuous. Let us show that the embedding $E_2' \subset E_1'$ is dense. Otherwise, by virtue of the corollary from the Hahn-Banach theorem, there exists the functional $u_1'' \in E_1''$ such that $u_1''(E_2') = 0$, but $u_1''(E_1') \ne 0$. Since E_1 is reflexive, then there exists $u_1 \in E_1$ such that

$$< u_1, u_1' > \; = \; < u_1', u_1'' >, \quad u_1' \in E_1'.$$

Hence,

$$< u_1, u_1' > \; = 0, \quad u_1' \in E_2', \tag{2.5}$$

$$< u_1, u_1' > \; \ne 0 \text{ for some } u_1' \in E_1'. \tag{2.6}$$

From (2.5) follows $u_1 = 0$ and this contradicts (2.6). ∎

1.2.3. Interpolation spaces. Let E_0 and E_1 be two Banach spaces, continuously embedded into the Banach space $E: \; E_0 \subset E, \; E_1 \subset E$. Two such spaces are called an *interpolation pair* $\{E_0, E_1\}$. Consider the Banach space

$$E_0 + E_1 = \{u | u \in E, \; \exists u_j \in E_j, \; j = 0, 1, \text{ where } u = u_0 + u_1\},$$

$$\|u\|_{E_0+E_1} = \inf_{\substack{u=u_0+u_1 \\ u_j \in E_j}} (\|u_0\|_{E_0} + \|u_1\|_{E_1}).$$

By virtue of [72, p.23] the functional

$$K(t, u) = \inf_{\substack{u=u_0+u_1 \\ u_j \in E_j}} (\|u_0\|_{E_0} + t\|u_1\|_{E_1}), \quad u \in E_0 + E_1,$$

is continuous on $(0, \infty)$ in t and the following estimate holds:

$$\min\{1, t\}\|u\|_{E_0+E_1} \le K(t, u) \le \max\{1, t\}\|u\|_{E_0+E_1}.$$

An *interpolation space* for $\{E_0, E_1\}$ by the K-method is defined as follows:

$$(E_0, E_1)_{\theta, p} = \{u | \; u \in E_0 + E_1, \; \|u\|_{(E_0, E_1)_{\theta, p}}$$

$$= \left(\int_0^\infty t^{-1-\theta p} K^p(t, u) \, dt \right)^{1/p} < \infty, \; 0 < \theta < 1, \; 1 \le p < \infty\},$$

and

$$(E_0, E_1)_{\theta, \infty} = \{u | u \in E_0 + E_1, \|u\|_{(E_0, E_1)_{\theta, \infty}}$$
$$= \sup_{t \in (0, \infty)} t^{-\theta} K(t, u) < \infty, \ 0 < \theta < 1\}.$$

In [72, p.25] it is proved that there exists a positive number $C_{\theta, p}$, $0 < \theta < 1$, $1 \leq p \leq \infty$, such that for all $u \in E_0 \cap E_1$

$$\|u\|_{(E_0, E_1)_{\theta, p}} \leq C_{\theta, p} \|u\|_{E_0}^{1-\theta} \|u\|_{E_1}^{\theta}. \tag{2.7}$$

Lemma 2.3. Let E_{j0} be a subspace of E_j, $j = 0, 1$. Then the embedding

$$(E_{00}, E_{10})_{\theta, p} \subset (E_0, E_1)_{\theta, p}, \qquad 0 < \theta < 1, \ 1 \leq p \leq \infty,$$

is continuous.

Proof. For all $u \in E_{00} + E_{10} \subset E_0 + E_1$ we have

$$K_0(t, u) = \inf_{\substack{u = u_{00} + u_{10} \\ u_{j0} \in E_{j0}}} (\|u_{00}\|_{E_0} + t\|u_{10}\|_{E_1})$$
$$\geq \inf_{\substack{u = u_0 + u_1 \\ u_j \in E_j}} (\|u_0\|_{E_0} + t\|u_1\|_{E_1}) \geq K(t, u).$$

Hence, for $1 \leq p < \infty$

$$\|u\|_{(E_0, E_1)_{\theta, p}} = (\int_0^\infty t^{-1-\theta p} K^p(t, u) \ dt)^{1/p}$$
$$\leq (\int_0^\infty t^{-1-\theta p} K_0^p(t, u) \ dt)^{1/p} = \|u\|_{(E_{00}, E_{10})_{\theta, p}}.$$

The case $p = \infty$ is proved in a similar way. ∎

If operator $-A$ generates a semigroup e^{-tA}, which is analytic for $t > 0$ and strongly continuous for $t \geq 0$, then

$$(E, E(A^n))_{\theta, p} = \{u | \ u \in E, \|u\|_{(E, E(A^n))_{\theta, p}}^*$$
$$= \int_0^\infty t^{-1+n(1-\theta)p} \|A^n e^{-tA} u\|^p \ dt)^{1/p} + \|u\| < \infty\},$$

moreover, the norm $\|u\|_{(E, E(A^n))_{\theta, p}}^*$ is equivalent to the norm $\|u\|_{(E, E(A^n))_{\theta, p}}$ [72, p.96].

There are other methods of interpolation as well. Many of them coincide with interpolation by the K-method up to the norm equivalency. But the complex method of interpolation generates new interpolation spaces.

Let S denote a stripe $S = \{z| \ z \in \mathbb{C}, \ 0 < \text{Re}z < 1\}$ of the complex plane, and $\overline{S}$ is its closure. Let $\{E_0, E_1\}$ be an interpolation pair. Then by definition [72, p.56] $F(E_0, E_1) = \{u(z)| \ u(z) \text{ is an } (E_0 + E_1)\text{-valued function, continuous in } \overline{S} \text{ and}$ analytic in S, $\sup_{z \in \overline{S}} \|u(z)\|_{E_0 + E_1} < \infty$, $u(j + it) \in E_j$, $j = 0, 1$, $t \in \mathbb{R}$, $u(j + it)$ is a continuous as E_j-valued function of t, $\|u\|_{F(E_0, E_1)} = \max_{j=0,1} \sup_{t \in \mathbb{R}} \|u(j + it)\|_{E_j} < \infty\}$.

Further, by definition under $0 < \theta < 1$ $[E_0, E_1]_\theta = \{u| \ u \in E_0 + E_1, \text{ for some}$ $u(\cdot) \in F(E_0, E_1)$, $u(\theta) = u$, $\|u\|_{[E_0,E_1]_\theta} = \inf_{u(\theta)=u} \|u(\cdot)\|_{F(E_0,E_1)}\}$.

In [72, p.59] it is proved that there exists a positive number C_θ, $0 < \theta < 1$, such that for all $u \in E_0 \cap E_1$

$$\|u\|_{[E_0,E_1]_\theta} \leq C_\theta \|u\|_{E_0}^{1-\theta} \|u\|_{E_1}^{\theta}. \tag{2.8}$$

Let the operator A in the Hilbert space H be selfadjoint and positive. Then [72, p.142] for $\alpha \geq 0$, $\beta \geq 0$, $0 < \theta < 1$

$$[H(A^\alpha), H(A^\beta)]_\theta = (H(A^\alpha), H(A^\beta))_{\theta,2} = H(A^{(1-\theta)\alpha+\theta\beta}).$$

1.2.4. Sobolev spaces and their interpolation. Let Ω be a domain in the n-dimensional real Euclidean space $\mathbb{R}^n$, $x = (x_1, \ldots, x_n)$.

$L_q(\Omega)$, $1 < q < \infty$, is a Banach space of functions $u(x)$, measurable on Ω for which the following norm is finite:

$$\|u\|_{L_q(\Omega)} = \left(\int_\Omega |u(x)|^q \, dx \right)^{1/q}.$$

$L_\infty(\Omega)$ is a Banach space of bounded almost everywhere on Ω functions $u(x)$ with the norm

$$\|u\|_{L_\infty(\Omega)} = \text{vrai} \max_{x \in \Omega} |u(x)|.$$

The generalized derivatives of the scalar functions are understood as they are asserted in most works on differential equations. Their definitions, different but equivalent, and basic properties can be found in [66], [73].

$W_q^m(\Omega)$ is a Banach space of functions $u(x)$ that have generalized derivatives on Ω up to the m-th order inclusive, for which the following norm is finite:

$$\|u\|_{W_q^m(\Omega)} = \left(\sum_{|\alpha| \leq m} \|D^\alpha u\|_{L_q(\Omega)}^q \right)^{1/q},$$

where $\alpha = (\alpha_1, \ldots, \alpha_n)$, $|\alpha| = \alpha_1 + \cdots + \alpha_n$, $D^\alpha = D_1^{\alpha_1} \cdots D_n^{\alpha_n}$, $D_k^{\alpha_k} = \partial^{\alpha_k}/\partial x_k^{\alpha_k}$. Let Ω be a bounded domain in $\mathbb{R}^n$ of class C^∞, i.e., the boundary Γ of the domain Ω is an $(n-1)$-dimensional infinitely differentiable manifold.

Let s_0 and s_1 be non-negative integers, $0 < \theta < 1$, $1 < p < \infty$, $1 \leq q \leq \infty$ and $s = (1 - \theta)s_0 + \theta s_1$. From [72, p.317, th.1, formula 2.4.2/16] it follows that if $s = (1 - \theta)s_0 + \theta s_1 = (1 - \theta')s_0' + \theta' s_1'$ then

$$(W_p^{s_0}(\Omega), W_p^{s_1}(\Omega))_{\theta,q} = (W_p^{s_0'}(\Omega), W_p^{s_1'}(\Omega))_{\theta',q}.$$

Consider the space

$$B_{p,q}^s(\Omega) = (W_p^{s_0}(\Omega), W_p^{s_1}(\Omega))_{\theta,q}, \tag{2.9}$$

where $0 \leq s_0, s_1$ are integers, $0 < \theta < 1$, $1 < p < \infty$, $1 \leq q \leq \infty$ and $s = (1 - \theta)s_0 + \theta s_1$. Set

$$W_p^s(\Omega) = B_{p,p}^s(\Omega) = (W_p^{s_0}(\Omega), W_p^{s_1}(\Omega))_{\theta,p}, \tag{2.10}$$

where $0 < s \neq$ integer.

If s_0, s_1, s are integers, $0 < \theta < 1$, $1 < p < \infty$, $s = (1-\theta)s_0 + \theta s_1$ then by virtue of [72, p.317, th. 1, formula 2.4.2/11]

$$W_p^s(\Omega) = [W_p^{s_0}(\Omega), W_p^{s_1}(\Omega)]_\theta. \tag{2.11}$$

Lemma 2.4. *Let $0 \leq s \leq \ell$, $1 < p < \infty$, $\lambda \in \mathbb{C}$. Then the following inequality holds:*

$$|\lambda|^{\ell-s}\|u\|_{W_p^s(\Omega)} \leq C(\|u\|_{W_p^\ell(\Omega)} + |\lambda|^\ell\|u\|_{L_p(\Omega)}).$$

Proof. Let $s \neq$ integer. Then by (2.10) we have

$$W_p^s(\Omega) = (L_p(\Omega), W_p^\ell(\Omega))_{s/\ell,p}.$$

Hence, by (2.7) we obtain

$$\|u\|_{W_p^s(\Omega)} \leq C\|u\|_{L_p(\Omega)}^{1-s/\ell}\|u\|_{W_p^\ell(\Omega)}^{s/\ell}. \tag{2.12}$$

Using the Young inequality, we have

$$|\lambda|^{\ell-s}\|u\|_{W_p^s(\Omega)} \leq C(|\lambda|^\ell\|u\|_{L_p(\Omega)})^{1-s/\ell}\|u\|_{W_p^\ell(\Omega)}^{s/\ell}$$
$$\leq C(|\lambda|^\ell\|u\|_{L_p(\Omega)} + \|u\|_{W_p^\ell(\Omega)}).$$

If s is an integer, then by virtue of (2.11)

$$W_p^s(\Omega) = [L_p(\Omega), W_p^\ell(\Omega)]_{s/\ell},$$

from which, by (2.8), we get (2.12). ∎

Lemma 2.5. *Let $k \geq 0$, $1 < q < \infty$, $1 \leq p < \infty$. Then for any $\varepsilon > 0$*

$$\|u\|_{W_p^k(\Omega)} \leq \varepsilon \|u\|_{W_q^{k+n/q}(\Omega)} + C(\varepsilon)\|u\|_{L_q(\Omega)}, \quad u \in W_q^{k+n/q}(\Omega).$$

Proof. Obviously, it is sufficient to prove the lemma for large p. By virtue of [72, p.350/12] the embedding $W_q^{k+n/q}(\Omega) \subset W_p^k(\Omega)$ is compact. By virtue of [72, p.328, th.4.6.2, (a)] for $p \geq q$ the embedding $W_p^k(\Omega) \subset L_q(\Omega)$ is continuous. Now it is enough to apply Lemma 2.1. ∎

1.2.5. Operators in interpolation spaces. Let $\{E_0, E_1\}$ and $\{F_0, F_1\}$ be interpolation pairs. By $B(\{E_0, E_1\}, \{F_0, F_1\})$ we denote a set of linear operators from $E_0 + E_1$ into $F_0 + F_1$ such that their restrictions on E_k, $k = 0, 1$, continuously map E_k into F_k. It is known [72, p.25, th.(a)] that for any $T \in B(\{E_0, E_1\}, \{F_0, F_1\})$ the following inequality is valid

$$\|T\|_{B((E_0,E_1)_{\theta,p},(F_0,F_1)_{\theta,p})} \leq C\|T\|_{B(E_0,F_0)}^{1-\theta}\|T\|_{B(E_1,F_1)}^\theta.$$

1.2.6. Inequalities. Finally let us state a number of well-known inequalities that are often used:

(1) Young inequality: for $1 < p < \infty$, $\frac{1}{p} + \frac{1}{p'} = 1$, $\varepsilon > 0$, $a, b > 0$,

$$ab \leq \frac{1}{p}(\varepsilon a)^p + \frac{1}{p'}\left(\frac{b}{\varepsilon}\right)^{p'}. \tag{2.13}$$

(2) Interpolation inequality for numbers:

$$a^k b^{m-k} \leq a^m + b^m, \quad a \geq 0, \ b \geq 0, \ m \geq 0, \ 0 \leq k \leq m.$$

If we set $a = xb$, then it turns out to be the obvious inequality $x^k \leq x^m + 1$, $x \geq 0$.

(3) Generalized Holder inequality for numbers

$$\sum_{k=1}^\infty |a_{k1} \cdots a_{kn}| \leq \left(\sum_{k=1}^\infty |a_{k1}|^{p_1}\right)^{1/p_1} \cdots \left(\sum_{k=1}^\infty |a_{kn}|^{p_n}\right)^{1/p_n},$$

when $\sum\limits_{k=1}^{n} \frac{1}{p_k} = 1$, $a_j = (a_{1j}, \ldots, a_{kj}, \ldots) \in \ell_{p_j}$.

(4) Interpolation inequality for functions – Lemma 2.4.

(5) Generalized Holder inequality for functions:

$$\int_\Omega \prod_{k=1}^{N} |u_k(x)| \; \mathrm{dx} \leq \prod_{k=1}^{N} \left(\int_\Omega |u_k(x)|^{p_k} \; \mathrm{dx} \right)^{1/p_k}, \qquad \sum_{k=1}^{N} \frac{1}{p_k} = 1,$$

where Ω is a bounded domain in $\mathbb{R}^n$, $u_k \in L_{p_k}(\Omega)$.

In the following chapters material used is briefly explained, namely that on spaces of vector-functions, fractional powers of operators, comparison of operators, interpolation of Banach spaces and operators, Fourier multipliers. For more detailed information the reader is referred to additional literature.

Chapter 2

Unbounded polynomial operator pencils

2.0. Introduction

In the theory of differential equations with constant operator coefficients, polynomial operator pencils play approximately the same role as characteristic polynomials play in the theory of differential equations with constant coefficients.

Consider, in a Banach space E, the following differential-operator equation

$$L(D)u = u^{(n)}(t) + A_1 u^{(n-1)}(t) + \cdots + A_n u(t) = 0, \qquad (0.1)$$

where A_k, $k = 1, \ldots, n$ are given, generally speaking, unbounded operators in E, $u(t)$ is an unknown function with values in E, $D = D_t = \frac{d}{dt}$, and the characteristic operator pencil

$$L(\lambda) = \lambda^n I + \lambda^{n-1} A_1 + \cdots + A_n. \qquad (0.2)$$

The main connection between equation (0.1) and pencil (0.2) is shown by the following lemma.

Lemma 0.1. *The function $u(t)$ of the form*

$$u(t) = e^{\lambda_0 t} \left(\frac{t^k}{k!} u_0 + \frac{t^{k-1}}{(k-1)!} u_1 + \cdots + u_k \right) \qquad (0.3)$$

where $u_j \in E$, $u_0 \neq 0$, is a solution to equation (0.1) if and only if the following correlations hold:

$$L(\lambda_0)u_p + \frac{1}{1!}L'(\lambda_0)u_{p-1} + \cdots + \frac{1}{p!}L^{(p)}(\lambda_0)u_0 = 0, \quad p = 0, \ldots, k. \qquad (0.4)$$

Proof. If function (0.3) is a solution to equation (0.1), then

$$L(D) \sum_{q=0}^{k} e^{\lambda_0 t} \frac{t^q}{q!} u_{k-q}$$

$$= \sum_{q=0}^{k} e^{\lambda_0 t} \left[L(\lambda_0) \frac{t^q}{q!} + \frac{L'(\lambda_0)}{1!} \frac{t^{q-1}}{(q-1)!} + \cdots + \frac{L^{(q)}(\lambda_0)}{q!} \right] u_{k-q}$$

$$= 0. \tag{0.5}$$

Here the Leibniz formula [24, Ch.1, formula 1.4.12] is used

$$L(D)(\varphi_1 \varphi_2) = \sum_{s} \frac{1}{s!} (L^{(s)}(D)\varphi_1) D^s \varphi_2.$$

Hence

$$L(\lambda_0)u_k + \frac{1}{1!}L'(\lambda_0)u_{k-1} + \cdots + \frac{1}{k!}L^{(k)}(\lambda_0)u_0 = 0,$$

. . . .

. . . .

. . . .

$$L(\lambda_0)u_0 = 0,$$

i.e., (0.4) is valid.

Conversely, if (0.4) holds, then (0.5) is true, i.e. function (0.3) is a solution to (0.1). ∎

Corollary 0.2. *If function (0.3) is a solution to equation (0.1), then functions*

$$u_q(t) = e^{\lambda_0 t} \left(\frac{t^q}{q!} u_0 + \frac{t^{q-1}}{(q-1)!} u_1 + \cdots + u_q \right), \quad q = 0, \ldots, k,$$

are also solutions to equation (0.1).

2.1. Completeness of root vectors of an operator from the class $\sigma_p(H)$

2.1.1. Notations and definitions.
E denotes a separable Banach space, and H denotes a separable Hilbert space.

The domain of definition of a linear operator A in E is denoted by $D(A)$, and the range of its values is denoted by $R(A)$.

The point λ of the complex plane is called a *regular point* of the operator A, if the operator $A - \lambda I$ in E is invertible.

The set $\rho(A)$ of all regular points of the operator A is called a *resolvent set* of the operator A, and the operator

$$R(\lambda, A) = (A - \lambda I)^{-1}$$

is called the *resolvent* of the operator A.

The resolvent set $\rho(A)$ is open. Indeed, if $\lambda_0 \in \rho(A)$ then from

$$A - \lambda I = A - \lambda_0 I + (\lambda_0 - \lambda)I = (A - \lambda_0 I)[I - (\lambda - \lambda_0)R(\lambda_0, A)]$$

it follows that in the circle

$$|\lambda - \lambda_0| < \|R(\lambda_0, A)\|^{-1}$$

there exists a resolvent, obtained by the formula

$$R(\lambda, A) = \sum_{k=0}^{\infty} (\lambda - \lambda_0)^k R^{k+1}(\lambda_0, A).$$

This equality is the *Taylor series expansion of the resolvent*. Hence, in particular for the derivatives of the resolvent, follows the formula

$$\frac{d^n R(\lambda, A)}{d\lambda^n} = n! R^{n+1}(\lambda, A).$$

At the same time we make sure that in the domain $\rho(A)$ the resolvent $R(\lambda, A)$ is a holomorphic operator function.

for any two points $\lambda, \mu \in \rho(A)$ the *Hilbert identity*:

$$R(\lambda, A) - R(\mu, A) = (\lambda - \mu)R(\lambda, A)R(\mu, A)$$

is directly verified.

The complement of the set $\rho(A)$ in the whole complex plane is called the *spectrum* $\sigma(A)$ of the operator A. Thus, the spectrum of an operator is a closed set.

The infinite point $\lambda = \infty$ is always attached to the resolvent set of a bounded operator A and to the spectrum of an unbounded operator A.

All *eigenvalues* of the operator A belong to the spectrum $\sigma(A)$, i.e., those numbers λ for which the equation $Au = \lambda u$ has at least one non-zero solution $u \in D(A)$. The element $u_0 \neq 0$ that satisfies the equation $Au_0 = \lambda_0 u_0$ is called an *eigenvector* of the operator A, corresponding to the *eignevalue* λ_0.

If the elements $u_0, u_1, \ldots, u_k$ correlate with

$$Au_p = \lambda_0 u_p + u_{p-1}, \qquad p = 1, \ldots, k, \tag{1.1}$$

then the element u_k is called an *associated vector of the k-th rank* to the eigenvector u_0.

The number $k + 1$ os called the *length of the chain* $u_0, u_1, \ldots, u_k$. The element u_0 is called an *eigenvector of the r-th rank*, if the longest chain corresponding to u_0 has length equal to r. The chain $u_0, u_1, \ldots, u_{r-1}$ is called the *Jordan chain*. The elements $u_0, u_1, \ldots, u_{r-1}$ are linearly independent.

The linear manifold

$$N^k = N^k_{\lambda_0} = \{u | u \in D(A^{k+1}), \ (A - \lambda_0 I)^{k+1} u = 0\}, \quad k = 0, \ldots, \infty,$$

is called a *root lineal of the k-th rank*. Obviously,

$$N^0 \subset N^1 \subset N^2 \subset \cdots$$

The linear manifold

$$N = N_{\lambda_0} = \bigcup_{k=0}^{\infty} N^k$$

is called a *root lineal*. The dimension $\gamma = \gamma(\lambda_0)$ of the lineal N is called an *algebraic multiplicity of the eigenvalue* λ_0. If $\gamma < \infty$, then the lineals N^k and N are closed, and in this case we call them a *root subspace of the k-th rank* and a *root subspace* respectively.

The root subspace of the k-th rank N^k is a subspace that spans over all eigenvectors and associated vectors of rank $\leq k$. Hence, $\dim N^k \geq k + 1$, when $k \leq \gamma$ and

$$N = N^{\gamma-1} = N^\gamma = \cdots,$$

but these equalities can arise much earlier.

In the sequel we shall deal with eigenvalues with finite algebraic multiplicity only.

A root subspace of zero rank is called a *proper subspace*.

A number r is called the *rank of the eigenvalue* λ_0, if the largest rank of the eigenvectors, corresponding to the number λ_0, equals r. Obviously

$$N = N^{r-1} = N^r = \cdots$$

and all the previous root subspaces are different. Consequently $r \leq \gamma$.

The eigenvectors and associated vectors are joined under the common name of *root vectors*.

In case of a linear pencil

$$L(\lambda) = \lambda I - A$$

correlations (0.4) and (1.1) coincide. So Lemma 0.1 yields the following:

Lemma 1.1. *The function $u(t)$ of the form*

$$u(t) = e^{\lambda_0 t}\left(\frac{t^k}{k!}u_0 + \frac{t^{k-1}}{(k-1)!}u_1 + \cdots + u_k\right),$$

where $u_j \in E$, $u_0 \neq 0$, is the solution to the equation

$$u'(t) - Au(t) = 0$$

if and only if the chain $u_0, u_1, \ldots, u_k$ is a chain of root vectors of the operator A, corresponding to the eigenvalue λ_0.

2.1.2. Invariant subspaces and restrictions of an operator; Projectors; Orthogonal projectors. Let A be an operator in a Banach space E. A subspace $E_1 \subset E$ is called an *invariant subspace* of the operator A, if from $u \in E_1 \cap D(A)$ follows $Au \in E_1$.

An operator A_1 is given in the space E_1 by the equalities

$$D(A_1) = E_1 \cap D(A), \quad A_1 u = Au, \quad u \in D(A_1),$$

is called the *restriction* of the operator A.

The definition of a root vector implies that the root subspace of any rank and root subspaces are invariant subspaces.

An eigenvalue of the restriction is also an eigenvalue of the operator itself, but root subspaces may not coincide.

In a Banach space E a bounded operator P, satisfying the condition $P^2 = P$, is called a *projector*. If P is a projector, then $P = I$ on the subspace $E_1 = PE$. Indeed, from $u_1 \in E_1$ follows that $u_1 = Pu$ under some $u \in E$. Hence $Pu_1 = P^2 u = Pu = u_1$.

If the projector P is commutative with a bounded operator A, then $E_1 = PE$ is an invariant subspace of A. Indeed, $Au_1 = APu = PAu \in E_1$. If P is a projector, then $I - P$ is also a projector. Hence, if P is a projector, commutative with a bounded operator A, then $E_1 = PE$ and $E_2 = (I - P)E$ are invariant subspaces of the operator A. Moreover, $PE_2 = 0$.

Let H_1 be a closed subspace in a Hilbert space H. The *orthogonal complement* of H_1 is denoted by $H_1^\perp$, i.e.,

$$H_1^\perp = \{u \mid u \in H, \; (u, v) = 0, \; v \in H_1\}.$$

It is known that

$$H = H_1 \oplus H_1^\perp,$$

i.e., an element $u \in H$ is uniquely representable in the form

$$u = u_1 + u_1^{\perp},$$

where $u_1 \in H_1$, $u_1^{\perp} \in H_1^{\perp}$. The operator

$$Pu = u_1$$

is called an operator of *orthogonal projection* onto H_1 and has the following properties: P is bounded in H, $P^2 = P$, $P^* = P$, $\|P\| = 1$, $Pu = u$, $u \in H_1$.

2.1.3. Laurent series expansion of the resolvent. Let λ_0 be an eigenvalue of a closed operator A, acting in a Banach space E that is an isolated point of the spectrum $\sigma(A)$ and a pole of the finite order of the resolvent

$$R(\lambda, A) = (A - \lambda I)^{-1}.$$

Let us expand $R(\lambda, A)$ in some neighborhood of λ_0 into the Laurent series:

$$R(\lambda, A) = \sum_{n=0}^{\infty} A_n(\lambda - \lambda_0)^n + \frac{A_{-1}}{\lambda - \lambda_0} + \cdots + \frac{A_{-q}}{(\lambda - \lambda_0)^q}, \qquad (1.2)$$

where A_k, $k = -q, \ldots, \infty$, are bounded operators and series (1.2) converges by the operator norm in E.

Theorem 1.2. *Let λ_0 be an eigenvalue of the operator A and a pole of the finite order of the resolvent $R(\lambda, A)$. Then the expansion coefficients of (1.2) are correlated as follows:*

(1) $A_{n+1}(A - \lambda_0 I) \subset (A - \lambda_0 I)A_{n+1} = A_n$, $\quad n \geq 0$;

(2) $A_{-p}(A - \lambda_0 I) \subset (A - \lambda_0 I)A_{-p} = A_{-(p+1)}$, $\quad p = 1, \ldots, q - 1$;

(3) $A_0(A - \lambda_0 I) \subset (A - \lambda_0 I)A_0 = A_{-1} + I$;

(4) $A_{-q}(A - \lambda_0 I) \subset (A - \lambda_0 I)A_{-q} = 0$.

Proof. Act from the left in (1.2) by the operator $A - \lambda I$. Then

$$I = (A - \lambda I) \sum_{n=-q}^{\infty} A_n(\lambda - \lambda_0)^n$$

$$= [(A - \lambda_0 I) - (\lambda - \lambda_0)I] \sum_{n=-q}^{\infty} A_n(\lambda - \lambda_0)^n.$$

Multiplying this identity by $(\lambda - \lambda_0)^q$, differentiating k times and passing to the limit under $\lambda \to \lambda_0$ we obtain

$$(A - \lambda_0 I)A_n - A_{n-1} = 0, \qquad n \neq 0, \ n \neq -q,$$
$$(A - \lambda_0 I)A_0 - A_{-1} = I,$$
$$(A - \lambda_0 I)A_{-q} = 0.$$

So, correlations 1–4 are partially proved. Now, act from the right in (1.2) by the operator $A - \lambda I$. Then for any $u \in D(A)$ we have

$$u = \sum_{n=-q}^{\infty} (\lambda - \lambda_0)^n A_n (A - \lambda I)u$$
$$= \sum_{n=-q}^{\infty} (\lambda - \lambda_0)^n A_n [(A - \lambda_0 I) - (\lambda - \lambda_0)I]u.$$

So, for $u \in D(A)$

$$A_n(A - \lambda_0 I)u - A_{n-1}u = 0, \qquad n \neq 0, \ n \neq q,$$
$$A_0(A - \lambda_0 I)u - A_{-1}u = u,$$
$$A_{-q}(A - \lambda_0 I)u = 0. \ \blacksquare$$

From the residue theory it is known that in the expansion (1.2) the operator A_{-1} plays a special role. So, let us establish some properties of the operator A_{-1}, which will be necessary in the sequel.

Lemma 1.3. *Let λ_0 be an eigenvalue of the operator A and a finite-order pole of the resolvent $R(\lambda, A)$. Then the coefficients of the principal part of the expansion (1.2) satisfy the following correlations*

$$R(A_{-q+k}) \subset N^k, \qquad k = 0, \ldots, q - 1;$$
$$R(A_{-1}) = N;$$

and

$$A_{-1}u = -u, \qquad u \in N. \tag{1.3}$$

Proof. From Theorem 1.2 it follows that

$$(A - \lambda_0 I)^{k+1} A_{-q+k} = 0, \qquad k = 0, \ldots, q - 1,$$

from which, in turn, follows $R(A_{-q+k}) \subset N^k$, $k = 0, \ldots, q - 1$. In particular

$$R(A_{-1}) \subset N^{q-1} \subset N. \tag{1.4}$$

If u_k is an associated vector of the k-th rank, then property 3 of Theorem 1.2 implies that

$$(A_{-1} + I)u_k = A_0(A - \lambda_0 I)u_k = A_0 u_{k-1}. \tag{1.5}$$

Property 1 of Theorem 1.2 implies that

$$A_0 u_{k-1} = A_k(A - \lambda_0 I)^k u_{k-1} = 0.$$

From this and (1.5) follows (1.3). From (1.3) in turn, follows $N \subset R(A_{-1})$. From this and (1.4) follows $R(A_{-1}) = N$. ∎

Corollary 1.4. *Let λ_0 be an eigenvalue of the operator A and a finite-order pole of the resolvent $R(\lambda, A)$. Then the rank of the eigenvalue coincides with the pole order of the resolvent, i.e., $r = q$.*

Proof. Indeed, from Lemma 1.3 and (1.4) follows

$$R(A_{-1}) = N^{q-1} = N = N^{r-1}.$$

Hence, $q \geq r$. On the other hand, $A_{-q} = (A - \lambda_0 I)^{q-1} A_{-1} \neq 0$, since λ_0 is a pole of order q. Hence, there exists $v_0 \in E$ such that $(A - \lambda_0 I)^{q-1} A_{-1} v_0 \neq 0$. Hence $A_{-1} v_0 \notin N^{q-2}$. So, $N^{q-2} \neq R(A_{-1}) = N = N^{r-1}$ from which follows $r \geq q$. ∎

In the sequel other notations will be also used for the operator A_{-1}:

$$A_{-1} = A_{-1}(\lambda_0) = A_{-1}(\lambda_0, A).$$

2.1.4. Infinite products. Let $\{z_j\}_{j=1}^{\infty}$ be a sequence of complex numbers. If the sequence of numbers

$$P_n = \prod_{j=1}^{n}(1 + z_j)$$

converges to the finite non-zero number p, then the infinite product

$$\prod_{j=1}^{\infty}(1 + z_j) \tag{1.6}$$

is called converging, and the number p is taken for the value of this product.

Lemma 1.5. *If the series $\sum_{j=1}^{\infty} z_j$, $z_j \geq 0$ converges, then product (1.6) converges too.*

2.1.5. The Phragmén-Lindelöf principle. Many proofs of completeness theorems of linear operator root vectors are based on the Phragmén-Lindelöf theorem. The Phragmén-Lindelöf principle is founded on the classical maximum principle of analytic functions (if an analytic function does not exceed some constant C on the boundary of a bounded domain, then it does not exceed the constant C inside the domain either) and represents a generalization of this property on the domain going to infinity.

A ray with origin at point $a \in \mathbb{C}$ and direction φ is denoted by $\ell(a, \varphi)$.

An angle between rays $\ell(a, \varphi_1)$ and $\ell(a, \varphi_2)$ is denoted by $G(a, \varphi_1, \varphi_2)$.

A circle with center a and radius r is denoted by $S(a, r)$. So,

$$\ell(a, \varphi) = \{\lambda | \lambda \in \mathbb{C}, \ \lambda = a + re^{i\varphi}, \ r \geq 0\},$$
$$G(a, \varphi_1, \varphi_2) = \{\lambda | \lambda \in \mathbb{C}, \ \lambda = a + re^{i\varphi}, \ r \geq 0, \ \varphi \in [\varphi_1, \varphi_2]\},$$
$$S(a, r) = \{\lambda | \lambda \in \mathbb{C}, \ |\lambda - a| = r\}.$$

When directions of the rays are not essential, then $\ell(a, \varphi)$ is denoted by $\ell(a)$ and $G(a, \varphi, \varphi + \theta)$ is denoted by $G(a, \theta)$.

Theorem 1.6. *Let the following conditions be satisfied:*

(1) *for some $p \in (0, \infty)$ a function $F(\lambda)$ is holomorphic in the angle $G(a, \theta)$, where $\theta < \pi/p$;*

(2) *there exist circles $S(a, r_k)$, $k = 1, \ldots, \infty$, with radii r_k going to infinity such that*

$$|F(\lambda)| \leq Ce^{\omega|\lambda|^p}, \qquad \lambda \in S(a, r_k) \cap G(a, \theta);$$

(3) *the function $F(\lambda)$ on the sides of angle $G(a, \theta)$ does not increase faster than a polynomial of power n.*

Then the function $F(\lambda)$ in the angle $G(a, \theta)$ does not increase faster than a polynomial of power n.

Proof. It is sufficient to prove the theorem for the case when the angle $G(a, \theta)$ coincides with the angle $G(0, -\theta/2, \theta/2)$ and the circles $S(a, r_k)$ coincide with the circles $S(0, r_k)$. The general case is proved by applying the theorem to the function

$$\Phi(\mu) = F(a + \mu e^{i(\varphi + \theta/2)})$$

in the angle $G(0, -\theta/2, \theta/2)$.

Let us find a number q that satisfies the inequality $\theta < \pi/q < \pi/p$. Consider the branch of the function

$$f_\varepsilon(\lambda) = e^{-\varepsilon\lambda^q}, \qquad |\arg \lambda| \le \theta/2 < \pi/2p, \quad \varepsilon > 0,$$

which takes the real values under real λ. Let us construct the function

$$F_\varepsilon(\lambda) = F(\lambda)f_\varepsilon(\lambda).$$

Since $q\theta/2 < \pi/2$, then $\cos q\theta/2 > 0$ and on the sides of angle

$$G(0, -\theta/2, \theta/2)$$

we have

$$|F_\varepsilon(\lambda)| \le |F(re^{\pm i\theta/2})||f_\varepsilon(re^{\pm i\theta/2})|$$
$$\le C(1 + r^n)e^{-\varepsilon r^q \cos q\theta/2} \le C(1 + r^n).$$

Since $\cos q\theta/2 > 0$, then on the arc $\lambda = r_k e^{i\psi}$, $|\psi| \le \theta/2$, we have

$$|F_\varepsilon(\lambda)| \le Ce^{\omega r_k^p - \varepsilon r_k^q \cos q\psi} \le Ce^{\omega r_k^p - \varepsilon r_k^q \cos q\theta/2}.$$

Since $q > p$, then for any $\varepsilon > 0$

$$\lim_{k \to \infty} e^{\omega r_k^p - \varepsilon r_k^q \cos q\theta/2} = 0,$$

therefore, on the contour Γ_k formed by segments of two rays $\ell(0, -\theta/2)$ and $\ell(0, \theta/2)$ and by the arc of the circle of radius r_k, the following estimate holds:

$$|F_\varepsilon(\lambda)| \le C(1 + |\lambda|^n),$$

with constant C independent of ε. Then, by the maximum principle of modulus of the holomorphic function, the estimate $|F_\varepsilon(\lambda)| \le C(1 + |\lambda|^n)$ is valid inside the contour Γ_k as well, and therefore, in the angle $G(0, -\theta/2, \theta/2)$ too. So, in the angle $G(0, -\theta/2, \theta/2)$

$$|F(\lambda)| \le |F_\varepsilon(\lambda)||f_\varepsilon^{-1}(\lambda)| \le C(1 + |\lambda|^n)e^{\varepsilon\lambda^q}.$$

When $\varepsilon \to 0$ we obtain

$$|F(\lambda)| \le C(1 + |\lambda|^n), \quad \lambda \in G(0, -\theta/2, \theta/2). \quad \blacksquare$$

2.1.6. Polar representation of a bounded operator. Let A from a Hilbert space H into a Hilbert space H_1 be bounded. Then A^* from H_1 into H is bounded and for $u \in H$, $u_1 \in H_1$ we have

$$(Au, u_1)_{H_1} = (u, A^* u_1)_H.$$

Consequently, the operator A^*A in H is selfadjoint and non-negative. Hence, there exists a unique non-negative operator $T = (A^*A)^{1/2}$ in H. Obviously, for $u \in H$ we will obtain

$$\|Au\|_{H_1}^2 = (Au, Au)_{H_1} = (A^*Au, u)_H = (T^2 u, u)_H = \|Tu\|_H^2.$$

The operator U that associates the vector Tu with the vector Au, maps $R(T) \subset H$ onto $R(A) \subset H_1$ isometrically. So, under $u \in H$

$$UTu = Au,$$

where U from $R(T)$ onto $R(A)$ is isometric. The representation

$$A = UT$$

is called a *polar representation* of the operator A. Since U maps $R(T)$ onto $R(A)$ isometrically, then

$$r(T) = \dim R(T) = \dim R(A) = r(A).$$

2.1.7. s-numbers of a compact operator. The following items of 2.1 introduce some essential information and results from the vast material that is expounded, for example, in [19], [13].

Let A from a Hilbert space H into a Hilbert space H_1 be compact. Then $T = (A^*A)^{1/2}$ in H is compact and non-negative.

The eigenvalues of the operator T are called *s-numbers* of the operator A and are denoted by $s_j(A; H, H_1)$.

Let us numerate s-numbers in decreasing order, taking into account their multiplicities, so that

$$s_j(A; H, H_1) = \lambda_j(T), \qquad j = 1, \ldots, \infty.$$

If $r(A) < \infty$, then $s_j(A; H, H_1) = 0$ under $j = r(A) + 1, \ldots, \infty$. Obviously,

$$s_1(A; H, H_1) = \|A\|,$$

and for any scalar λ

$$s_j(\lambda A; H, H_1) = |\lambda| s_j(A; H, H_1).$$

If A in H is selfadjoint, then

$$s_j(A; H, H) = |\lambda_j(A)|.$$

2.1.8. Schmidt expansion for a compact operator. Let A from H into H_1 be compact and $A = UT$ be its polar representation.

Denote by $u_j \in H$, $j = 1, \ldots, r(T)$ an orthonormal system of eigenvectors of the operator T, complete in $R(T)$. Then

$$T = \sum_{j=1}^{r(T)} s_j(A; H, H_1)(\cdot, u_j)u_j, \tag{1.7}$$

where the series converges by the operator norm in H. Applying the operator U to both sides of (1.7), we obtain

$$A = \sum_{j=1}^{r(A)} s_j(A; H, H_1)(\cdot, u_j)Uu_j.$$

Since $u_j \in R(T)$, then the system $v_j = Uu_j$, $j = 1, \ldots, r(A)$ is orthonormal in H_1. Thus, if A from H into H_1 is compact, then the *Schmidt expansion*

$$A = \sum_{j=1}^{r(A)} s_j(A; H, H_1)(\cdot, u_j)v_j, \tag{1.8}$$

is true, where u_j and v_j are some systems of vectors, orthonormal in H and H_1 respectively, and the series converges by the norm in $B(H, H_1)$.

2.1.9. Elementary properties of s-numbers.

Property 1.7. *If A from H into H_1 is compact, then*

$$s_j(A; H, H_1) = s_j(A^*; H_1, H). \tag{1.9}$$

Proof. From (1.8) follows the Schmidt expansion for the conjugate operator

$$A^* = \sum_{j=1}^{r(A)} s_j(A; H, H_1)(\cdot, v_j)_{H_1}u_j. \tag{1.10}$$

From (1.8) and (1.10) we obtain

$$A^*Au_j = s_j^2(A; H, H_1)u_j,$$
$$AA^*v_j = s_j^2(A; H, H_1)v_j.$$

From this (1.9) follows. ∎

Property 1.8. *If A from H into H_1 is compact and B from H_1 into H is bounded, then*

$$s_j(BA; H, H) \leq \|B\|_{B(H_1,H)} s_j(A; H, H_1), \tag{1.11}$$

$$s_j(AB; H_1, H_1) \leq \|B\|_{B(H_1,H)} s_j(A; H, H_1), \tag{1.12}$$

Proof. For $u \in H$ we have

$$(A^* B^* B A u, u) = \|BAu\|^2 \leq \|B\|^2_{B(H_1,H)} \|Au\|^2_{H_1} \leq \|B\|^2_{B(H_1,H)}(A^* A u, u),$$

i.e., $A^* B^* B A \leq \|B\|^2_{B(H_1,H)} A^* A$. Hence,

$$\lambda_j(A^* B^* B A) \leq \|B\|^2_{B(H_1,H)} \lambda_j(A^* A). \tag{1.13}$$

On the other hand, by the definition

$$s_j^2(BA; H, H) = \lambda_j(A^* B^* B A), \qquad s_j^2(A; H, H_1) = \lambda_j(A^* A). \tag{1.14}$$

From (1.13) and (1.14) follows (1.11).

Since, by virtue of (1.9)

$$s_j(AB; H_1, H_1) = s_j(B^* A^*; H_1, H_1)$$

and by the above proved

$$s_j(B^* A^*; H_1, H_1) \leq \|B^*\|_{B(H,H_1)} s_j(A^*; H_1, H)$$
$$= \|B\|_{B(H_1,H)} s_j(A; H, H_1),$$

then correlations (1.12) are also proved. ∎

Property 1.9. *[19, corollary II.2.2]. If operators A and B in H are compact, then*

$$s_{m+n-1}(A + B) \leq s_m(A) + s_n(B),$$
$$s_{m+n-1}(AB) \leq s_m(A) s_n(B).$$

2.1.10. Weyl's inequality: a correlation between the eigenvalues and the s-numbers. For brevity, s-numbers of the compact operator A in H are denoted by s_j, i.e., $s_j = s_j(A) = s_j(A; H, H)$.

Lemma 1.10. *[19, lemma II.3.1]. Let A in H be compact. Then for any system of the elements u_k, $k = 1, \ldots, n$, the following correlation is valid:*

$$\det((Au_j, Au_k))_1^n \leq s_1^2 s_2^2 \cdots s_n^2 \det((u_j, u_k))_1^n.$$

The sum of algebraic multiplicities of all non-zero eigenvalues of the compact operator A in H will be denoted by $\gamma(A)$ and the closure of the linear span of all root vectors, corresponding to the non-zero eigenvalues of such operator A, will be denoted by $(spA)_{-0}$ (span). Hence $\dim(spA)_{-0} = \gamma(A)$.

Lemma 1.11. *[19, lemma I.4.1]. Let A in H be compact. Then there exists an orthonormal basis $\{u_j\}_1^{\gamma(A)}$ of the subspace $(spA)_{-0}$ such that*

$$Au_j = a_{j1}u_1 + a_{j2}u_2 + \cdots + a_{jj}u_j,$$

moreover,

$$a_{jj} = (Au_j, u_j) = \lambda_j(A).$$

Corollary 1.12. *[19, remark I.4.1]. Let A be an operator in the n-dimensional Hilbert space H_n. Then there exists an orthonormal basis $\{e_j\}_1^n$ of the space H_n such that*

$$Ae_j = a_{j1}e_1 + a_{j2}e_2 + \cdots + a_{jj}e_j,$$
$$a_{jj} = (Ae_j, e_j) = \lambda_j(A).$$

Lemma 1.13. *[19, lemma II.3.3]. Let A in H be compact. Then*

$$|\lambda_1(A)\lambda_2(A)\cdots\lambda_n(A)| \leq s_1(A)s_2(A)\cdots s_n(A), \quad n = 1, \ldots, \gamma(A). \tag{1.15}$$

Lemma 1.14. *[19, corollary II.3.1]. Let A in H be compact. Then*

$$\sum_{j=1}^n |\lambda_j(A)|^p \leq \sum_{j=1}^n s_j^p(A), \quad p > 0, \ n = 1, \ldots, \gamma(A).$$

2.1.11. Classes $\sigma_p(H, H_1)$. By $\sigma_p(H, H_1)$, $p > 0$, we denote the space of operators A that act from H into H_1 compactly and for which

$$\|A\|_p^p = \sum_{j=1}^\infty s_j^p(A; H, H_1) < \infty.$$

For brevity denote $\sigma_p(H, H) = \sigma_p(H)$.

Let us show that $\sigma_p(H)$ under $p \geq 1$ is a linear manifold. Let $A \in \sigma_p(H)$, $B \in \sigma_p(H)$. Represent the natural number m uniquely in the form

$$m = 2k + \beta, \qquad k = 0, \ldots, \infty, \quad \beta = 0, 1.$$

By virtue of Property 1.9 we have

$$s_{2k}(A + B) \leq s_{k+1}(A) + s_k(B), \qquad k = 1, \ldots, \infty,$$

and

$$s_{2k+1}(A + B) \leq s_{k+1}(A) + s_{k+1}(B), \qquad k = 0, \ldots, \infty.$$

Hence,

$$\|A + B\|_p = \left(\sum_{j=1}^{\infty} s_j^p (A + B) \right)^{1/p}$$

$$\leq \left(2 \sum_{j=1}^{\infty} (s_j(A) + s_j(B))^p \right)^{1/p} \leq 2^{1/p}(\|A\|_p + \|B\|_p).$$

So, $A + B \in \sigma_p(H)$. From the property $s_m(\lambda A) = |\lambda| s_m(A)$ it follows that $\lambda A \in \sigma_p(H)$, if λ is a scalar and the operator $A \in \sigma_p(H)$. ∎

Let us state some properties of classes $\sigma_p(H)$.

Property 1.15. *[19, III.7.2.1°]. Let $0 < p_1 < p_2 < \infty$ and $A \in \sigma_{p_1}(H)$, then $A \in \sigma_{p_2}(H)$ and*

$$\|A\|_{p_2} \leq \|A\|_{p_1}.$$

Property 1.16. *If operators $A_j \in \sigma_{p_j}(H)$, $p_j > 0$, $j = 1, \ldots, n$, then the operator $A_1 A_2 \cdots A_n \in \sigma_p(H)$, where $p^{-1} = \sum_{j=1}^{n} p_j^{-1}$ and*

$$\|A_1 A_2 \cdots A_n\|_p \leq n^{1/p} \|A_1\|_{p_1} \|A_2\|_{p_2} \cdots \|A_n\|_{p_n}.$$

Proof. Let us represent the natural number m uniquely as

$$m = nk + \beta, \qquad k = 0, \ldots, \infty, \quad \beta = 1, \ldots, n.$$

By virtue of Property 1.9, we have

$$s_{nk+\beta}(A_1 A_2 \cdots A_n) \leq s_{k+1}(A_1) s_{(n-1)k+\beta}(A_2 \cdots A_n) \leq \cdots$$

$$\leq s_{k+1}(A_1) s_{k+1}(A_2) \cdots s_{k+\beta}(A_n),$$

$$k = 0, \cdots, \infty, \qquad \beta = 1, \cdots, n.$$

then applying the generalized Holder inequality, we obtain

$$
\|A_1 A_2 \cdots A_n\|_p = \left(\sum_{k=0}^{\infty} \sum_{\beta=1}^{n} s_{nk+\beta}^p (A_1 A_2 \cdots A_n) \right)^{1/p}
$$

$$
\leq \left(\sum_{k=0}^{\infty} \sum_{\beta=1}^{n} s_{k+1}^p (A_1) s_{k+1}^p (A_2) \cdots s_{k+\beta}^p (A_n) \right)^{1/p}
$$

$$
\leq \left(n \sum_{k=1}^{\infty} s_k^p (A_1) s_k^p (A_2) \cdots s_k^p (A_n) \right)^{1/p}
$$

$$
\leq n^{1/p} \left(\sum_{k=1}^{\infty} s_k^{p_1} (A_1) \right)^{1/p_1} \cdots \left(\sum_{k=1}^{\infty} s_k^{p_n} (A_n) \right)^{1/p_n}
$$

$$
\leq n^{1/p} \|A_1\|_{p_1} \|A_2\|_{p_2} \cdots \|A_n\|_{p_n}. \quad \blacksquare
$$

In fact, this inequality holds without the multiplier $n^{1/p}$ [19, Ch.III, §7.2.2].

Property 1.17. *If the operator* $A \in \sigma_p(H)$, $p > 0$, *then for any natural* n *the operator* $A^n \in \sigma_{p/n}(H)$ *and moreover*

$$
\|A^n\|_{p/n} \leq n^{n/p} \|A\|_p^n.
$$

Proof. It follows from Property 1.16 under $A_1 = \cdots = A_n = A$. $\blacksquare$

Property 1.18. *[19, theorem III.7.1]. If the operator* $A \in \sigma_p(H)$, $p > 0$, *and an operator* B *in* H *is bounded, then* $AB \in \sigma_p(H)$ *and* $BA \in \sigma_p(H)$.

2.1.12. A growth estimate for the resolvent of a Volterra operator from the class $\sigma_p(H)$. First, let us prove an estimate for an invertible operator, acting in a finite-dimensional Hilbert space.

Lemma 1.19. *Let an operator* C *in the* n-*dimensional Hilbert space* H_n *be invertible. Then*

$$
\|C^{-1}\| \leq \frac{\displaystyle\prod_{j=1}^{n-1} s_j(C)}{\left| \displaystyle\prod_{j=1}^{n} \lambda_j(C) \right|}.
\tag{1.16}
$$

Proof. Consider a polar representation of the operator C, i.e., the representation

$$
C = UT,
$$

where $T = (C^*C)^{1/2}$, and U is a unitary operator in H_n. Then from

$$\|C^{-1}\| \leq \|T^{-1}\| \leq \lambda_n^{-1}(T) = s_n^{-1}(C)$$

and the Weyl inequality (1.15) follows the inequality

$$\left| \prod_{j=1}^{n} \lambda_j(C) \right| \|C^{-1}\| \leq \prod_{j=1}^{n} s_j(C) s_n^{-1}(C) = \prod_{j=1}^{n-1} s_j(C),$$

i.e., (1.16). ∎

Let C be an operator in the n-dimensional Hilbert space H_n. Consider the Schmidt expansion

$$C = \sum_{j=1}^{n} s_j(C)(\cdot, u_j)v_j,$$

where $\{u_j\}_1^n$, $\{v_j\}_1^n$ are some orthonormal bases in H_n.

Lemma 1.20. *[19, Ch.III, §8.5)]. Let C be an operator in the n-dimensional Hilbert space H_n. Then the following equality holds:*

$$\sum_{j=1}^{n} \lambda_j(C) = \sum_{j=1}^{n} (Cv_j, v_j) = \sum_{j=1}^{n} s_j(C)(v_j, u_j).$$

Lemma 1.21. *[19, Ch.I, §1.3°]. Let the following conditions be satisfied:*

(1) *the operator C in H is bounded;*
(2) *F is a closed set of the complex plane, belonging to $\rho(C)$.*

Then there exists $\delta > 0$ such that for all bounded operators B satisfying the condition $\|B - C\| < \delta$, the set F also belongs to $\rho(B)$.

The operator A in H is a *Volterra operator*, if it is compact in H and does not have any non-zero eigenvalues.

Lemma 1.22. *Let the following conditions be satisfied:*

(1) *operators B_n, $n = 1, \ldots, \infty$ in H are compact;*
(2) *an operator B in H is a Volterra operator;*
(3) *$\lim_{n \to \infty} \|B_n - B\| = 0$.*

Then

$$\lim_{n \to \infty} \lambda_1(B_n) = 0,$$

where $\lambda_1(B_n)$ is the largest, in modulus, eigenvalue of the operator B_n.

Proof. Under any $\varepsilon > 0$ the closed set $F = \{\lambda | \lambda \in \mathbb{C}, \ |\lambda| \geq \varepsilon\}$ belongs to $\rho(B)$. Then, by Lemma 1.21 there exists a number $n(\varepsilon)$ such that under $n > n(\varepsilon)$ all eigenvalues of the operators B_n are in the circle $|\lambda| < \varepsilon$. ∎

Theorem 1.23. [13, Ch.XI, §9.27]. Let an operator B in H be a Volterra operator and under some $p > 0$ $B \in \sigma_p(H)$. Then for some $\omega > 0$ the following inequality holds:

$$\|(I - \lambda B)^{-1}\| \leq C e^{\omega |\lambda|^p}, \quad \lambda \in \mathbb{C}.$$

2.1.13. Completeness of root vectors of an operator from the class $\sigma_p(H)$. Here we prove a theorem from which follows the main theorem of the next paragraph 2.2. First, let us derive some well-known results of functional analysis.

Lemma 1.24. [28, theorem III.5.30]. Let an operator B in H be invertible and the domain of definition $D(B)$ be dense in H. Then the operator B^* is invertible and

$$(B^*)^{-1} = (B^{-1})^*.$$

Corollary 1.25. If an operator B in H is closed and the domain of definition $D(B)$ is dense in H, then $\rho(B^*) = \overline{\rho(B)}$ and

$$(R(\lambda, B))^* = R(\overline{\lambda}, B^*), \quad \lambda \in \rho(B).$$

Lemma 1.26. Let an operator B in a Banach space E be compact and λ_0 be its non-zero point of the spectrum. Then λ_0 is the eigenvalue of the finite algebraic multiplicity of the operator B and the finite-order pole of the resolvent $R(\lambda, B)$.

Proof. For any $\lambda \in \rho(B)$ we have

$$R(\lambda, B)(B - \lambda I) = I.$$

Hence

$$R(\lambda, B) = -\lambda^{-1}(I - R(\lambda, B)B).$$

Since the non-zero spectrum of the compact operator B consists of isolated eigenvalues [31, Ch.4, §6.3], then under small enough $\varepsilon > 0$ inside the circle $|\lambda - \lambda_0| = \varepsilon$

there are no eigenvalues of the operator B other than λ_0 (zero is also outside the circle $|\lambda - \lambda_0| = \varepsilon$). Consider the operator

$$P = -\frac{1}{2\pi i} \int\limits_{|\lambda-\lambda_0|=\varepsilon} R(\lambda, B) \, d\lambda$$

$$= -\frac{1}{2\pi i} \int\limits_{|\lambda-\lambda_0|=\varepsilon} \frac{d\lambda}{\lambda} I - \frac{1}{2\pi i} \int\limits_{|\lambda-\lambda_0|=\varepsilon} \frac{R(\lambda, B)B}{\lambda} \, d\lambda$$

$$= -\frac{1}{2\pi i} \int\limits_{|\lambda-\lambda_0|=\varepsilon} \frac{R(\lambda, B)B}{\lambda} \, d\lambda. \tag{1.17}$$

The integral sums, corresponding to (1.17), are compact operators in E. Since the limit of compact operators in the operator norm is compact, then the operator P in E is compact. On the other hand, P is a projector, commutative with B. Indeed, choosing $\varepsilon_1 > \varepsilon$ so that

$$P^2 = \frac{1}{(2\pi i)^2} \int\limits_{|\lambda-\lambda_0|=\varepsilon} \int\limits_{|\mu-\lambda_0|=\varepsilon_1} R(\lambda, B)R(\mu, B) \, d\lambda d\mu$$

$$= \frac{1}{(2\pi i)^2} \int\limits_{|\lambda-\lambda_0|=\varepsilon} \int\limits_{|\mu-\lambda_0|=\varepsilon_1} \frac{R(\lambda, B) - R(\mu, B)}{\lambda - \mu} \, d\lambda d\mu$$

$$= \frac{1}{(2\pi i)^2} \int\limits_{|\lambda-\lambda_0|=\varepsilon} R(\lambda, B) \, d\lambda \int\limits_{|\mu-\lambda_0|=\varepsilon_1} \frac{d\mu}{\lambda - \mu}$$

$$- \frac{1}{(2\pi i)^2} \int\limits_{|\mu-\lambda_0|=\varepsilon_1} R(\mu, B) \, d\mu \int\limits_{|\lambda-\lambda_0|=\varepsilon} \frac{d\lambda}{\lambda - \mu},$$

and observing that

$$\int\limits_{|\mu-\lambda_0|=\varepsilon_1} \frac{d\mu}{\lambda - \mu} = -2\pi i, \qquad \int\limits_{|\lambda-\lambda_0|=\varepsilon} \frac{d\lambda}{\lambda - \mu} = 0,$$

we obtain

$$P^2 = -\frac{1}{2\pi i} \int\limits_{|\lambda-\lambda_0|=\varepsilon} R(\lambda, B) \, d\lambda = P.$$

So, P is a projector. Commutativity of P with B follows from commutativity of B with $R(\lambda, B)$. Since a compact projector is finite-dimensional, then P is a finite-dimensional projector. So, $E = E_1 + E_2$, where $E_1 = PE$ is a finite-dimensional invariant subspace of the operator B, and $E_2 = (I - P)E$ is an infinite-dimensional invariant subspace of the operator B.

Denote by B_k the restriction of the operator B on E_k. Let us show that $\sigma(B_1) = \{\lambda_0\}$, $\sigma(B_2) = \sigma(B) \setminus \{\lambda_0\}$. Indeed, from the obvious correlation

$$(B - \mu I)R(\lambda, B) = I + (\lambda - \mu)R(\lambda, B)$$

follows

$$(B - \mu I)\frac{1}{2\pi i} \int_{|\lambda - \lambda_0| = \varepsilon} \frac{R(\lambda, B)}{\lambda - \mu} \, \mathrm{d}\lambda = \frac{1}{2\pi i} \int_{|\lambda - \lambda_0| = \varepsilon} \frac{\mathrm{d}\lambda}{\lambda - \mu} I$$

$$+ \frac{1}{2\pi i} \int_{|\lambda - \lambda_0| = \varepsilon} R(\lambda, B) \, \mathrm{d}\lambda$$

$$= \begin{cases} 0 \cdot I - P = -P & \text{if } \mu \text{ is outside the circle } |\lambda - \lambda_0| = \varepsilon, \\ 1 \cdot I - P = I - P & \text{if } \mu \text{ is inside the circle } |\lambda - \lambda_0| = \varepsilon. \end{cases}$$

Since $P = I$ on E_1 and $P = 0$ on E_2, then the operator $R(\mu, B_1)$ exists and coincides with the restriction of the operator $\frac{1}{2\pi i} \int_{|\lambda - \lambda_0| = \varepsilon} \frac{R(\lambda, B)}{\lambda - \mu} \mathrm{d}\lambda$ on E_1, if μ is outside the circle $|\lambda - \lambda_0| = \varepsilon$; and the operator $R(\mu, B_2)$ exists and coincides with the restriction of the operator $\frac{1}{2\pi i} \int_{|\lambda - \lambda_0| = \varepsilon} \frac{R(\lambda, B)}{\lambda - \mu} \mathrm{d}\lambda$ on E_2, if μ is inside the circle $|\lambda - \lambda_0| = \varepsilon$.

From $\sigma(B_1) = \{\lambda_0\}$ and the Jordan representation of the operator B_1 follows

$$(B_1 - \lambda_0 I)^\gamma = 0 \tag{1.18}$$

where $\gamma = \dim E_1$. So, $(B - \lambda_0 I)^\gamma E_1 = 0$ and E_1 coincides with the root subspace $N_{\lambda_0}(B)$. Otherwise, there exists $u \in N_{\lambda_0}(B)$ such that $u \notin E_1$. Hence, under some n we have $(B - \lambda_0 I)^n u = 0$, but $u_2 = (I - P)u \neq 0$. For the vector $u_2 = (I - P)u \in E_2$, in turn, we have $(B - \lambda_0 I)^n u_2 = 0$. Hence, by virtue of the invertibility of $B - \lambda_0 I$ in E_2 it follows that $u_2 = 0$. Thus, the contradiction obtained shows that $N = E_1$. So, the algebraic multiplicity of λ_0 is equal to $\gamma = \dim E_1 < \infty$.

From (1.18) it follows that there exists the smallest integer r such that $(B_1 - \lambda_0 I)^r = 0$. Assuming $A_1 = B_1 - \lambda_0 I$ we obtain

$$-(\lambda - \lambda_0)^r I = A_1^r - (\lambda - \lambda_0)^r I = (B_1 - \lambda I)[(\lambda - \lambda_0)^{r-1} I$$
$$+ (\lambda - \lambda_0)^{r-2} A_1 + \cdots + A_1^{r-1}].$$

Hence,

$$-(B_1 - \lambda I)^{-1} = (\lambda - \lambda_0)^{-1} I + \sum_{j=1}^{r-1} (\lambda - \lambda_0)^{-j-1} A_1^j, \quad \lambda \neq \lambda_0.$$

Then, for $0 < |\lambda - \lambda_0| < \varepsilon$ we have

$$(B - \lambda I)^{-1} = (B_1 - \lambda I)^{-1} P + (B_2 - \lambda I)^{-1} (I - P)$$
$$= (\lambda - \lambda_0)^{-1} P - (\lambda - \lambda_0)^{-2} A_1 P + \cdots + (\lambda - \lambda_0)^{-r} A_1^{r-1} P$$
$$+ (B_2 - \lambda I)^{-1} (I - P).$$

So, λ_0 is a pole of order r of the resolvent $R(\lambda, B)$. ∎

Let λ_0 be a non-zero eigenvalue of the compact operator B in H. Let us expand the resolvent $R(\lambda, B)$ in some neighborhood of λ_0 in the Laurent series

$$R(\lambda, B) = \sum_{n=0}^{\infty} B_n (\lambda - \lambda_0)^n + \frac{B_{-1}}{\lambda - \lambda_0} + \cdots + \frac{B_{-r}}{(\lambda - \lambda_0)^r}, \qquad (1.19)$$

where B_k, $k = -r, \ldots, \infty$ are bounded operators and series (1.19) converges by the operator norm in H. For the operator $B_{-1} = B_{-1}(\lambda_0, B)$ the following will be proved.

Lemma 1.27. *Let the operator B in H be compact and λ_0 be its non-zero eigenvalue. Then*

$$(B_{-1}(\lambda_0, B))^* = B_{-1}(\overline{\lambda_0}, B^*).$$

Proof. From (1.19), by virtue of the residue theory, follows

$$B_{-1}(\lambda_0, B) = \frac{1}{2\pi i} \int\limits_{|\lambda - \lambda_0| = \varepsilon} R(\lambda, B) \, d\lambda,$$

where ε is chosen so small that inside the circle $|\lambda - \lambda_0| = \varepsilon$ there are no eigenvalues, other than λ_0. From Corollary 1.25 and the compactness of B in H it follows that $\overline{\lambda_0}$ is the eigenvalue of B^* and

$$(B_{-1}(\lambda_0, B))^* = \left(\frac{1}{2\pi i} \int\limits_{|\lambda - \lambda_0| = \varepsilon} R(\lambda, B) \, d\lambda \right)^*$$

$$= \frac{1}{2\pi i} \int\limits_{|s - \overline{\lambda_0}| = \varepsilon} R(s, B^*) \, ds = B_{-1}(\overline{\lambda_0}, B^*). \quad ∎$$

As above, by $(\mathrm{sp}B)_{-0}$ we denote the closure of the linear span of all root vectors, corresponfing to the non-zero eigenvalues of the operator B. By $(\mathrm{sp}B)_{-0}^{\perp}$ we denote the orthogonal complement to $(\mathrm{sp}B)_{-0}$.

Lemma 1.28. *Let an operator B in H be compact. Then $(\mathrm{sp}B)_{-0}^{\perp}$ is an invariant subspace of the operator B^* and the restriction of B^* on $(\mathrm{sp}B)_{-0}^{\perp}$ is a Volterra operator.*

Proof. If $u \in (\mathrm{sp}B)_{-0}^{\perp}$, then under all $v \in (\mathrm{sp}B)_{-0}$ we have $(u, v) = 0$. Since from $v \in (\mathrm{sp}B)_{-0}$ it follows that $Bv \in (\mathrm{sp}B)_{-0}^{\perp}$, then

$$(B^*u, v) = (u, Bv) = 0, \qquad v \in (\mathrm{sp}B)_{-0}.$$

Hence, $B^*u \in (\mathrm{sp}B)_{-0}^{\perp}$.

Denote the restriction of B^* onto $(\mathrm{sp}B)_{-0}^{\perp}$ by B_1^*. Suppose B_1^* is not a Volterra operator. Then there exists a non-zero number μ and a non-zero vector u_0 in $(\mathrm{sp}B)_{-0}^{\perp}$ such that

$$B_1^*u_0 = \mu u_0.$$

By Lemma 1.27 $B_{-1}(\mu, B^*) = (B_{-1}(\overline{\mu}, B))^*$. By Lemma 1.3 $B_{-1}(\overline{\mu}, B)H \subset (\mathrm{sp}B)_{-0}$. Hence,

$$((\mathrm{sp}B)_{-0}^{\perp}, B_{-1}(\overline{\mu}, B)H) = 0.$$

So,

$$(B_{-1}(\mu, B^*)(\mathrm{sp}B)_{-0}^{\perp}, H) = 0. \tag{1.20}$$

On the other hand, $u_0 \in (\mathrm{sp}B)_{-0}^{\perp}$ and is an eigenvector of the operator B^*, corresponding to the eigenvalue μ. By Lemma 1.3 $u_0 \in B_{-1}(\mu, B^*)(\mathrm{sp}B)_{-0}^{\perp}$. Hence, $B_{-1}(\mu, B^*)(\mathrm{sp}B)_{-0}^{\perp} \neq \{0\}$, and this contradicts (1.20). ∎

Let E be a Banch space and $\{u_k\}_1^{\infty}$ be a system of elements from E. The closure of the linear span of the system $\{u_k\}_1^{\infty}$ will be denoted by $\mathrm{sp}\{u_k\}$. The system $\{u_k\}_1^{\infty}$ is called *complete* in the set F of the space E, if $F \subset \mathrm{sp}\{u_k\}$. The system $\{u_k\}$ is called *complete* in the space E, if $\mathrm{sp}\{u_k\} = E$.

If a functional $u' \in E'$ orthogonal to $\mathrm{sp}\{u_k\}$ is orthogonal to the set F as well, then the system $\{u_k\}$ is complete in the set F of the space E. Indeed, if there is no completeness, then there exists $u \in F$ such that $u \notin \mathrm{sp}\{u_k\}$. Then by the Hahn-Banach theorem, there exists $u' \in E'$ such that $< \mathrm{sp}\{u_k\}, u' > = 0$ and $< u, u' > = 1$, and this contradicts the assumption.

If B is an operator in E, then $(\mathrm{sp}B)_{-0} = \mathrm{sp}\{u_k$, where u_k is a root vector, corresponding to the non-zero eigenvalue of the operator $B\}$.

Denote

$$\mathrm{sp}B = \mathrm{sp}\{u_k, \text{ where } u_k \text{ is a root vector of the operator } B\}.$$

Theorem 1.29. *Let the following conditions be satisfied:*

(1) $B \in \sigma_p(H)$ *for some* $p > 0$;

(2) *there exist rays* $\ell_k(a)$ *with angles between the neighboring rays less than* π/p *and an integer* $n \geq -1$ *such that*

$$\|(I - \lambda B)^{-1}\| \leq C|\lambda|^n, \qquad \lambda \in \ell_k(a), \ |\lambda| \to \infty.$$

Then a system of root vectors, corresponding to the non-zero eigenvalues of the operator B, *is complete in the set* $R(B^{n+1})$ *of the space* H.

Proof. Let a vector u be orthogonal to $(\mathrm{sp}B)_{-0}$, i.e., $((\mathrm{sp}B)_{-0}, u) = 0$. Let $v \in H$. Consider the following function

$$F(\lambda) = ((I - \lambda B)^{-1}v, u), \qquad \lambda \neq \lambda_k^{-1} = \lambda_k^{-1}(B).$$

Show that points $\lambda_k^{-1}(B)$ are removable singular points of $F(\lambda)$. By (1.19) and the equality $\lambda^{-1} - \lambda_k = -\lambda^{-1}\lambda_k(\lambda - \lambda_k^{-1})$, in some neighborhood of λ_k^{-1} we have

$$F(\lambda) = -\lambda^{-1}((B - \lambda^{-1}I)^{-1}v, u) = \lambda^{-1}\sum_{n=-q_k}^{\infty}(\lambda^{-1} - \lambda_k)^n(B_n v, u)$$

$$= \sum_{n=-q_k}^{\infty}(-1)^{n+1}\lambda^{-n-1}\lambda_k^n(\lambda - \lambda_k^{-1})^n(B_n v, u).$$

By Lemma 1.3 we have $B_n v \in N_{\lambda_k} \subset (\mathrm{sp}B)_{-0}$, $n = -q_k, \ldots, -1$. Hence, the main part of the Laurent series expansion for $F(\lambda)$ at the point λ_k^{-1} is equal to zero.

Denote the restriction of B^* onto $(\mathrm{sp}B)_{-0}^{\perp}$ by B_1^*. Since $u \in (\mathrm{sp}B)_{-0}^{\perp}$, then

$$F(\lambda) = (v, (I - \bar{\lambda}B^*)^{-1}u) = (v, (I - \bar{\lambda}B_1^*)^{-1}u).$$

Denote the operator of the orthogonal projection onto $(\mathrm{sp}B)_{-0}^{\perp}$ by Q. Then on $(\mathrm{sp}B)_{-0}^{\perp}$ we have $B_1^* = B^*Q$ and

$$F(\lambda) = (v, (I - \bar{\lambda}B^*Q)^{-1}u).$$

By Lemma 1.28 B_1^* is a Volterra operator, hence the B^*Q is a Volterra operator in H. By condition 1 and Properties 1.7, 1.18 we have $B^*Q \in \sigma_p(H)$. Then, by Theorem 1.23

$$|F(\lambda)| \leq \|v\| \, \|(I - \bar{\lambda}B^*Q)^{-1}u\| \leq C\|u\| \, \|v\|e^{\omega|\lambda|^p}, \qquad \lambda \in \mathbb{C}.$$

From condition 2 follows the estimate

$$|F(\lambda)| \leq C\|u\| \, \|v\| \, |\lambda|^n, \qquad \lambda \in \ell_k(a), \, |\lambda| \to \infty.$$

Applying Theorem 1.6 in each of the angles less than π/p with a vertex at a, into which the compex plane is divided be rays $\ell_k(a)$, we establish that the function $F(\lambda)$ over the whole plane increases not faster than a polynomial of power n. Then, by the Liouville theorem, we obtain that under $n \geq 0$

$$F(\lambda) = \alpha_0 + \alpha_1\lambda + \cdots + \alpha_n\lambda^n, \tag{1.21}$$

and under $n = -1$ $\quad F(\lambda) = 0$.

On the other hand, under $|\lambda| \, \|B\| < 1$ we have

$$F(\lambda) = (v, u) + \lambda(Bv, u) + \cdots + \lambda^n(B^nv, u) + \cdots.$$

From this and (1.21) follows

$$(B^{n+1}v, u) = 0. \quad \blacksquare$$

2.2. Completeness of root vectors of an unbounded operator

Let an operator A in a Banach space E be closed. The spectrum $\sigma(A)$ of the operator A is called *discrete*, if $\sigma(A)$ consists of isolated eigenvalues with finite algebraic multiplicities and infinity is the only limit point of $\sigma(A)$.

Obviously, infinity is the unique limit point of the set $\sigma(A)$. It is well known that if A has a compact resolvent[2] then, generally speaking, the spectrum of the operator A is not discrete. In this case, by virtue of Lemma 1.26, the spectrum consists of isolated eigenvalues with finite algebraic multiplicities, but it is possible that infinity is not a limit point of $\sigma(A)$. The case when $\sigma(A) = \{\infty\}$ is also possible.

---

[2] From the Hilbert identity $R(\lambda, A) - R(\lambda_0, A) = (\lambda - \lambda_0)R(\lambda, A)R(\lambda_0, A)$ it follows that if for one $\lambda_0 \in \rho(A)$ the operator $R(\lambda_0, A)$ is compact, then under any $\lambda \in \rho(A)$ the operattor $R(\lambda, A)$ is compact.

2.2.1. Completeness of root vectors of an operator with a compact resolvent. Before we pass on to the proof of the main theorem on completeness of root vectors of an unbounded operator, let us formulate and prove auxiliary theorems both on completeness and decomposition.

Theorem 2.1. *Let the following conditions be satisfied:*

(1) *an operator A in a Banach space E has a compact resolvent;*

(2) *there exist circles $S(a, r_k)$, $k = 1, \ldots, \infty$, with radii r_k going to ∞ and numbers $\omega \geq 0$, $p > 0$ such that*

$$\|R(\lambda, A)\| \leq Ce^{\omega|\lambda|^p}, \qquad |\lambda - a| = r_k;$$

(3) *there exist rays rays $\ell_k(A)$ with angles between the neighboring rays less than π/p and an integer $n \geq -1$ such that*

$$\|R(\lambda, A)\| \leq C|\lambda|^n, \qquad \lambda \in \ell_k(a), \ |\lambda| \to \infty.$$

Then the spectrum of the operator A is discrete and the system of root vectors of the operator A is complete in the set $D(A^{n+2})$ of the space E.

Proof. Let $\lambda_0 \in \rho(A)$. Then from the identity

$$A - \lambda I = (A - \lambda_0 I) - (\lambda - \lambda_0)I$$
$$= (\lambda - \lambda_0)(A - \lambda_0 I)[(\lambda - \lambda_0)^{-1}I - R(\lambda_0, A)]$$

and condition 1, by virtue of Lemma 1.26, it follows that the spectrum of the operator A consists of isolated eigenvalues $\lambda_k(A) = \lambda_k^{-1}(R(\lambda_0, A)) + \lambda_0$ with finite algebraic multiplicities. Since under $\lambda \neq \lambda_k^{-1}(R(\lambda_0, A)) + \lambda_0$ we have

$$R(\lambda, A) = -(\lambda - \lambda_0)^{-1}R(\lambda_0, A)[R(\lambda_0, A) - (\lambda - \lambda_0)^{-1}I]^{-1},$$

then, by virtue of the same Lemma 1.26, the eigenvalues $\lambda_k(A)$ are finite-order poles of the resolvent $R(\lambda, A)$.

Let $u \in E$. Then by Lemma 1.3, in some neighborhood of $\lambda_k = \lambda_k(A)$ the Laurent expansion holds:

$$R(\lambda, A)u = \sum_{n=0}^{\infty} (\lambda - \lambda_k)^n A_n u + \frac{A_{-1}u}{\lambda - \lambda_k} + \cdots + \frac{A_{-q_k}u}{(\lambda - \lambda_k)^{q_k}}, \qquad (2.1)$$

where $A_{-j}u \in N_{\lambda_k}^{q_k - j}$, $j = 1, \ldots, q_k$.

Let the functional $u' \in E'$ be orthogonal to sp$A$. Consider the function

$$F(\lambda) = < R(\lambda, A)u, u' >,$$

where $< u, u' >$ denotes the value of the linear functional $u' \in E'$ at the point $u \in E$. The function $F(\lambda)$ is analytic under $\lambda \neq \lambda_k$. Since $u' \perp$ sp$A$, then from (2.1) it follows that the principal part of the Laurent series expansion for $F(\lambda)$ at the points λ_k equals zero. Thus, $F(\lambda)$ has a removable singularity at the points $\lambda = \lambda_k$. So, $F(\lambda)$ is an analytic function. From condition 3 follows the estimate

$$|F(\lambda)| \leq C|\lambda|^n, \qquad \lambda \in \ell_k(a), \ |\lambda| \to \infty, \tag{2.2}$$

and from condition 2 follows the estimate

$$|F(\lambda)| \leq Ce^{\omega|\lambda|^p}, \qquad |\lambda - a| = r_k.$$

Let us now apply Theorem 1.6 in each of the angles less than π/p with a vertex at a, into which the complex plane is divided by rays $\ell_k(a)$. Then the function $F(\lambda)$ satisfies condition (2.2) over all the plane. By the Liouville theorem, we have that under $n \geq 0$

$$F(\lambda) = \alpha_0 + \alpha_1 \lambda + \cdots + \alpha_n \lambda^n, \tag{2.3}$$

and under $n = -1$ $\quad F(\lambda) = 0$. Let us expand $R(\lambda, A)$ in some neighborhood of λ_0 in the Taylor series:

$$R(\lambda, A) = R(\lambda_0, A) + (\lambda - \lambda_0)R^2(\lambda_0, A) + \cdots + (\lambda - \lambda_0)^n R^{n+1}(\lambda_0, A) + \cdots .$$

Then,

$$F(\lambda) = < R(\lambda, A)u, u' > \ = \ < R(\lambda_0, A)u, u' > + \cdots$$
$$+ (\lambda - \lambda_0)^n < R^{n+1}(\lambda_0, A)u, u' > + \cdots .$$

From this and (2.3) follows

$$< R^{n+2}(\lambda_0, A)u, u' > \ = 0, \qquad u \in E.$$

So, a system of root vectors of the operator A is complete in $D((A - \lambda_0 I)^{n+2})$. Since the operator $R^{n+2}(\lambda_0, A)$ maps E onto $D((A - \lambda_0 I)^{n+2})$ and has the inverse operator, then $\dim D((A - \lambda_0 I)^{n+2}) = \infty$. From this and the completeness of the root vectors it follows that the set of different eigenvalues of the operator A is countable, which, in turn, implies the discreteness of the spectrum of the operator A. ∎

2.2.2. Expansion of a smooth vector in the root subspaces of an unbounded operator. Let the operator A in the Banach space E have a compact resolvent. Enumerate the eigenvalues of such an operator (if they are exist) in the nondecreasing order of their moduli and without taking into account their multiplicities. If some eigenvalues have equal moduli, then the enumeration is arbitrary. For $u \in E$ we consider the formal Fourier series

$$-\sum_j A_{-1}(\lambda_j)u, \tag{2.4}$$

where $|\lambda_1| \leq |\lambda_2| \leq \cdots$, $\lambda_j \neq \lambda_m$ under $j \neq m$,

$$A_{-1}(\lambda_j) = \frac{1}{2\pi i} \int\limits_{|\lambda-\lambda_j|=\varepsilon} R(\lambda, A)\, d\lambda.$$

Radius ε_j is chosen so small that inside the circle $|\lambda - \lambda_j| = \varepsilon_j$ there are no eigenvalues, other than λ_j. By Lemma 1.3, the operator $A_{-1}(\lambda_j)$ maps the whole space E into the root subspace N_{λ_j}.

Theorem 2.2. *Let the following conditions be satisfied:*

(1) *an operator A in a Banach space E has a compact resolvent;*

(2) *there exist circles $S(a, r_k)$, $k = 1, \ldots, \infty$ with radii r_k going to ∞ and an integer $n \geq -1$ such that*

$$\|R(\lambda, A)\| \leq C|\lambda|^n, \qquad |\lambda - a| = r_k;$$

Then the spectrum of the operator A is discrete and for any $u \in D(A^{n+2})$ there exists a subsequence of partial sums of series (2.4), converging to u in the sense of E.

Proof. Denote $v = (A - \lambda_0 I)^{n+2}u$, where $\lambda_0 \in \rho(A)$. Then $u = R^{n+2}(\lambda_0, A)v$. Using the Hilbert identity, we obtain

$$R(\lambda, A)u = R(\lambda, A)R^{n+2}(\lambda_0, A)v = \frac{R(\lambda, A) - R(\lambda_0, A)}{\lambda - \lambda_0} R^{n+1}(\lambda_0, A)v$$

$$= -\frac{R^{n+2}(\lambda_0, A)v}{\lambda - \lambda_0} + \frac{1}{\lambda - \lambda_0} R(\lambda, A)R^{n+1}(\lambda_0, A)v = \cdots$$

$$= -\frac{u}{\lambda - \lambda_0} - \frac{1}{(\lambda - \lambda_0)^2} R^{n+1}(\lambda_0, A)v - \cdots$$

$$-\frac{1}{(\lambda - \lambda_0)^{n+2}} R(\lambda_0, A)v + \frac{1}{(\lambda - \lambda_0)^{n+2}} R(\lambda, A)v.$$

Hence,

$$\frac{1}{2\pi i} \int\limits_{S(a,r_k)} R(\lambda, A)u \ d\lambda = -\frac{1}{2\pi i} \int\limits_{S(a,r_k)} \frac{u}{\lambda - \lambda_0} \ d\lambda$$

$$+ \frac{1}{2\pi i} \int\limits_{S(a,r_k)} \frac{1}{(\lambda - \lambda_0)^{n+2}} R(\lambda, A)v \ d\lambda.$$

Denote by n_k a number of the eigenvalues of the operator A, situated inside the circle $S(a, r_k)$. Using the residues theory for $|\lambda - a| < r_k$ we have

$$u + \sum_{j=1}^{n_k} A_{-1}(\lambda_j)u = \frac{1}{2\pi i} \int\limits_{S(a,r_k)} \frac{1}{(\lambda - \lambda_0)^{n+2}} R(\lambda, A)v \ d\lambda.$$

Then,

$$\left\| u + \sum_{j=1}^{n_k} A_{-1}(\lambda_j)u \right\| \le C \int\limits_{S(a,r_k)} \frac{\|R(\lambda, A)\| \ \|v\|}{|\lambda|^{n+2}} \ d\lambda \le C r_k^{-1},$$

from which follows the theorem statement. ∎

2.2.3. Completeness of root vectors of an operator with the resolvent from the class $\sigma_p(H)$. In this item we prove the main theorem of section 2.1.

Theorem 2.3. *Let the following conditions be satisfied:*

(1) *an operator A in H has a dense domain of definition $D(A)$;*

(2) *$R(\lambda_0, A) \in \sigma_p(H)$ under some $p > 0$ and $\lambda_0 \in \rho(A)$;*

(3) *there exist rays $\ell_k(a)$ with angles between the neighboring rays less than π/p and an integer $n \ge -1$ such that*

$$\|R(\lambda, A)\| \le C|\lambda|^n, \qquad \lambda \in \ell_k(a), \ |\lambda| \to \infty.$$

Then the spectrum of the operator A is discrete and a system of root vectors of the operator A is complete in the space $H(A^k)$, $k = 0, \dots, \infty$.

Proof. For $\lambda_0 \in \rho(A)$ we have

$$A - \lambda I = (A - \lambda_0 I) - (\lambda - \lambda_0)I = [I - (\lambda - \lambda_0)R(\lambda_0, A)](A - \lambda_0 I).$$

Hence, under $\lambda \in \rho(A)$

$$[I - (\lambda - \lambda_0)R(\lambda_0, A)]^{-1} = (A - \lambda_0 I)(A - \lambda I)^{-1}. \tag{2.5}$$

From the identity

$$A(A - \lambda I)^{-1} = I + \lambda(A - \lambda I)^{-1}, \qquad \lambda \in \rho(A),$$

and the conditions of the theorem the following estimate follows

$$\|A(A - \lambda I)^{-1}\| \le C|\lambda|^{n+1}, \qquad \lambda \in \ell_k(a), \ |\lambda| \to \infty.$$

Taking this into account in (2.5), we obtain the inequality

$$\|(I - (\lambda - \lambda_0)R(\lambda_0, A))^{-1}\| \le C|\lambda|^{n+1}, \qquad \lambda \in \ell_k(a), \ |\lambda| \to \infty.$$

So, the operator $B = R(\lambda_0, A)$, on rays $\ell_k(-\lambda_0 + a) = -\lambda_0 + \ell_k(a)$, satisfies condition 2 of Theorem 1.29. Hence, Theorem 1.29 is applicable to the operator $B = R(\lambda_0, A)$, from which it follows that a system of root vectors of the operator A is complete in the set $R(B^{n+2}) = R(R^{n+2}(\lambda_0, A))$ of the space H. From

$$H = \overline{R(R^{n+2}(\lambda_0, A))} \oplus N_0^0((R^{n+2}(\lambda_0, A))^*)$$

and

$$N_0^0((R^{n+2}(\lambda_0, A))^*) = N_0^0(R^{n+2}(\overline{\lambda_0}, A^*)) = 0$$

follows

$$\overline{R(R^{n+2}(\lambda_0, A))} = H.$$

So, the system of root vectors of the operator A is complete in the space H.

Let $u \in H(A^k)$ and $\varepsilon > 0$. Then there exist root vectors v_j, $j = 1, \ldots, m$ of the operator A and numbers C_j, such that

$$\left\| (A - \lambda_0 I)^k u - \sum_{j=1}^m C_j v_j \right\| < \varepsilon.$$

Since the vectors $(A - \lambda_0 I)^{-k} v_j$, $j = 1, \ldots, m$ are also root vectors of the operator A, then the last inequality implies

$$\overline{\mathrm{sp}A}|_{H(A^k)} = H(A^k). \quad \blacksquare$$

Remark 2.4. *From the Hilbert identity*

$$R(\lambda, A) - R(\lambda_0, A) = (\lambda - \lambda_0)R(\lambda, A)R(\lambda_0, A)$$

and Property 1.18 it follows that if for one $\lambda_0 \in \rho(A)$ the operator $R(\lambda_0, A) \in \sigma_p(H)$, then under any $\lambda \in \rho(A)$ the operator $R(\lambda, A) \in \sigma_p(H)$.

2.2.4. Comparison of operators. Let an operator A in a Banach space E be closed.

The operator B is called A-compact with order $\eta \in [0,1]$, if $D(B) \supset D(A)$ and for any $\varepsilon > 0$

$$\|Bu\| \leq \varepsilon \|Au\|^{\eta} \|u\|^{1-\eta} + C(\varepsilon)\|u\|, \qquad u \in D(A).$$

The operator B is called A-compact, if B is A-compact with unit order.

Lemma 2.5. *Let an operator A in H has a dense domain of definition $D(A)$[3] and at least one regular point. Then for the operator B to be A-compact it is sufficient that $D(B) \supset D(A)$ and the operator $BR(\lambda, A)$ in H is compact; besides if the operator $R(\lambda, A)$ in H is compact, then it is necessary.*

Proof. Let B be A-compact, i.e., $D(B) \supset D(A)$ and for any $\varepsilon > 0$

$$\|Bu\| \leq \varepsilon \|Au\| + C(\varepsilon)\|u\|, \qquad u \in D(A).$$

Then under any $v \in H$, $\lambda_0 \in \rho(A)$, $\varepsilon > 0$ we have

$$\|BR(\lambda_0, A)v\| \leq \varepsilon \|AR(\lambda_0, A)v\| + C(\varepsilon)\|R(\lambda_0, A)v\|$$
$$\leq C\varepsilon\|v\| + C(\varepsilon)\|R(\lambda_0, A)v\|.$$

Let $v_n \in H$ and $\|v_n\| \leq M$. Since $R(\lambda_0, A)$ in H is compact, then there exists a subsequence $\{v_{n_k}\}_{k=1}^{\infty}$ such that $R(\lambda_0, A)v_{n_k}$ converges in H. Then from

$$\|BR(\lambda_0, A)v_{n_k} - BR(\lambda_0, A)v_{n_m}\|$$
$$\leq C\varepsilon\|v_{n_k} - v_{n_m}\| + C\varepsilon\|R(\lambda_0, A)v_{n_k} - R(\lambda_0, A)v_{n_m}\|$$

it follows that under any $\varepsilon > 0$ one can choose $N(\varepsilon)$ so that under $k > N(\varepsilon)$, $m > N(\varepsilon)$

$$\|BR(\lambda_0, A)v_{n_k} - BR(\lambda_0, A)v_{n_m}\| \leq 2MC\varepsilon + \varepsilon = (2MC + 1)\varepsilon.$$

Conversely, let $D(B) \supset D(A)$ and the operator $T = BR(\lambda_0, A)$ in H be compact. Under the given $\varepsilon > 0$ let us construct a finite-dimensional operator $Q = \sum_{k=1}^{n} (\cdot, u_k) v_k$ such that $\|T - Q\| < \varepsilon/2$. From the lemma condition it follows that the domain of definition of the conjugate operator A^* is dense in H. Then the elements u_k can be approximated by elements $w_k \in D(A^*)$ and the operator

---

[3]The density of $D(A)$ in H is used only when we prove sufficiency.

$P = \sum_{k=1}^{n} (\cdot, w_k) v_k$ can be constructed so that $\|Q - P\| \leq \varepsilon/2$. Then under $u \in D(A)$ and $\lambda_0 \in \rho(A)$ we have

$$\|Bu\| = \|T(A - \lambda_0 I)u\| \leq \|P(A - \lambda_0 I)u\| + \|(T - P)(A - \lambda_0 I)u\|$$

$$\leq \sum_{k=1}^{n} |((A - \lambda_0 I)u, w_k)| \, \|v_k\| + \varepsilon(\|Au\| + |\lambda_0|\|u\|)$$

$$\leq \sum_{k=1}^{n} (|(u, A^* w_k)| + |\lambda_0| \, |(u, w_k)|) \|v_k\| + \varepsilon(\|Au\| + |\lambda_0| \, \|u\|)$$

$$\leq \varepsilon\|Au\| + C(\varepsilon)\|u\|. \quad \blacksquare$$

Let A and B be operators in a Banach space E.

The operator B is said to be *compact with respect to A*, if $D(B) \supset D(A)$ and the operator $BR(\lambda, A)$ in E is compact.

By virtue of Lemma 2.5 if $R(\lambda, A)$ in H is compact, then we have: if the operator B is A-compact then the operator B is compact with respect to A.

Lemma 2.6. *Let the following conditions be satisfied:*

(1) *a Banach space E is compactly embedded into a Banach space G.*

(2) *an operator B from E into a Banach space F is bounded and for any $\varepsilon > 0$*

$$\|Bu\|_F \leq \varepsilon\|u\|_E + C(\varepsilon)\|u\|_G, \qquad u \in E. \tag{2.6}$$

Then the operator B from E into F is compact.

Proof. Let $u_n \in E$ and $\|u_n\|_E \leq M$, $n = 1, \ldots, \infty$. Since the embedding $E \subset G$ is compact, then from the sequence $\{u_n\}_1^\infty$ one can choose a subsequence $\{u_{n_k}\}_{k=1}^\infty$ converging in G. From (2.6) under any $\varepsilon > 0$ follows

$$\|Bu_{n_k} - Bu_{n_m}\|_F \leq \varepsilon\|u_{n_k} - u_{n_m}\|_E + C(\varepsilon)\|u_{n_k} - u_{n_m}\|_G.$$

Hence, for any $\varepsilon > 0$ one can choose $N(\varepsilon)$ so that if $k > N(\varepsilon)$ and $m > N(\varepsilon)$ then

$$\|Bu_{n_k} - Bu_{n_m}\|_F \leq (2M + 1)\varepsilon. \quad \blacksquare$$

This lemma is almost invertible, to be more precise, the following lemma holds:

Lemma 2.7. *Let the following conditions be satisfied:*

(1) *E and F are Banach spaces with bases and E is reflexive;*

(2) *the embedding $E \subset F$ is continuous and dense;*

(3) *an operator B from E into F is compact.*

Then for any $\varepsilon > 0$

$$\|Bu\|_F \le \varepsilon\|u\|_E + C(\varepsilon)\|u\|_F, \qquad u \in E.$$

Proof. For the given $\varepsilon > 0$ let us construct a finite-dimensional operator $Q = \sum_{k=1}^{n} < \cdot, u'_k > v_k$, where $u'_k \in E'$, $v_k \in F$ such that $\|B - Q\|_{B(E,F)} < \varepsilon/2$. By Lemma 1.2.2, the embedding $F' \subset E'$ is dense. Consequently, one can find functionals $v'_k \in F'$ such that $\|u'_k - v'_k\|_{E'} < \varepsilon/2nM$, $k = 1, \ldots, n$, where $\|v_k\|_F < M$. Then for $u \in E$ we have

$$\|Bu\|_F \le \|(B - Q)u\|_F + \|Qu\|_F \le \|B - Q\|_{B(E,F)}\|u\|_E$$
$$+ \sum_{k=1}^{n} \|u'_k - v'_k\|_{E'}\|u\|_E\|v_k\|_F + \sum_{k=1}^{n} \|v'_k\|_{F'}\|u\|_F\|v_k\|_F$$
$$\le \varepsilon\|u\|_E + C(\varepsilon)\|u\|_F. \qquad \blacksquare$$

2.2.5. Perturbation of an operator with a decreasing resolvent. Let us prove some lemmas, which turn to be useful for estimating the resolvent of the perturbed operator.

Lemma 2.8. *Let the following conditions be satisfied:*

(1) *A is an operator in a Banach space E and under some $\eta \in (0, 1]$*

$$\|R(\lambda, A)\| \le C|\lambda|^{-\eta}, \qquad \lambda \in \Gamma, \ |\lambda| \to \infty,$$

where Γ is an unbounded set of the complex plane;

(2) *B is an operator in E, $D(B) \supset D(A)$ and for any $\varepsilon > 0$*

$$\|Bu\| \le \varepsilon\|Au\|^{\eta}\|u\|^{1-\eta} + C(\varepsilon)\|u\|, \qquad u \in D(A).$$

Then,

$$\|R(\lambda, A + B)\| \le C|\lambda|^{-\eta}, \qquad \lambda \in \Gamma, \ |\lambda| \to \infty.$$

Proof. Since when $\lambda \in \rho(A)$

$$AR(\lambda, A) = [(A - \lambda I) + \lambda I]R(\lambda, A) = I + \lambda R(\lambda, A),$$

then for $\lambda \in \Gamma$, $|\lambda| \to \infty$ we have

$$\|AR(\lambda, A)\| \le C|\lambda|^{1-\eta}.$$

Therefore, by condition 2 of the lemma, for any $v \in E$

$$\|BR(\lambda, A)v\| \leq \varepsilon\|AR(\lambda, A)v\|^{\eta}\|R(\lambda, A)v\|^{1-\eta} + C(\varepsilon)\|R(\lambda, A)v\|$$
$$\leq (C\varepsilon + C(\varepsilon)|\lambda|^{-\eta})\|v\|, \qquad \lambda \in \Gamma, \ |\lambda| \to \infty.$$

Hence,

$$\|BR(\lambda, A)\| \leq q < 1, \qquad \lambda \in \Gamma, \ |\lambda| \to \infty.$$

Then, by the Neyman identity

$$R(\lambda, A + B) = R(\lambda, A) \sum_{k=0}^{\infty} (-BR(\lambda, A))^k \qquad (2.7)$$

we obtain

$$\|R(\lambda, A + B)\| \leq C\|R(\lambda, A)\| \leq C|\lambda|^{-\eta}, \qquad \lambda \in \Gamma, \ |\lambda| \to \infty. \ \blacksquare$$

Lemma 2.9. *Let the following conditions be satisfied:*

(1) *A is an operator in a Banach space E and under some $\eta \in (0, 1]$*

$$\|R(\lambda, A)\| \leq C|\lambda|^{-\eta}, \quad \lambda \in \Gamma, \ |\lambda| \to \infty,$$

where Γ is an unbounded set of the complex plane;

(2) *B is an operator in E, $D(B) \supset D(A)$ and under some $\mu \in [0, \eta)$*

$$\|Bu\| \leq C(\|Au\|^{\mu}\|u\|^{1-\mu} + \|u\|), \qquad u \in D(A).$$

Then,

$$\|R(\lambda, A + B)\| \leq C|\lambda|^{-\eta}, \qquad \lambda \in \Gamma, \ |\lambda| \to \infty.$$

Proof. By the Young inequality (1.2.13) we have

$$\|Au\|^{\mu}\|u\|^{1-\mu} = \|u\|^{1-\eta}\|Au\|^{\mu}\|u\|^{\eta-\mu}$$
$$\leq \|u\|^{1-\eta}(\varepsilon\|Au\|^{\eta}\|u\|^{1-\eta} + C(\varepsilon)\|u\|)$$
$$\leq \varepsilon\|Au\|^{\eta}\|u\|^{1-\eta} + C(\varepsilon)\|u\|,$$

which implies that Lemma 2.9 follows from Lemma 2.8. $\blacksquare$

2.2.6. Completeness of root vectors of a perturbed unbounded operator. We mainly apply Theorem 2.10 when we investigate differential operators:

Theorem 2.10. *Let the following conditions be satisfied:*

(1) *an operator A in H has a dense domain of definition $D(A)$;*

(2) *$R(\lambda_0, A) \in \sigma_p(H)$ for some $p > 0$ and $\lambda_0 \in \rho(A)$;*

(3) *there exist rays $\ell_k(a)$ with angles between the neighboring rays less than π/p and a number $\eta \in (0, 1]$ such that*

$$\|R(\lambda, A)\| \leq C|\lambda|^{-\eta}, \qquad \lambda \in \ell_k(a), \ |\lambda| \to \infty;$$

(4) *B is an operator in H, $D(B) \supset D(A)$ and for any $\varepsilon > 0$*

$$\|Bu\| \leq \varepsilon \|Au\|^\eta \|u\|^{1-\eta} + C(\varepsilon)\|u\|, \qquad u \in D(A).$$

Then the spectrum of the operator $A + B$ is discrete and a system of root vectors of the operator $A + B$ is complete in the space $H(A)$.

Proof. From (2.7), by condition 2 of the Theorem and by Property 1.18, it follows that $R(\lambda, A + B) \in \sigma_p(H)$, i.e., the operator $A + B$ satisfies condition 2 of Theorem 2.3. From conditions 3 and 4 and by Lemma 2.8, it follows that the operator $A + B$ satisfies condition 3 of Theorem 2.3. So, Theorem 2.3 is applicable to the operator $A + B$, from which it follows that the system of root vectors of the operator $A + B$ is complete in the space $H(A + B) = H(A)$. ∎

2.2.7. Completeness of root vectors of a perturbed unbounded self-adjoint operator. Since for a selfadjoint operator any non-real ray is a ray of maximum decrease of the resolvent, then the completeness theorem, in the case when the operator is principally selfadjoint, has a more simplified formulation.

Theorem 2.11. *Let the following conditions be satisfied:*

(1) *an operator A in H is selfadjoint;*

(2) *$R(\lambda_0, A) \in \sigma_p(H)$ for some $p > 0$ and $\lambda_0 \in \rho(A)$;*

(3) *B is an operator in H, $D(B) \supset D(A)$ and the operator $BR(\lambda, A)$ in H is compact.*

Then the spectrum of the operator $A + B$ is discrete; for any $\varepsilon > 0$ outside the angles $|\arg \lambda| < \varepsilon$ and $|\arg \lambda - \pi| < \varepsilon$ there is a finite number of eigenvalues, and a system of root vectors of the operator $A + B$ is complete in the space $H(A)$.

Proof. Since for any $\lambda \neq \lambda_k(A)$, $k = 1, \ldots, \infty$, the expansion

$$R(\lambda, A) = \sum_{k=1}^{\infty} \frac{1}{\lambda_k - \lambda}(\cdot, u_k)u_k \tag{2.8}$$

holds, where $\{u_k\}_1^\infty$ is a complete orthonormal system of the eigenvectors of operator A then under $\varphi \neq 0$, $\varphi \neq \pi$

$$\|R(\lambda, A)\| \leq \frac{1}{|\mathrm{Im}\lambda|} \leq \frac{C(\varphi)}{|\lambda|}, \qquad \lambda \in \ell(0, \varphi),$$

i.e., the operator A satisfies condition 3 of Theorem 2.10 under $\eta = 1$ on any non-real ray. From condition 3, by virtue of Lemma 2.5, it follows that the operator B satisfies condition 4 of Theorem 2.10 under $\eta = 1$. So, Theorem 2.10 is applicable to the operator $A + B$, from which the statement of Theorem 2.11 follows. ■

2.2.8. Expansion of a smooth vector in the root subspaces of the perturbed unbounded operator. In the theorem given below, condition 2 imposed on the principal term of the operator seems to be stringent. However, for ordinary differential operators, generated by both regular and some irregular boundary value conditions, this condition holds.

In Theorem 2.2 it was shown that if $A + B$ is a closed operator in a Banach space E with a compact resolvent, then the terms of the series

$$\sum_k P(\lambda_k, A + B)u, \qquad u \in E, \tag{2.9}$$

where $|\lambda_1| \leq |\lambda_2| \leq \cdots$, $\lambda_k \neq \lambda_m$ if $k \neq m$,

$$P(\lambda_k, A + B) = -\frac{1}{2\pi i} \int\limits_{|\lambda - \lambda_k| = \varepsilon_k} R(\lambda, A + B) \, d\lambda,$$

belong to the root subspaces $N_{\lambda_k}(A + B)$, when radius ε_k is chosen so small that inside the circle $|\lambda - \lambda_k| = \varepsilon_k$ there are no eigenvalues, other than $\lambda_k = \lambda_k(A + B)$.

Theorem 2.12. *Let the following conditions be satisfied:*

(1) *an operator A in a Banach space E has a compact resolvent;*

(2) *there exist circles $S(a, r_k)$, $k = 1, \ldots, \infty$ with radii r_k going to ∞ and a number $\eta \in (0, 1]$ such that*

$$\|R(\lambda, A)\| \leq C|\lambda|^{-\eta}, \qquad |\lambda - a| = r_k;$$

(3) *B is an operator in E, $D(B) \supset D(A)$ and for any $\varepsilon > 0$*

$$\|Bu\| \leq \varepsilon\|Au\|^\eta \|u\|^{1-\eta} + C(\varepsilon)\|u\|, \qquad u \in D(A).$$

*Then the spectrum of the operator $A + B$ is discrete and for any $u \in D((A+B)^2)$[4]
there exists a subsequence of partial sums of series (2.9), converging to u in the
sense of E.*

Proof. From (2.7), by virtue of condition 1 of the Theorem, it follows that the
resolvent of the operator $A + B$ is compact in E, i.e., the operator $A + B$ satisfies
condition 1 of Theorem 2.2. From conditions 2 and 3 and Lemma 2.8, it follows that
the operator $A + B$ also satisfies condition 2 of Theorem 2.2. So, Theorem 2.2 is
applicable to the operator $A + B$, from which follows the Theorem 2.12 statement.
∎

**2.2.9. Expansion of a smooth vector in the root subspaces of the
perturbed unbounded selfadjoint operator.** Let us show that, in one case
condition 2 of Theorem 2.12 can be changed to a condition that is more easily
verified.

Denote by λ_n the eigenvalues of an unbounded selfadjoint operator A in the
nondecreasing order of their moduli, with their multiplicities taken into account.

Lemma 2.13. *Let eigenvalues of a selfadjoint operator A under some $q \geq p \geq 1$
satisfy the condition*

$$C_1 n^p \leq |\lambda_n| \leq C_2 n^q.$$

Then there exist circles $S(0, r_k)$, $k = 1, \ldots, \infty$, with radii r_k going to ∞, such that

$$\|R(\lambda, A)\| \leq C|\lambda|^{-\frac{p-1}{q}}, \qquad |\lambda| = r_k.$$

Proof. First, let us show that there exists a number $C > 0$ and a sequence of
numbers n_k such that

$$|\lambda_{n_k+1}| - |\lambda_{n_k}| \geq Cn_k^{p-1}. \tag{2.10}$$

Suppose the opposite, i.e., under any $\varepsilon > 0$ there exists a number n_ε such that for
all $n \geq n_\varepsilon$

$$|\lambda_{n+1}| - |\lambda_n| < \varepsilon n^{p-1}.$$

Then for any $\varepsilon > 0$ and any m we have

$$|\lambda_{n_\varepsilon+m}| - |\lambda_{n_\varepsilon}| < \varepsilon[n_\varepsilon^{p-1} + (n_\varepsilon + 1)^{p-1} + \cdots + (n_\varepsilon + m - 1)^{p-1}]$$
$$< \varepsilon m(n_\varepsilon + m)^{p-1} < \varepsilon(n_\varepsilon + m)^p.$$

Dividing both sides of the obtained inequality by $(n_\varepsilon + m)^p$ we obtain

$$\frac{|\lambda_{n_\varepsilon+m}|}{(n_\varepsilon + m)^p} - \frac{|\lambda_{n_\varepsilon}|}{(n_\varepsilon + m)^p} < \varepsilon.$$

---

[4] If $\eta = 1$ then the theorem statement is true for any $u \in D(A)$.

Hence, passing to the limit under $m \to \infty$ we obtain $C_1 \leq \varepsilon$, which contradicts the arbitrariness of ε. Thus, (2.10) holds for some $C > 0$ and some sequence of numbers n_k.

Consider now the circles $S(0, r_k)$ with radii

$$r_k = \frac{|\lambda_{n_k+1}| + |\lambda_{n_k}|}{2} = |\lambda_{n_k}| + \frac{|\lambda_{n_k+1}| - |\lambda_{n_k}|}{2}.$$

From (2.8) follows the equality

$$\|R(\lambda, A)\| = \frac{1}{\rho(\lambda, \sigma(A))}, \qquad \lambda \in \rho(A),$$

where $\rho(\lambda, \sigma(A))$ is the distance between the point λ and the spectrum of the operator A. Then

$$\|R(\lambda, A)\| = \frac{1}{\rho(\lambda, \sigma(A))} \leq \frac{2}{|\lambda_{n_k+1}| - |\lambda_{n_k}|}, \qquad |\lambda| = r_k. \tag{2.11}$$

On the other hand,

$$r_k = \frac{|\lambda_{n_k+1}| + |\lambda_{n_k}|}{2} \leq C_2 \frac{(n_k+1)^q + n_k^q}{2} \leq C n_k^q.$$

Then, taking into account (2.10), we obtain

$$(|\lambda_{n_k+1}| - |\lambda_{n_k}|)^{-1} \leq C n_k^{-(p-1)} = C(n_k^q)^{-\frac{p-1}{q}} = C r_k^{-\frac{p-1}{q}}.$$

From this and (2.11) follows the Lemma 2.13 statement. ∎

Theorem 2.14. *Let the following conditions be satisfied:*

(1) *the eigenvalues of a selfadjoint operator A in H under some $q \geq p \geq 1$ satisfy the condition*

$$C_1 n^p \leq |\lambda_n| \leq C_2 n^q;$$

(2) *B is an operator in H, $D(B) \supset D(A)$ and for any $\varepsilon > 0$*

$$\|Bu\| \leq \varepsilon \|Au\|^{\frac{p-1}{q}} \|u\|^{1-\frac{p-1}{q}} + C(\varepsilon)\|u\|, \qquad u \in D(A).$$

Then the spectrum of the operator $A + B$ is discrete; under any $\varepsilon > 0$ outside the angles $|\arg \lambda| < \varepsilon$ and $|\arg \lambda - \pi| < \varepsilon$ there is a finite number of eigenvalues and for $u \in D((A + B)^2)$ there exists a subsequence of partial sums of series (2.9), converging to u in the sense of H.

Proof. By virtue of Lemma 2.13 the statement of Theorem 2.14 follows from Theorem 2.12. ∎

2.3.   n-fold completeness of root vectors of a system of unbounded polynomial operator pencils

The result of this paragraph is used as a basis of subsequent work in this chapter and has many applications in the theory of differential equations.

2.3.1. A system of unbounded polynomial operator pencils. The results of M. V. Keldysh [30] about completeness of root vectors of an operator pencil and of J. T. Schwartz [13, Ch.XI, §9.31] about completeness for one operator were improved and generalized in the works of S. Ya. Yakubov [89, 91]. This problem for a system of operator pencils was investigated for the first time in [91].

If an operator C in a Hilbert space is invertible, then $H(C) = \{u|u \in D(C),$ with the scalar product $(u,v)_{H(C)} = (Cu, Cv)\}$ which is a Hilbert space.

In the sequel we will repeatedly use the s-numbers of operators that act complactly from one Hilbert space into another.

Lemma 3.1. Let operators C_1 and C_2 in a Hilbert space H be invertible. Then

$$s_j(A; H(C_1), H(C_2)) = s_j(C_2 A C_1^{-1}; H, H), \qquad j = 1, \ldots, \infty.$$

Proof. For any $u \in H(C_1)$, $v \in H(C_2)$ we have

$$(Au, v)_{H(C_2)} = (u, A^*v)_{H(C_1)},$$

i.e.,

$$(C_2 Au, C_2 v) = (C_1 u, C_1 A^* v).$$

Hence, for all $g \in H$, $w \in H$ we have

$$(C_2 A C_1^{-1} g, w) = (g, C_1 A^* C_2^{-1} w),$$

from which follows

$$(C_2 A C_1^{-1})^* = C_1 A^* C_2^{-1}.$$

Then,

$$(C_2 A C_1^{-1})^* C_2 A C_1^{-1} = C_1 A^* A C_1^{-1}.$$

On the other hand, it is easy to see that eigenvalues of the operator $A^* A$ coincide with eigenvalues of the operator $C_1 A^* A C_1^{-1}$. ∎

Theorem 3.2. Let an operator A from H into H_1 be compact. Then,

$$s_{j+1}(A; H, H_1) = \min_{\substack{\dim R(K) \le j \\ K \in B(H, H_1)}} \|A - K\|_{B(H, H_1)}, \qquad j = 0, \ldots, \infty.$$

'roof. Let K be a j-dimensional operator from H into H_1, i.e., there exist ems of elements u_m and v_m, $m = 1, \ldots, j$, linearly independent in H and H_1 'ectively, such that

$$K = \sum_{m=1}^{j} (\cdot, u_m) v_m.$$

'iously, $R(K) = \mathrm{sp}\{v_m\}$, $N_0^0(K) = (\mathrm{sp}\{u_m\})^{\perp}$. Then, by virtue of minimax 'erties of eigenvalues we have

$$s_{j+1}^2(A; H, H_1) = \lambda_{j+1}(A^*A) \leq \max_{u \in N(K)} \frac{(A^*Au, u)}{(u, u)} = \max_{u \in N(K)} \frac{\|Au\|_{H_1}^2}{\|u\|_H^2}.$$

:e for all $u \in N_0^0(K)$

$$\|Au\|_{H_1} = \|(A - K)u\|_{H_1} \leq \|A - K\|_{B(H,H_1)} \|u\|_H,$$

ι for each j-dimensional operator K we have

$$s_{j+1}(A; H, H_1) \leq \|A - K\|_{B(H,H_1)}.$$

refore, for any p-dimensional, $p = 0, \ldots, j$, operator K we have

$$s_{j+1}(A; H, H_1) \leq s_{p+1}(A; H, H_1) \leq \|A - K\|_{B(H,H_1)}.$$

sider the j-th segment of Schmidt expansion (1.8) for A

$$K_j = \sum_{m=1}^{j} s_m(A; H, H_1)(\cdot, u_m) v_m.$$

iously,

$$\|A - K_j\|_{B(H,H_1)} = s_{j+1}(A; H, H_1). \quad \blacksquare$$

emma 3.3. *Let H_1, H_2, H_3 be Hilbert spaces with continuous embedding* $\sqsubset H_2$. *Then,*

$$s_j(A; H_1, H_3) \leq C s_j(A; H_2, H_3), \qquad j = 1, \ldots, \infty.$$

'roof. Since for all $u \in H_1$

$$\|u\|_{H_2} \leq C \|u\|_{H_1},$$

then for $T \in B(H_2, H_3)$ and $u \in H_1$ we have

$$\|Tu\|_{H_3} \leq \|T\|_{B(H_2,H_3)} \|u\|_{H_2} \leq C\|T\|_{B(H_2,H_3)} \|u\|_{H_1}.$$

Therefore,

$$\|T\|_{B(H_1,H_3)} \|u\|_{H_2} \leq C\|T\|_{B(H_2,H_3)}.$$

By virtue of Theorem 3.2

$$s_{j+1}(A; H_1, H_3) = \min_{\substack{\dim R(K) \leq j \\ K \in B(H_1, H_3)}} \|A - K\|_{B(H_1, H_3)}$$

$$\leq C \min_{\substack{\dim R(K) \leq j \\ K \in B(H_2, H_3)}} \|A - K\|_{B(H_2, H_3)}$$

$$= C s_{j+1}(A; H_2, H_3). \ \blacksquare$$

Let H and H^p, $p = 1, \ldots, m$ be Hilbert spaces. Consider a problem for a system of polynomial operator pencils in H

$$L(\lambda)u = \lambda^n u + \lambda^{n-1} A_1 u + \cdots + A_n u = 0,$$
$$L_p(\lambda)u = \lambda^{n_p} A_{p0} u + \lambda^{n_p - 1} A_{p1} u + \cdots + A_{pn_p} u = 0, \quad p = 1, \ldots, m, \tag{3.1}$$

where $n \geq 1$, $0 \leq n_p \leq n - 1$, $m \geq 0$, and A_k are, generally speaking, unbounded operators in H and A_{pk}, $k = 0, \ldots, n_p$ are, generally speaking, unbounded operators from H into H^p.

It is obvious that for $\lambda \neq 0$ $D(L(\lambda)) = \bigcap_{k=1}^{n} D(A_k)$.

A number λ_0 is called an *eigenvalue* of problem (3.1) if the problem

$$L(\lambda_0) = 0, \qquad L_p(\lambda_0)u = 0, \quad p = 1, \ldots, m$$

has a nontrivial solution. The nontrivial solution u_0 is called an *eigenvector* of problem (3.1) corresponding to the eigenvalue λ_0. A solution of the problem

$$L(\lambda_0)u_p + \frac{1}{1!}L'(\lambda_0)u_{p-1} + \cdots + \frac{1}{p!}L^{(p)}(\lambda_0)u_0 = 0,$$
$$L_k(\lambda_0)u_p + \frac{1}{1!}L'_k(\lambda_0)u_{p-1} + \cdots + \frac{1}{p!}L_k^{(p)}(\lambda_0)u_0 = 0, \qquad k = 1, \ldots, m,$$

u_p is called an *associated vector* of the p-th rank to the eigenvector u_0 of problem (3.1).

Eigenvectors and associated vectors of problem (3.1) are combined under the general name *root vectors* of problem (3.1).

A complex number λ is called a *regular point* of problem (3.1) or of the pencil $\mathbb{L}(\lambda) = (L(\lambda), L_1(\lambda), \ldots, L_m(\lambda))$ acting from H into $H \oplus H^1 \oplus \cdots \oplus H^m$, if the problem

$$L(\lambda)u = f, \qquad L_p(\lambda)u = f_p, \quad p = 1, \ldots, m$$

for any $f \in H$, $f_p \in H^p$, has a unique solution and the estimate

$$\|u\| \le C(\lambda) \left(\|f\| + \sum_{p=1}^{m} \|f_p\|_{H^p} \right)$$

is satisfied.

The complement of the regular point set in the complex plane is called the *spectrum* of problem (3.1) or of the pencil $\mathbb{L}(\lambda) = (L(\lambda), L_1(\lambda), \ldots, L_m(\lambda))$.

The spectrum of problem (3.1) is called *discrete*, if:

a) all points λ, not coinciding with the eigenvalues of problem (3.1), are regular points of problem (3.1);

b) the eigenvalues are isolated and have finite algebraic multiplicities;

c) infinity is the only limit point of the set of the eigenvalues of problem (3.1).

Consider a system of differential equations

$$\begin{aligned}
L(D_t)u &= u^{(n)}(t) + A_1 u^{(n-1)}(t) + \cdots + A_n u(t) = 0, \\
L_p(D_t)u &= A_{p0} u^{(n_p)}(t) + \cdots + A_{pn_p} u(t) = 0, \quad p = 1, \ldots, m,
\end{aligned} \tag{3.2}$$

$$u^{(k)}(0) = v_{k+1}, \qquad k = 0, \ldots, n - 1, \tag{3.3}$$

where v_{k+1} are given elements of H, $D_t = \frac{\partial}{\partial t}$, $t \ge 0$.

By virtue of Lemma 2.0.1 a function of the form

$$u(t) = e^{\lambda_0 t} \left(\frac{t^k}{k!} u_0 + \frac{t^{k-1}}{(k-1)!} u_1 + \cdots + u_k \right) \tag{3.4}$$

is a solution to system (3.2), if and only if the system of vectors $u_0, u_1, \cdots, u_k$ is a chain of root vectors of problem (3.1), corresponding to the eigenvalue λ_0.

A solution of the form (3.4) is called an *elementary solution* to system (3.2).

The inclination to approximate a solution to the Cauchy problem (3.2)–(3.3), by linear combinations of the elementary solutions, suggests that the vector $(v_1, v_2, \ldots, v_n)$ should be approximated by linear combinations of vectors of the form

$$(u(0), u'(0), \ldots, u^{(n-1)}(0)) \tag{3.5}$$

where $u(t)$ is an elementary solution of the form (3.4).

Let $\mathcal{H}$ be a Hilbert space, continuously embedded in $\overset{n}{\bigoplus} H$.

A system of root vectors of the problem (3.1) is called *n-fold complete* in $\mathcal{H}$ if the system of vectors (3.5) is complete in the space $\mathcal{H}$.

Theorem 3.4. *Let the following conditions be satisfied:*

(1) *there exist Hilbert spaces H_k, $k = 0, \ldots, n$, for which the compact embeddings $H_n \subset H_{n-1} \subset \cdots \subset H_0 = H$ take place and $\overline{H_n} = H$;*

(2) *for some $p > 0$ $J \in \sigma_p(H_k, H_{k-1})$, $k = 1, \ldots, n$;*

(3) *the operators A_k, $k = 1, \ldots, n$, from H_k into H act boundedly;*

(4) *the operators A_{pk}, $k = 0, \ldots, n_p$, $p = 1, \ldots, m$, from H_{n-n_p+k} into H^p act boundedly;*

(5) *there exist Hilbert spaces H_0^p such that continuous embeddings $H^p \subset H_0^p$, $p = 1, \ldots, m$ hold, and the linear manifold*

$$\mathcal{H}_1 = \{v| \ v = (v_1, \ldots, v_n) \in \overset{n-1}{\underset{k=0}{\bigoplus}} H_{n-k}, \ \overset{n_p}{\underset{k=0}{\sum}} A_{pk} v_{n_p - k + s} = 0,$$

$$s = 1, \ldots, n - n_p, \ p = 1, \ldots, m;$$

$$A_{pk} \in B(H_{n+1-n_p+k-s}, H_0^p), \ k = 0, \ldots, n_p\}$$

is dense in the Hilbert space

$$\mathcal{H} = \{v| \ v = (v_1, \ldots, v_n) \in \overset{n-1}{\underset{k=0}{\bigoplus}} H_{n-k-1}, \ \overset{n_p}{\underset{k=0}{\sum}} A_{pk} v_{n_p - k + s} = 0,$$

$$s = 1, \ldots, n - n_p - 1, \ p = 1, \ldots, m;$$

$$A_{pk} \in B(H_{n-n_p+k-s}, H_0^p), \ k = 0, \ldots, n_p\};$$

(6) *there exist rays ℓ_k with the angles between the neighboring rays less than π/p and a number q such, that*

$$\|L^{-1}(\lambda)\|_{B(H \oplus H^1 \oplus \cdots \oplus H^m, H_n)} \leq C|\lambda|^q, \quad \lambda \in \ell_k, \ |\lambda| \to \infty.$$

Then the spectrum of problem (3.1) is discrete and a system of root vectors of problem (3.1) is n-fold complete in the spaces $\mathcal{H}$ and $\mathcal{H}_1$.

Proof. By the substitution $v_k = \lambda^{k-1} u$, $k = 1, \ldots, n$, the system

$$L(\lambda)u = 0, \qquad L_p(\lambda)u = 0, \quad p = 1, \ldots, m$$

is reduced to the equivalent system

$$\lambda v = \underline{A}v,$$

where $\underline{A}$ is an operator in the Hilbert space $\mathcal{H}$ (see condition 5) given by the equalities

$$D(\underline{A}) = \mathcal{H}_1, \qquad \underline{A}(v_1, \ldots, v_n) = (v_2, \ldots, v_n, -A_n v_1 - \cdots - A_1 v_n).$$

If $u(t)$ is a solution to system (3.2), then

$$v(t) = (u(t), \ldots, u^{(n-1)}(t))$$

is a solution to the system

$$v'(t) = \underline{A}v(t). \tag{3.6}$$

Conversely, if $v(t) = (v_1(t), \ldots, v_n(t))$ is a solution to system (3.6), then $u(t) = v_1(t)$ is a solution to system (3.2). Since the set of root vectors coincides with the set of values of elementary solutions in zero, then the set of root vectors of the operator $\underline{A}$ coincides with the set

$$\{v(0)\} = \{(u(0), \ldots, u^{(n-1)}(0))\},$$

where $u(t)$ are elementary solutions (3.4) to system (3.2).

Let us apply Theorem 2.3 to the operator $\underline{A}$.

By virtue of condition 5, $D(\underline{A}) = \mathcal{H}_1$ is dense in $\mathcal{H}$, i.e., condition 1 of Theorem 2.3 is satisfied. To show that the resolvent of operator $\underline{A}$ is compact we instead of the equation

$$(\underline{A} - \lambda I)v = F$$

solve the system

$$v_{k+1} - \lambda v_k = f_k, \quad k = 1, \ldots, n - 1,$$

$$-\sum_{k=1}^{n} A_{n-k+1} v_k - \lambda v_n = f_n, \tag{3.7}$$

$$A_{p0} v_{n_p+1} + A_{p1} v_{n_p} + \cdots + A_{pn_p} v_1 = 0, \qquad p = 1, \ldots, m,$$

which is equivalent to it in the space $\mathcal{H}$. Let us show that if $v = (v_1, \ldots, v_n)$ is a solution to problem (3.7) and $F = (f_1, \ldots, f_n) \in \mathcal{H}$, then v satisfies all conditions of connection in $\mathcal{H}_1$. Let operators A_{pk}, $k = 0, \ldots, n_p$, from H_{n+1-n_p+k-s} into H_0^p be bounded for some $s = r$, $r = 2, \ldots, n - n_p$. Then, by virtue of the continuity of the embeddings $H_{k+1} \subset H_k$, $k = 1, \ldots, n - 1$, they are bounded for all $s = 2, \ldots, r - 1$. Further, if under some s

$$\sum_{k=0}^{n_p} A_{pk} v_{n_p-k+s-1} = 0 \tag{3.8}$$

then, since $F \in \mathcal{H}$, i.e.

$$\sum_{k=0}^{n_p} A_{pk} f_{n_p-k+s-1} = 0, \qquad s = 2, \ldots, p$$

from the first $n-1$ equations of the system (3.7), (3.8) and the last equality we get

$$\sum_{k=0}^{n_p} A_{pk} v_{n_p-k+s} = \lambda \sum_{k=0}^{n_p} A_{pk} v_{n_p-k+s-1} + \sum_{k=0}^{n_p} A_{pk} f_{n_p-k+s-1} = 0. \tag{3.9}$$

In turn, if $s = 2$ then (3.8) holds. It follows from the last m equations of system (3.7). So (3.9) is true for $s = 2$. This means that (3.8) is true for $s = 3$. Continuing these considerations we find that (3.9) is also true for $s = r$, i.e., v satisfies all conditions of connections in $\mathcal{H}_1$.

From the first $n - 1$ equations of system (3.7) we successively find

$$v_k = \lambda^{k-1} v_1 + \sum_{j=1}^{k-1} \lambda^{k-1-j} f_j, \qquad k = 2, \ldots, n. \tag{3.10}$$

Substituting these values of v_k into other equations of system (3.7), we obtain

$$-L(\lambda) v_1 - \sum_{k=2}^{n} A_{n-k+1} \sum_{j=1}^{k-1} \lambda^{k-1-j} f_j - \sum_{j=1}^{n-1} \lambda^{n-j} f_j = f_n,$$

$$L_p(\lambda) v_1 + \sum_{k=0}^{n_p-1} A_{pk} \sum_{j=1}^{n_p-k} \lambda^{n_p-k-j} f_j = 0, \qquad p = 1, \ldots, m. \tag{3.11}$$

If λ is a regular point of the pencil $\mathbb{L}(\lambda) = (L(\lambda), L_1(\lambda), \ldots, L_m(\lambda))$ acting from H into $H \oplus H^1 \oplus \cdots \oplus H^m$ then by virtue of conditions 1, 3 and 4 for any $(f_1, \ldots, f_n) \in \mathcal{H}_{n-1} \oplus \cdots \oplus \mathcal{H}_0$ the problem (3.11) has a unique solution

$$v_1 = \mathbb{L}^{-1}(\lambda) \Big(-\sum_{k=2}^{n} A_{n-k+1} \sum_{j=1}^{k-1} \lambda^{k-1-j} f_j - \sum_{j=1}^{n-1} \lambda^{n-j} f_j,$$

$$-\sum_{k=0}^{n_1-1} A_{1k} \sum_{j=1}^{n_1-k} \lambda^{n_1-k-j} f_j, \ldots, -\sum_{k=0}^{n_m-1} A_{mk} \sum_{j=1}^{n_m-k} \lambda^{n_m-k-j} f_j \Big). \tag{3.12}$$

From conditions 3, 4, 6 and formulas (3.12), and (3.10) it follows that $v_k \in H_{n-k+1}$, $k = 1, \ldots, n$. From condition 6 it follows that the operator $\mathbb{L}^{-1}(\lambda)$ from $H \oplus H^1 \oplus \cdots \oplus H^m$ into H_n acts boundedly, but by virtue of condition 1 the embeddings $H_{n-k+1} \subset H_{n-k}$, $k = 1, \ldots, n$ are compact. Hence, from (3.10) and

(3.12) it follows that $R(\lambda, \underline{A})$ is compact. From conditions 3, 4, 6 and equalities (3.12), and (3.10) it follows that there exists a number r such that

$$\|R(\lambda, \underline{A})\| \le C|\lambda|^r, \quad \lambda \in \ell_k, \ |\lambda| \to \infty,$$

i.e., condition 3 of Theorem 2.3 is satisfied.

That $R(\lambda, \underline{A}) \in \sigma_p(\mathcal{H})$ remains to be shown. Consider selfadjoint positively defined operators C_k in H such that $H(C_k) = H_k$, $k = 0, \ldots, n$ [44, Ch.1, §2.1]. Consider in the space

$$\mathcal{H}_c = \{v| \ v = (v_1, \ldots, v_n) \in \bigoplus_{}^{n} H, \ \sum_{k=0}^{n_p} A_{pk} C_{n-n_p+k-s}^{-1} v_{n_p-k+s} = 0,$$
$$s = 1, \ldots, n - n_p - 1, \ p = 1, \ldots, m;$$
$$A_{pk} \in B(H_{n-n_p+k-s}, H_0^p), \ k = 0, \ldots, n_p\}$$

an operator C is defined by the equalities

$$D(C) = \mathcal{H}, \quad C = \mathrm{diag}(C_{n-1}, \ldots, C_0).$$

The operator C from $\mathcal{H}$ into $\mathcal{H}_c$ acts boundedly, is invertible and $\mathcal{H}_c(C) = \mathcal{H}$. Then, by virtue of Lemma 3.1, $R(\lambda, \underline{A}) \in \sigma_p(\mathcal{H})$ if and only if $CR(\lambda, \underline{A})C^{-1} \in \sigma_p(\mathcal{H}_c)$.

Consider in the space $\mathcal{H}_c$ an unbounded operator $\underline{A}_c$ defined by the equalities

$$D(\underline{A}_c) = \{v| \ v \in D(A_n C_{n-1}^{-1}) \oplus D(C_{n-1} C_{n-2}^{-1} \oplus \cdots \oplus D(C_1 C_0^{-1}),$$
$$\sum_{k=0}^{n_p} A_{pk} C_{n-n_p+k-s}^{-1} v_{n_p-k+s} = 0, s = 1, \ldots, n - n_p, \ p = 1, \ldots, m;$$
$$A_{pk} \in B(H_{n-n_p+k-s+1}, H_0^p), \ k = 0, \ldots, n_p\},$$

$$\underline{A}_c(v_1, \ldots, v_n) = C\underline{A}C^{-1}(v_1, \ldots, v_n)$$
$$= (C_{n-1} C_{n-2}^{-1} v_2, \ldots, C_1 C_0^{-1} v_n, -A_n C_{n-1}^{-1} v_1 - \cdots - A_1 C_0^{-1} v_n).$$

Let us show, if λ is a regular point of the pencil $\mathbb{L}(\lambda)$, that $\underline{A}_c - \lambda I$ is invertible. To find $(\underline{A}_c - \lambda I)^{-1}$ consider the equation $(\underline{A}_c - \lambda I)v = F$, which is equivalent to the system

$$C_{n-k} C_{n-k-1}^{-1} v_{k+1} - \lambda v_k = f_k, \qquad k = 1, \ldots, n - 1$$
$$-\sum_{k=1}^{n} A_{n-k+1} C_{n-k}^{-1} v_k - \lambda v_n = f_n, \tag{3.13}$$
$$A_{p0} C_{n-n_p-1}^{-1} v_{n_p+1} + \cdots + A_{pn_p} C_{n-1}^{-1} v_1 = 0, \qquad p = 1, \ldots, m$$

in the space $\mathcal{H}_c$.

From the first $n-1$ equations of system (3.13) we successively find

$$v_k = \lambda^{k-1}C_{n-k}C_{n-1}^{-1}v_1 + C_{n-k}\sum_{j=1}^{k-1}\lambda^{k-1-j}C_{n-j}^{-1}f_j, \qquad k = 2,\ldots,n. \qquad (3.14)$$

Substituting these values of v_k into the other equations of system (3.13) and taking into account that $C_0 = I$ we find

$$-L(\lambda)C_{n-1}^{-1}v_1 - \sum_{k=2}^{n}A_{n-k+1}\sum_{j=1}^{k-1}\lambda^{k-1-j}C_{n-j}^{-1}f_j = \sum_{j=1}^{n-1}\lambda^{n-j}C_{n-j}^{-1}f_j,$$

$$L_p(\lambda)C_{n-1}^{-1}v_1 + \sum_{k=0}^{n_p-1}A_{pk}\sum_{j=1}^{n_p-k}\lambda^{n_p-k-j}C_{n-j}^{-1}f_j = 0, \qquad p = 1,\ldots,m.$$

Hence,

$$v_1 = C_{n-1}\mathbb{L}^{-1}(\lambda)(-\sum_{k=2}^{n}A_{n-k+1}\sum_{j=1}^{k-1}\lambda^{k-1-j}C_{n-j}^{-1}f_j - \sum_{j=1}^{n-1}\lambda^{n-j}C_{n-j}^{-1}f_j,$$

$$-\sum_{k=0}^{n_1-1}A_{1k}\sum_{j=1}^{n_1-k}\lambda^{n_1-k-j}C_{n-j}^{-1}f_j,\ldots,-\sum_{k=0}^{n_m-1}A_{mk}\sum_{j=1}^{n_m-k}\lambda^{n_m-k-j}C_{n-j}^{-1}f_j).$$

Taking this expresion into account in (3.14) we find

$$v_p = C_{n-p}\mathbb{L}^{-1}(\lambda)(-\sum_{k=2}^{n}A_{n-k+1}\sum_{j=1}^{k-1}\lambda^{p+k-2-j}C_{n-j}^{-1}f_j$$

$$-\sum_{j=1}^{n-1}\lambda^{n+p-j-1}C_{n-j}^{-1}f_j, -\sum_{k=0}^{n_1-1}A_{1k}\sum_{j=1}^{n_1-k}\lambda^{n_1-k-j}C_{n-j}^{-1}f_j,\ldots,$$

$$-\sum_{k=0}^{n_m-1}A_{mk}\sum_{j=1}^{n_m-k}\lambda^{n_m-k-j}C_{n-j}^{-1}f_j)$$

$$+C_{n-k}\sum_{j=1}^{k-1}\lambda^{k-1-j}C_{n-j}^{-1}f_j, \qquad k = 2,\ldots,n.$$

By virtue of Lemma 3.1, condition 2 of the theorem is equivalent to the condition $C_{k-1}C_k^{-1} \in \sigma_p(H)$, $k = 1,\ldots,n$. Then, from

$$C_{n-k}\mathbb{L}^{-1}(\lambda) = C_{n-k}C_{n-k+1}^{-1} \cdot C_{n-k+1}\mathbb{L}^{-1}(\lambda), \qquad k = 1,\ldots,n,$$

$$C_{n-k}C_{n-k+1}^{-1} \in \sigma_p(H), \qquad k = 1,\ldots,n,$$

$$C_{n-k}C_{n-j}^{-1} = C_{n-k}C_{n-k+1}^{-1} \cdot C_{n-k+1}C_{n-j}^{-1} \in \sigma_p(H),$$

$$j = 1,\ldots,k-1,\ k = 1,\ldots,n,$$

and from boundedness of the operators

$$A_{n-k+1}C_{n-j}^{-1} = A_{n-k+1}C_{n-k+1}^{-1} \cdot C_{n-k+1}C_{n-j}^{-1},$$

$$j = 1, \ldots, k-1, \quad k = 1, \ldots, n,$$

and

$$A_{pk}C_{n-j}^{-1} = A_{pk}C_{n-n_p+k}^{-1} \cdot C_{n-n_p+k}C_{n-j}^{-1},$$

$$j = 1, \ldots, n_p - k, \quad k = 0, \ldots, n_p,$$

it follows that all elements of the matrix $\overline{R(\lambda, \underline{A}_c)}^5$ are operators, belonging to $\sigma_p(H)$. Hence $\overline{R(\lambda, \underline{A}_c)} \in \sigma_p(\overset{n}{\bigoplus} H)$. Indeed, if $B = (B_{ij})_1^n$ and all the $B_{ij} \in \sigma_p(H)$, then $B \in \sigma_p(\overset{n}{\bigoplus} H)$, since the operator-matrix B is represented as a sum of matrices, consisting of the operators B_{ij} at the intersection of the i-th row and j-th column, and all the remaining elements are zero operators. It is evident that any such operator-matrix belongs to $\sigma_p(\overset{n}{\bigoplus} H)$. So, $\overline{R(\lambda, \underline{A}_c)} \in \sigma_p(\overset{n}{\bigoplus} H)$. On the other hand, since $\mathcal{H}_c$ is a subspace $\overset{n}{\bigoplus} H$ then

$$s_j(R(\lambda, \underline{A}_c); \mathcal{H}_c) \le s_j(\overline{R(\lambda, \underline{A}_c)}; \overset{n}{\bigoplus} H), \quad j = 1, \ldots, \infty.$$

Indeed, if $\mathcal{H}_k^0$ is a subspace of $\mathcal{H}_k$, $k = 1, 2$, then

$$s_j^2(A; \mathcal{H}_1^0, \mathcal{H}_2^0) = \lambda_j(A^*A; \mathcal{H}_1^0, \mathcal{H}_2^0)$$

$$\le \lambda_j(A^*A; \mathcal{H}_1, \mathcal{H}_2) = s_j^2(A; \mathcal{H}, \mathcal{H}), \quad j = 1, \ldots, \infty.$$

Hence $R(\lambda, \underline{A}_c) \in \sigma_p(\mathcal{H}_c)$, i.e., condition 2 of Theorem 2.3 is satisfied. ∎

2.3.2. A system of operator pencils that are partially polynomially dependent on the parameter. Here we will give the formulation of the theorem for differential equations when boundary conditions do not depend on the spectral parameter.

Let H and H^p, $p = 1, \ldots, m$ be Hilbert spaces. Consider in H the problem

$$L(\lambda)u = \lambda^n u + \lambda^{n-1}A_1 u + \cdots + A_n u = 0,$$

$$L_p u = 0, \quad p = 1, \ldots, m. \tag{3.15}$$

From Theorem 3.4, in particular, follows

---

[5] $\overline{R(\lambda, \underline{A}_c)}$ is the extension of $R(\lambda, \underline{A}_c)$ onto $\overset{n}{\bigoplus} H$.

Theorem 3.5. *Let the following conditions be satisfied:*

(1) *there exist Hilbert spaces H_k, $k = 0, \ldots, n$, for which the compact embeddings $H_n \subset H_{n-1} \subset \cdots \subset H_0 = H$ take place and $\overline{H_n} = H$;*

(2) *for some $p > 0$ $J \in \sigma_p(H_k, H_{k-1})$, $k = 1, \ldots, n$;*

(3) *the operators A_k, $k = 1, \ldots, n$, from H_k into H, act boundedly;*

(4) *the operators L_p, $p = 1, \ldots, m$, from H_n into H^p, act boundedly;*

(5) *there exist Hilbert spaces H_0^p such that continuous embeddings $H^p \subset H_0^p$, $p = 1, \ldots, m$ hold and the linear manifold*

$$\{u \mid u \in H_k, \ L_p u = 0, \ L_p \in B(H_k, H_0^p), \ p = 1, \ldots, m\}$$

is dense in the Hilbert space

$$\{u \mid u \in H_{k-1}, \ L_p u = 0, \ L_p \in B(H_{k-1}, H_0^p), \ p = 1, \ldots, m\}$$

for $k = 2, \ldots, n$, and the linear manifold

$$\{u \mid u \in H_1, \ L_p u = 0, \ L_p \in B(H_1, H_0^p), \ p = 1, \ldots, m\}$$

is dense in the space H;

(6) *there exist rays ℓ_k with the angles between the neighboring rays less than π/p and a number q such, that*

$$\|L^{-1}(\lambda)\|_{B(H \oplus H^1 \oplus \cdots \oplus H^m, H_n)} \leq C|\lambda|^q, \qquad \lambda \in \ell_k, \ |\lambda| \to \infty.$$

Then the spectrum of problem (3.15) is discrete and a system of root vectors of problem (3.15) is n-fold complete in the spaces

$$\mathcal{H} = \bigoplus_{k=0}^{n-1} \{u \mid u \in H_{n-k-1}, \ L_p u = 0, \ L_p \in B(H_{n-k-1}, H_0^p), \ p = 1, \ldots, m\},$$

and

$$\mathcal{H}_1 = \bigoplus_{k=0}^{n-1} \{u \mid u \in H_{n-k}, \ L_p u = 0, \ L_p \in B(H_{n-k}, H_0^p), \ p = 1, \ldots, m\}.$$

Proof. For $n_p = 0$ conditions 1–4 and 6 coincide with the corresponding conditions of Theorem 3.4, and condition 5 of Theorem 3.4 for $n_p = 0$ is transformed into the following condition: the linear manifold

$$\mathcal{H}_1 = \{v \mid v = (v_1, \ldots, v_n), \ v_k \in H_{n+1-k}, \ L_p v_s = 0,$$
$$L_p \in B(H_{n+1-s}, H_0^p), \ s = 1, \ldots, n, \ p = 1, \ldots, m\}$$

is dense in the Hilbert space

$$\mathcal{H} = \{v \mid v = (v_1, \ldots, v_n), \ v_k \in H_{n-k}, \ L_p v_s = 0,$$
$$L_p \in B(H_{n-s}, H_0^p), \ s = 1, \ldots, n-1, \ p = 1, \ldots, m\}.$$

And this coincides with condition 5. ∎

Let us now give the other formulation of Theorem 3.5. Consider, in a Hilbert space H, the pencil

$$L(\lambda)u = \lambda^n u + \lambda^{n-1} A_1 u + \cdots + A_n u. \tag{3.16}$$

Theorem 3.6. *Let the following conditions be satisfied:*

(1) *there exist Hilbert spaces H_k, $k = 0, \ldots, n$, for which the compact embeddings $H_n \subset H_{n-1} \subset \cdots \subset H_0 = H$ take place and $\overline{H_k}|_{H_{k-1}} = H_{k-1}$, $k = 1, \ldots, n$;*

(2) *$J \in \sigma_p(H_k, H_{k-1})$, $k = 1, \ldots, n$, for some $p > 0$;*

(3) *the operators A_k from H_k into H act boundedly;*

(4) *there exist rays ℓ_k with the angles between the neighboring rays less than π/p and a number q such, that*

$$\|L^{-1}(\lambda)\|_{B(H, H_n)} \le C|\lambda|^q, \qquad \lambda \in \ell_k, \ |\lambda| \to \infty.$$

Then the spectrum of pencil (3.16) is discrete and the system of root vectors of pencil (3.16) is n-fold complete in the spaces $\mathcal{H} = H_{n-1} \oplus \cdots \oplus H_0$ and $\mathcal{H}_1 = H_n \oplus \cdots \oplus H_1$.

Proof. Repeat the proof of Theorem 3.4 with new spaces $\mathcal{H} = H_{n-1} \oplus \cdots \oplus H_0$ and $\mathcal{H}_1 = H_n \oplus \cdots \oplus H_1$. ∎

2.3.3. *n*-fold completeness of root vectors of a perturbed coercive operator pencil with a defect. In the theory of regular boundary value problems for ordinary differential equations and elliptic partial differential equations, *coercive operator pencils* (3.16) appear, i.e., operator pencils for which on some rays ℓ_k the estimate

$$\sum_{k=0}^{n} |\lambda|^{n-k} \|L^{-1}(\lambda)\|_{B(E, E_k)} \le C \tag{3.17}$$

holds.

However, in the theory of irregular boundary value problems *coercive operator pencils with a defect* appear, i.e., pencils for which the estimate (3.17) holds with some loss $\eta \in [0, n)$ in λ. Let us show that for such pencils it is possible to prove the perturbation theorem on the completeness of root vectors.

Consider, in a Banach space E, the following unbounded operator pencil

$$L(\lambda) = \lambda^n I + \lambda^{n-1}(A_1 + B_1) + \cdots + (A_n + B_n). \qquad (3.18)$$

Denote

$$\begin{aligned} L_0(\lambda) &= \lambda^n I + \lambda^{n-1} A_1 + \cdots + A_n, \\ L_1(\lambda) &= \lambda^{n-1} B_1 + \cdots + B_n. \end{aligned} \qquad (3.19)$$

Lemma 3.7. *Let the following conditions be satisfied:*

(1) A_j *are operators in E; there exist Banach spaces E_k, $k = 1, \ldots, n$, continuously embedded into $E_0 = E$, and a number $\eta \geq 0$ such that $D(L_0(\lambda)) = D(L(\lambda)) = \bigcap\limits_{k=1}^{n} E_k$ and*

$$\sum_{k=0}^{n} |\lambda|^{n-k-\eta} \|L_0^{-1}(\lambda)\|_{B(E, E_k)} \leq C, \qquad \lambda \in S, \; |\lambda| \to \infty,$$

where S is an unbounded set of the complex plane;

(2) B_k *are operators in E; under $k = 1, \ldots, [\eta]$ $B_k = 0$ and under $k = [\eta] + 1, \ldots, n$ $D(B) \supset \bigcap\limits_{k=1}^{n} E_k$ and for any $\varepsilon > 0$*

$$\|B_k u\| \leq \varepsilon \|u\|_{E_k}^{1-\eta/k} \|u\|^{\eta/k} + C(\varepsilon)\|u\|, \qquad u \in \bigcap_{k=1}^{n} E_k,$$

where $\eta = [\eta] + \{\eta\}$, $0 \leq \{\eta\} < 1$.

Then there exists $R > 0$ such that all complex numbers $\lambda \in S$, for which $|\lambda| > R$, are regular points of pencil (3.18) and the estimate

$$\sum_{k=0}^{n} |\lambda|^{n-k-\eta} \|L^{-1}(\lambda)\|_{B(E, E_k)} \leq C, \qquad \lambda \in S, \; |\lambda| \to \infty$$

holds.

Proof. Let us use the identity

$$L(\lambda) = [I + L_1(\lambda)L_0^{-1}(\lambda)]L_0(\lambda), \qquad \lambda \in S, \; |\lambda| \to \infty, \qquad (3.20)$$

where $L_0(\lambda)$, $L_1(\lambda)$ are defined by formulas (3.19). By virtue of conditions 1 and 2 for $\lambda \in S$, $|\lambda| \to \infty$ and for any $\varepsilon > 0$ we have

$$\sum_{k=[\eta]+1}^{n} |\lambda|^{n-k}\|B_k L_0^{-1}(\lambda)\| \leq \sum_{k=[\eta]+1}^{n} |\lambda|^{n-k}(\varepsilon \|L_0^{-1}(\lambda)\|_{B(E,E_k)}^{1-\eta/k}\|L_0^{-1}(\lambda)\|^{\eta/k}$$
$$+ C(\varepsilon)\|L_0^{-1}(\lambda)\|) \leq C\varepsilon + C(\varepsilon)|\lambda|^{\{\eta\}-1}.$$

Thus, for $\lambda \in S$, $|\lambda| \to \infty$

$$\|L_1(\lambda)L_0^{-1}(\lambda)\| \leq \sum_{k=[\eta]+1}^{n} |\lambda|^{n-k}\|B_k L_0^{-1}(\lambda)\| \leq q < 1. \tag{3.21}$$

Then from (3.20) is follows that for $\lambda \in S$, $|\lambda| \to \infty$ the operator $L(\lambda)$ is invertible in E and

$$L^{-1}(\lambda) = L_0^{-1}(\lambda)[I + L_1(\lambda)L_0^{-1}(\lambda)]^{-1}.$$

Hence, by virtue of condition 1 and (3.21) we have

$$\sum_{k=0}^{n} |\lambda|^{n-k-\eta}\|L^{-1}(\lambda)\|_{B(E,E_k)} \leq \sum_{k=0}^{n} |\lambda|^{n-k-\eta}\|L_0^{-1}(\lambda)\|_{B(E,E_k)}$$
$$\times \|[I + L(\lambda)L_0^{-1}(\lambda)]^{-1}\|_{B(E,E_k)} \leq C, \quad \lambda \in S, \ |\lambda| \to \infty. \ \blacksquare$$

For ideal pencils, i.e., for pencils in which the action law of subordinated terms has interpolation character, the following becomes useful.

Lemma 3.8. *Let the following conditions be satisfied:*

(1) A_k *are operators in* E; *there exist Banach spaces* E_k, $k = 1,\ldots,n$, *continuously embedded into* $E_0 = E$ *and a number* $\eta \geq 0$ *such that* $D(L_0(\lambda)) = D(L(\lambda)) = \bigcap_{k=1}^{n} E_k$ *and*

$$\sum_{k=0}^{n} |\lambda|^{n-k-\eta}\|L_0^{-1}(\lambda)\|_{B(E,E_k)} \leq C, \quad \lambda \in S, \ |\lambda| \to \infty,$$

where S *is an unbounded set of the complex plane;*

(2) B_k *are operators in* E; *under* $k = 1,\ldots,[\eta]$ $B_k = 0$ *and under* $k = [\eta]+1,\ldots,n$ $D(B_k) \supset \bigcap_{k=1}^{n} E_k$ *and for any* $\varepsilon > 0$

$$\|B_k u\| \leq \varepsilon \|u\|_{E_n}^{(k-\eta)/n}\|u\|^{1-(k-\eta)/n} + C(\varepsilon)\|u\|, \quad u \in \bigcap_{k=1}^{n} E_k.$$

Then there exists $R > 0$ such that all complex numbers $\lambda \in S$, for which $|\lambda| > R$, are regular points of the pencil (3.18) and the estimate

$$\sum_{k=0}^{n} |\lambda|^{n-k-\eta} \|L^{-1}(\lambda)\|_{B(E,E_k)} \leq C, \quad \lambda \in S, \ |\lambda| \to \infty$$

holds.

Proof. Under $\lambda \in S$, $|\lambda| \to \infty$ and $\varepsilon > 0$ we have

$$\sum_{k=[\eta]+1}^{n} |\lambda|^{n-k} \|B_k L_0^{-1}(\lambda)\| \leq \sum_{k=[\eta]+1}^{n} |\lambda|^{n-k} \left(\varepsilon \|L_0^{-1}(\lambda)\|_{B(E,E_k)}^{(k-\eta)/n} \|L_0^{-1}(\lambda)\|^{1-(k-\eta)/n} \right.$$
$$+ C(\varepsilon) \|L_0^{-1}(\lambda)\| \Big) \leq C\varepsilon + C(\varepsilon) |\lambda|^{\{\eta\}-1}.$$

The rest of the proof coincides with the proof of the above Lemma 3.7. ∎

Theorem 3.9. *Let the following conditions be satisfied:*

(1) *there exist Hilbert spaces H_k, $k = 0, \ldots, n$, for which the compact embeddings $H_n \subset H_{n-1} \subset \cdots \subset H_0 = H$ take place and $\overline{H_k}|_{H_{k-1}} = H_{k-1}$, $k = 1, \ldots, n$;*

(2) *$J \in \sigma_p(H_k, H_{k-1})$, $k = 1, \ldots, n$, for some $p > 0$;*

(3) *the operators A_k from H_k into H act boundedly;*

(4) *there exist rays ℓ_k with the angles between the neighboring rays less than π/p and a number $\eta \geq 0$ such that*

$$\sum_{k=0}^{n} |\lambda|^{n-k-\eta} \|L_0^{-1}(\lambda)\|_{B(H,H_k)} \leq C, \quad \lambda \in \ell_k, \ |\lambda| \to \infty;$$

(5) *B_k are operators in H; under $k = 1, \ldots, [\eta]$ $B_k = 0$ and under $k = [\eta] + 1, \ldots, n$ $D(B_k) \supset H_k$ and for any $\varepsilon > 0$*

$$\|B_k u\| \leq \varepsilon \|u\|_{H_k}^{1-\eta/k} \|u\|^{\eta/k} + C(\varepsilon) \|u\|, \quad u \in H_k.$$

Then the spectrum of operator pencil (3.18) is discrete and a system of root vectors of pencil (3.18) is n-fold complete in the spaces $H_n \oplus \cdots \oplus H_1$ and $H_{n-1} \oplus \cdots \oplus H_0$.

Proof. By virtue of Lemma 3.7, Theorem 3.6 is applicable to pencil (3.18), from which follows the Theorem 3.9 statement. ∎

In Theorems 3.4 – 3.9, the choices of the spaces H_k, $k = 1, \ldots, n$ are somewhat arbitrary. The spaces should be chosen to yield the minimal p, since the number of rays ℓ_k then also becomes minimal.

When applying these theorems to differential pencils, the spaces H_k, $k = 1, \ldots, n$ become optimal if they are chosen to the following principle: let A_k be a differential operator of order m_k, then

$$H_k = W_2^{m_n k/n}, \qquad m_k \leq m_n k/n,$$
$$H_k = W_2^{m_k}, \qquad m_k > m_n k/n.$$

Let us call the lower bound of numbers η, which satisfies the 4-th condition of Theorem 3.9, the *defect of coerciveness* or the *order of noncoerciveness* of the pencil $L_0(\lambda)$. From Theorem 3.9 it follows that all terms of the coercive pencil, i.e., the pencil $L_0(\lambda)$ that satisfies condition 4 of Theorem 3.9 when $\eta = 0$, admit the perturbation (the most possible) since, by virtue of Lemma 2.6, condition 5 is transformed into the condition of compactness of the operator B_k acting from H_k into H. The greater the defect of coerciveness η, the fewer terms that admit the perturbation, and the order of the perturbation operators becomes less. When the defect of the coerciveness is too high, i.e., $\eta > n$, then no term admits the perturbation.

Also note that the order of the perturbation operator may be more than the order of the perturbed operator. This happens when $m_k < m_n k/n$.

The differential pencils of normal type have

$$m_1 < m_2 < \cdots < m_n.$$

2.3.4. n-fold completeness of root vectors of an operator pencil with a weight. Consider in a Hilbert space H the unbounded operator pencil (3.16) with weight A, i.e. the pencil, for which there exists an unbounded invertible operator A in H such that operators $A_k A^{-k}$, $k = 1, \ldots, n$ in H act boundedly. The class of pencils with a weight is a subclass of the normal pencil class and does not coincide with the latter. However, as a rule, polynomial operator pencils with a weight appear in the theory of differential equations.

Lemma 3.10. *Let an operator A in a Banach space E be invertible and have a dense domain of definition $D(A)$. Then,*

$$\overline{E(A^k)}|_{E(A^{k-1})} = E(A^{k-1}), \qquad k = 1, \ldots, \infty.$$

Proof. Let $u \in E(A^{k-1})$. Denote $v = A^{k-1}u$. Since $\overline{D(A)} = E$ then there exists a sequence $v_n \in D(A)$ such that $\lim\limits_{n \to \infty} \|v_n - v\| = 0$. It is easy to see that $u_n = A^{-(k-1)}v_n \in D(A^k)$ and

$$\lim_{n \to \infty} \|A^{k-1}u_n - A^{k-1}u\| = \lim_{n \to \infty} \|v_n - v\| = 0. \qquad \blacksquare$$

Corollary 3.11. *Let the condition of Lemma 3.10 hold. Then*

$$\overline{E(A^k)}|_E = E, \qquad k = 1, \ldots, \infty.$$

The simple corollary of Theorem 3.9 is the following

Theorem 3.12. *Let the following conditions be satisfied:*

(1) *an operator A in H is invertible and has a dense domain of definition $D(A)$;*

(2) $J \in \sigma_p(H(A), H)$ *for some $p > 0$;*

(3) A_k *are operators in H, $D(A_k) \supset D(A^k)$, $D(A_n) = D(A^n)$ and operators $A_k A^{-k}$, $k = 1, \ldots, n$ in H act boundedly;*

(4) *there exist rays ℓ_k with the angles between the neighboring rays less than π/p and an integer m such that*

$$\|A^n L^{-1}(\lambda)\| \leq C|\lambda|^m, \qquad \lambda \in \ell_k, \ |\lambda| \to \infty,$$

where $L(\lambda) = \lambda^n I + \lambda^{n-1} A_1 + \cdots + A_n$.

Then the spectrum of operator pencil (3.16) is discrete and a system of root vectors of pencil (3.16) is n-fold complete in the spaces $H(A^n) \oplus \cdots \oplus H(A)$ and $H(A^{n-1}) \oplus \cdots \oplus H$.

Proof. It is obvious from Lemma 3.10 that Hilbert spaces $H_k = H(A^k)$, $k = 1, \ldots, n$ satisfy condition 1 of Theorem 3.6. By virtue of Lemma 3.1 the condition $J \in \sigma_p(H(A^k), H(A^{k-1}))$ is equivalent to the condition $J \in \sigma_p(H(A), H)$. Hence, from condition 2 follows condition 2 of Theorem 3.6. Conditions 3 and 4 of Theorem 3.6 coincide with conditions 3 and 4 of Theorem 3.12. ∎

Similarly, from Theorem 3.9 follows the completeness theorem of root vectors of a perturbed operator pencil with a weight.

Theorem 3.13. *Let the following conditions be satisfied:*

(1) *an operator A in H is invertible and has a dense domain of definition $D(A)$;*

(2) $J \in \sigma_p(H(A), H)$ *for some $p > 0$;*

(3) A_k *are operators in H, $D(A_k) \supset D(A^k)$, $D(A_n) = D(A^n)$ and operators $A_k A^{-k}$, $k = 1, \ldots, n$ in H act boundedly;*

(4) *there exist rays ℓ_k with the angles between the neighboring rays less than π/p and a number $\eta \geq 0$ such that*

$$\sum_{k=0}^{n} |\lambda|^{n-k-\eta} \|A^k L_0^{-1}(\lambda)\| \leq C, \qquad \lambda \in \ell_k, \ |\lambda| \to \infty,$$

where $L_0(\lambda) = \lambda^n I + \lambda^{n-1} A_1 + \cdots + A_n$;

(5) B_k are operators in H; when $k = 1, \ldots, [\eta]$ $B_k = 0$ and when $k = [\eta] + 1, \ldots, n$ $D(B_k) \supset D(A^k)$ and for any $\varepsilon > 0$

$$\|B_k u\| \leq \varepsilon \|A^k u\|^{1 - \eta/k} \|u\|^{\eta/k} + C(\varepsilon) \|u\|, \qquad u \in D(A^k).$$

Then the spectrum of operator pencil (3.18) is discrete and a system of root vectors of pencil (3.18) is n-fold complete in the spaces $H(A^n) \oplus \cdots \oplus H(A)$ and $H(A^{n-1}) \oplus \cdots \oplus H$.

2.3.5. n-fold completeness of root vectors for the Keldysh operator pencil. Consider, in a Hilbert space H, the operator pencil

$$L(\lambda) = \lambda^n I + \lambda^{n-1}(a_1 A + B_1) + \cdots + (a_n A^n + B_n), \qquad (3.22)$$

where A, B_k, $k = 1, \ldots, n$ are operators in H, a_k are complex numbers, $a_n \neq 0$. Let us show that for such pencils the formulation of completeness theorems of root vectors becomes simpler.

Denote by ω_k, $k = 1, \ldots, n$ roots of the characteristic equation

$$\omega^n + a_1 \omega^{n-1} + \cdots + a_n = 0.$$

Lemma 3.14. Let for some $\eta \geq 0$

$$|\lambda|^{1-\eta} \|R(\lambda, A)\| \leq C, \qquad \lambda \in \ell(a\omega_k^{-1}, \varphi - \arg \omega_k), \ |\lambda| \to \infty.$$

Then for the operator pencil

$$L_0(\lambda) = \lambda^n I + \lambda^{n-1} a_1 A + \cdots + a_n A^n$$

the estimate

$$\sum_{k=0}^{n} |\lambda|^{n-k-n\eta} \|A^k L_0^{-1}(\lambda)\| \leq C, \qquad \lambda \in \ell(a, \varphi), \ |\lambda| \to \infty$$

is valid.

Proof. Let us use the relation

$$L_0(\lambda) = \prod_{k=1}^{n} (\lambda I - \omega_k A) = \prod_{k=1}^{n} \omega_k (\lambda \omega_k^{-1} I - A), \qquad (3.23)$$

valid for any $\lambda \in \mathbb{C}$. Since for $k = 1, \ldots, n$

$$|\lambda|^{1-\eta}\|R(\lambda\omega_k^{-1}, A)\| \leq C, \qquad \lambda \in \ell(a, \varphi), \ |\lambda| \to \infty, \tag{3.24}$$

then from

$$(A - \lambda\omega_k^{-1}I)R(\lambda\omega_k^{-1}, A) = I$$

it follows that for $k = 1, \ldots, n$, $\lambda \in \ell(a, \varphi)$, $|\lambda| \to \infty$

$$|\lambda|^{-\eta}\|AR(\lambda\omega_k^{-1}, A)\| \leq C\left(|\lambda|^{-\eta} + |\lambda|^{1-\eta}\|R(\lambda\omega_k^{-1}, A)\|\right) \leq C. \tag{3.25}$$

From (3.23)–(3.25) it follows that for $k = 0, \ldots, n$, $\lambda \in \ell(a, \varphi)$, $|\lambda| \to \infty$ the estimate

$$|\lambda|^{n-k-n\eta}\|A^k L_0^{-1}(\lambda)\| \leq |\lambda|^{-k\eta} \prod_{j=1}^{k} |\omega_j| \ \|A(\lambda\omega_j^{-1}I - A)^{-1}\|$$

$$\times |\lambda|^{(n-k)(1-\eta)} \prod_{j=k+1}^{n} |\omega_j| \ \|(\lambda\omega_j^{-1}I - A)^{-1}\| \leq C$$

is valid. ■

Theorem 3.15. *Let the following conditions be satisfied:*

(1) *an operator A in H is invertible and has a dense domain of definition $D(A)$;*

(2) *$J \in \sigma_p(H(A), H)$ for some $p > 0$;*

(3) *there exist rays $\ell_k(a, \varphi_k)$ with the angles between the neighboring rays less than π/p and a number $\eta \geq 0$ such that*

$$|\lambda|^{1-\eta}\|R(\lambda, A)\| \leq C, \qquad \lambda \in \ell_{kj}(a\omega_j^{-1}, \varphi_k - \arg\omega_j), \ |\lambda| \to \infty;$$

(4) *B_k are operators in H; when $k = 1, \ldots, [n\eta]$ $B_k = 0$ and when $k = [n\eta] + 1, \ldots, n$ $D(B_k) \supset D(A^k)$ and for any $\varepsilon > 0$*

$$\|B_k u\| \leq \varepsilon\|A^k u\|^{1-n\eta/k}\|u\|^{n\eta/k} + C(\varepsilon)\|u\|, \qquad u \in D(A^k).$$

Then the spectrum of operator pencil (3.22) is discrete and a system of root vectors of pencil (3.22) is n-fold complete in the spaces $H(A^n) \oplus \cdots \oplus H(A)$ and $H(A^{n-1}) \oplus \cdots \oplus H$.

Proof. By virtue of Lemma 3.14, Theorem 3.15 is a special case of Theorem 3.13. ■

Theorem 3.16. *Let the following conditions be satisfied:*

(1) *an operator A in H is invertible and selfadjoint;*

(2) *$J \in \sigma_p(H(A), H)$ for some $p > 0$;*

(3) *B_k are operators in H and operators $B_k A^{-k}$, $k = 1, \ldots, n$, in H, are compact.*

Then the spectrum of operator pencil (3.22) is discrete, under any $\varepsilon > 0$ outside the angles $|\arg\lambda - \arg\omega_k| < \varepsilon$, $|\arg\lambda - \arg\omega_k - \pi| < \varepsilon$, $k = 1,\ldots,n$ there exists a finite number of eigenvalues, and a system of root vectors of pencil (3.22) is n-fold complete in the spaces $H(A^n) \oplus \cdots \oplus H(A)$ and $H(A^{n-1}) \oplus \cdots \oplus H$.

Proof. The estimate

$$|\lambda|\,\|R(\lambda,A)\| \le C, \qquad \lambda \in \ell(0,\varphi),\ \varphi \ne 0,\ \varphi \ne \pi$$

was proved in Theorem 2.11 for a selfadjoint operator A. Hence the operator A satisfies condition 3 of Theorem 3.15 when $\eta = 0$. Then, from Lemma 2.5 it follows that condition 3 of our Theorem and condition 4 of Theorem 3.15 are equivalent. So, Theorem 3.15 is applicable to the pencil (3.22). The fact that outside the angles $|\arg\lambda - \arg\omega_k| < \varepsilon$, $|\arg\lambda - \arg\omega_k - \pi| < \varepsilon$, $k = 1,\ldots,n$ there is a finite number of eigenvalues follows from Lemmas 3.14 and 3.7. ∎

2.3.6. $(u(0), u''(0))$**-completeness of root vectors, corresponding to a part of the operator pencil spectrum.** When nonstationary equations are solved, the question about completeness of root vectors, corresponding to the whole spectrum, arises. However, in case of stationary equations, the question applies to the completeness of root vectors, corresponding to some part of the spectrum.

Let us consider, in a Hilbert space H, the following unbounded operator pencil

$$L(\lambda) = \lambda^4 I + \lambda^2 A_2 + A_4. \tag{3.26}$$

Let H_1, H_2 be Hilbert spaces continuously embedded in H. A system of root vectors of the pencil (3.26), corresponding to the eigenvalues λ_i with $\mathrm{Re}\lambda_i < 0$, is said to be $(u(0), u''(0))$-complete in the space $H_1 \oplus H_2$, if the system of vectors

$$(u_i(0), u_i''(0)),$$

where $u_i(t)$ are elementary solutions

$$u_i(t) = e^{\lambda_i t}\Big(\frac{t^{k_i}}{k_i!}u_{i0} + \frac{t^{k_i-1}}{(k_i-1)!}u_{i1} + \cdots + u_{ik_i}\Big)$$

of the equation

$$u''''(t) + A_2 u''(t) + A_4 u(t) = 0, \quad t > 0, \tag{3.27}$$

corresponding to the eigenvalues λ_i with $\mathrm{Re}\lambda_i < 0$, is complete in the space $H_1 \oplus H_2$.

Theorem 3.17. *Let the following conditions be satisfied:*

(1) *there exist Hilbert spaces H, H_2, H_4 for which the compact embeddings $H_4 \subset H_2 \subset H_0 = H$ take place; $\overline{H_2}|_H = H$ and $\overline{H_4}|_{H_2} = H_2$;*

(2) *operators A_k, $k = 2, 4$, from H_k into H, act boundedly;*

(3) *$J \in \sigma_p(H_4, H_2)$, $J \in \sigma_p(H_2, H)$ for some $p > 0$;*

(4) *there exist rays ℓ_k with angles between the neighboring rays less than $\pi/2p$ and a number q such, that*

$$\|L^{-1}(\lambda)\|_{B(H, H_4)} \le C|\lambda|^q, \quad \lambda \in \ell_k, \ |\lambda| \to \infty;$$

Then the spectrum of pencil (3.26) is discrete and a system of root vectors of pencil (3.26), corresponding to the eigenvalues λ_i with $\operatorname{Re}\lambda_i < 0$, is $(u(0), u''(0))$-complete in the spaces $H_4 \oplus H_2$ and $H_2 \oplus H$.

Proof. If the function $u(t)$ is an elementary solution to equation (3.27), then the function

$$v(t) = \begin{pmatrix} v_1(t) \\ v_2(t) \end{pmatrix} = \begin{pmatrix} u(t) \\ u''(t) \end{pmatrix} \tag{3.28}$$

is an elementary solution of the equation

$$v''(t) - Gv(t) = 0, \tag{3.29}$$

where

$$G = \begin{pmatrix} 0 & I \\ -A_4 & -A_2 \end{pmatrix}.$$

Conversely, if the function $v(t) = (v_1(t), v_2(t))$ is an elementary solution to equation (3.29), then the function $u(t) = v_1(t)$ is an elementary solution to equation (3.27).

Equation (3.29) and the operator G are considered in the space $\mathcal{H} = H_2 \oplus H$. Let $D(G) = H_4 \oplus H_2$.

It is easy to see that the set of root vectors coincides with the set of values of the elementary solutions in zero. Therefore, the set of root vectors of the operator G coincides with the set

$$\{v(0)\} = \{(v_1(0), v_2(0))\}, \tag{3.30}$$

where $\{v(t)\}$ is the set of elementary solutions to equation (3.29). According to (3.28) the set (3.30) has the form

$$\{v(0)\} = \{(u(0), u''(0))\},$$

where $\{u(t)\}$ is the set of elementary solutions to equation (3.27).

Let us apply Thoerem 2.3 to the operator G. By virtue of condition 1, $D(G) = H_4 \oplus H_2$ dense in $\mathcal{H} = H_2 \oplus H$, i.e., condition 1 of Theorem 2.3 is satisfied. We will show that the resolvent of operator G is compact. Let $F = (f_1, f_2) \in \mathcal{H} = H_2 \oplus H$. From the first equation of the system

$$(\lambda^2 I - G)v = F \qquad (3.31)$$

we find

$$v_2 = \lambda^2 v_1 - f_1.$$

Substituting this expression into the second equation of system (3.31) we have

$$\lambda^2(\lambda^2 v_1 - f_1) = -A_4 v_1 - A_2(\lambda^2 v_1 - f_1) + f_2.$$

Hence,

$$L(\lambda)v_1 = \lambda^2 f_1 + A_2 f_1 + f_2,$$

i.e.,

$$v_1 = \lambda^2 L^{-1}(\lambda)f_1 + L^{-1}(\lambda)A_2 f_1 + L^{-1}(\lambda)f_2. \qquad (3.32)$$

Consequently,

$$v_2 = \lambda^4 L^{-1}(\lambda)f_1 + \lambda^2 L^{-1}(\lambda)A_2 f_1 - f_1 + \lambda^2 L^{-1}(\lambda)f_2. \qquad (3.33)$$

These formulas and condition 1 imply compactness of the resolvent

$$R(\lambda^2, G) = - \begin{pmatrix} \lambda^2 L^{-1}(\lambda) + L^{-1}(\lambda)A_2 & L^{-1}(\lambda) \\ \\ \lambda^4 L^{-1}(\lambda) + \lambda^2 L^{-1}(\lambda)A_2 - I & \lambda^2 L^{-1}(\lambda) \end{pmatrix}.$$

From (3.32), (3.33) and conditions 2 and 4 it follows that under some r we have

$$\|R(\lambda^2, G)F\| \le C\left(\|A_2 v_1\| + \|v_1\| + \|v_2\|\right) \le C|\lambda|^r \|F\|, \qquad \lambda \in \ell_k, \ |\lambda| \to \infty,$$

i.e., condition 3 of Theorem 2.3 is satisfied.

It remains to be shown only that $R(\lambda^2, G) \in \sigma_p(\mathcal{H})$. Consider a selfadjoint positively-defined operator C in H such that $H(C_k) = H_{2k}$, $k = 1, 2$, [44, Ch.1, §2.1]. Consider, in the space $H^2 = H \oplus H$, an operator S defined by the equalities

$$D(S) = H_2 \oplus H, \qquad S = \mathrm{diag}(C_1, I).$$

The operator S from $\mathcal{H} = H_2 \oplus H$ into $H^2 = H \oplus H$ acts boundedly, is invertible and $H^2(S) = \mathcal{H}$. Then, by virtue of Lemma 3.1 $R(\lambda, G) \in \sigma_p(\mathcal{H})$ if and only if $SR(\lambda, G)S^{-1} \in \sigma_p(H^2)$. Obviously,

$$SR(\lambda^2, G)S^{-1} = - \begin{pmatrix} C_1(\lambda^2 L^{-1}(\lambda) + L^{-1}(\lambda)A_2)C_1^{-1} & C_1 L^{-1}(\lambda) \\ \\ (\lambda^4 L^{-1}(\lambda) + \lambda^2 L^{-1}(\lambda)A_2 - I)C_1^{-1} & \lambda^2 L^{-1}(\lambda) \end{pmatrix} \quad (3.34)$$

Since $H(C_k) = H_{2k}$, then by virtue of Lemma 3.1 the condition $J \in \sigma_p(H_4, H_2) = \sigma_p(H(C_2), H(C_1))$ is equivalent to the condition $C_1 C_2^{-1} \in \sigma_p(H)$ and the condition $J \in \sigma_p(H_2, H) = \sigma_p(H(C_1), H)$ is equivalent to the condition $C_1^{-1} \in \sigma_p(H)$. Then,

$$L^{-1}(\lambda) = C_1^{-1} \cdot C_1 C_2^{-1} \cdot C_2 L^{-1}(\lambda) \in \sigma_p(H),$$
$$C_1 L^{-1}(\lambda) = C_1 C_2^{-1} \cdot C_2 L^{-1}(\lambda) \in \sigma_p(H),$$
$$(\lambda^4 L^{-1}(\lambda) + \lambda^2 L^{-1}(\lambda)A_2 - I)C_1^{-1} \in \sigma_p(H)$$

and by virtue of (3.34)
$$SR(\lambda^2, G)S^{-1} \in \sigma_p(H^2),$$

i.e., condition 2 of Theorem 2.3 is satisfied. ∎

2.4. Completeness of root vectors of a system of unbounded linear operator pencils

In 2.3 systems of operator pencils were discussed, in which the powers of the parameter in the "boundary value conditions" are less than the parameter power in the "equation". Now let us discuss the case in which the powers of the parameter in the "boundary value conditions" and in the "equations" can be equal.

Let H and H^p, $p = 1, \ldots, m$, be Hilbert spaces. Consider, in H, the problem for a system of linear operator pencils

$$\begin{aligned} L(\lambda)u &= \lambda u + Au = 0, \\ L_p(\lambda)u &= \lambda A_{p0}u + A_{p1}u = 0, \quad p = 1, \ldots, m, \end{aligned} \quad (4.1)$$

where A is an unbounded operator in H and A_{pk}, $k = 0, 1$, are, generally speaking, unbounded operators from H into H^p.

Elements $u_0, u_1, \ldots, u_k$ of H, satisfying the relations

$$\begin{aligned} (\lambda_0 I + A)u_q + u_{q-1} &= 0, \\ (\lambda_0 A_{p0} + A_{p1})u_q + A_{p0}u_{q-1} &= 0, \quad p = 1, \ldots, m, \end{aligned} \quad (4.2)$$

where $q = 0, \ldots, k$ and $u_{-1} = 0$, are called *root vectors* of problem (4.1), corresponding to the eigenvalue λ_0. The vector $u_0 \neq 0$ is called an *eigenvector* of problem (4.1).

Let $s = 0, \ldots, m$ be such an integer that $A_{p0} \neq 0$ for $p = 1, \ldots, s$ and $A_{p0} = 0$ for $p = s + 1, \ldots, m$. It is obvious that, with the help of the new enumeration, it can be always done.

Definitions of regular points, spectrum of problem (4.1) and discreteness of the spectrum are analogous to the ones in 3.1.

Theorem 4.1. *Let the following conditions be satisfied:*

(1) *there exists a Hilbert space H_1 such that the embedding $H_1 \subset H_0 = H$ is compact and dense; the operator A from H_1 into H acts boundedly;*

(2) *there exist Hilbert spaces H_0^p, $p = 1, \ldots, s$, such that the embeddings $H_0^p \subset H^p$, $p = 1, \ldots, s$ are compact and dense;*

(3) *for some $p > 0$ $J \in \sigma_p(H_1, H)$, $J_k \in \sigma_p(H_0^k, H^k)$, $k = 1, \ldots, s$;*

(4) *the operators A_{p0}, $p = 1, \ldots, s$, from H_1 into H_0^p, act boundedly;*

(5) *the operators A_{p1}, $p = 1, \ldots, m$, from H_1 into H^p, act boundedly;*

(6) *the linear manifold*

$$\{v \mid v = (u, A_{10}u, \ldots, A_{s0}u), \ u \in H_1, \ A_{p1}u = 0, \ p = s + 1, \ldots, m\},$$

is dense in the Hilbert space $H \oplus H^1 \oplus \cdots \oplus H^s$;

(7) *there exist rays ℓ_k with the angles between the neighboring rays less than π/p and a number q such that*

$$\|L^{-1}(\lambda)\|_{B(H \underset{p=1}{\overset{s}{\oplus}} H^p \underset{s+1}{\overset{m}{\oplus}} \{0\}, H_1)} \leq C|\lambda|^q, \qquad \lambda \in \ell_k, \ |\lambda| \to \infty,$$

where the pencil $\mathbb{L}(\lambda) = (L(\lambda), L_1(\lambda), \ldots, L_m(\lambda))$ acts from H into $H \oplus H^1 \oplus \cdots \oplus H^m$.

Then the spectrum of problem (4.1) is discrete and a system of vectors $(u_k, A_{10}u_k, \ldots, A_{s0}u_k)$, where u_k are root vectors of problem (4.1), is complete in the space $H \oplus H^1 \oplus \cdots \oplus H^s$.

Proof. Consider, in the Hilbert space $\mathcal{H} = H \oplus H^1 \oplus \cdots \oplus H^s$, the operator $\underline{A}$, given by the equalities

$$D(\underline{A}) = \{v \mid v = (u, A_{10}u, \ldots, A_{s0}u), \ u \in H_1,$$
$$A_{p1}u = 0, \ p = s + 1, \ldots, m\},$$
$$\underline{A}(u, A_{10}u, \ldots, A_{s0}u) = (-Au, -A_{11}u, \ldots, -A_{s1}u).$$

It is easy to see that if $u_0, u_1, \ldots, u_k$ are root vectors of problem (4.1), i.e., they satisfy relations (4.2), then the vectors $v_p = (u_p, A_{10}u_p, \ldots, A_{s0}u_p)$ are root vectors of the problem

$$(\lambda I - \underline{A})v = 0,$$

i.e., they satisfy the relations

$$(\lambda I - \underline{A})v_p + v_{p-1} = 0, \qquad p = 0, \ldots, k,$$

where $v_{-1} = 0$, and conversely.

Let us apply Theorem 2.3 to the operator $\underline{A}$. By virtue of condition 6, $D(\underline{A})$ is dense in $\mathcal{H}$. We will show that the resolvent of operator $\underline{A}$ is compact. For this instead of the equation

$$(\underline{A} - \lambda I)v = F \tag{4.3}$$

we solve the system

$$\begin{aligned}
-Au - \lambda u &= f, \\
-A_{p1}u - \lambda A_{p0}u &= f_p, \qquad p = 1, \ldots, s, \\
A_{p1}u &= 0, \qquad p = s+1, \ldots, m,
\end{aligned} \tag{4.4}$$

which is equivalent to (4.3) in the space H_1.

If λ is a regular point of the pencil $\mathbb{L}(\lambda) = (L(\lambda), L_1(\lambda), \ldots, L_m(\lambda))$, acting from H into $H \oplus H^1 \oplus \cdots \oplus H^m$, then by virtue of condition 7, problem (4.4) has a unique solution

$$u = \mathbb{L}^{-1}(\lambda)(-f, -f_1, \ldots, -f_s, 0, \ldots, 0),$$

moreover $u \in H$. So, a solution to (4.3) has the following form

$$\begin{aligned}
v &= (\mathbb{L}^{-1}(\lambda)(-f, -f_1, \ldots, -f_s, 0, \ldots, 0), \\
&\quad A_{p0}\mathbb{L}^{-1}(\lambda)(-f, -f_1, \ldots, -f_s, 0, \ldots, 0), \ p = 1, \ldots, s).
\end{aligned} \tag{4.5}$$

Then, by virtue of conditions 1, 2 and 4 the operator $R(\lambda, \underline{A})$ is compact.

From conditions 2, 4 and 7 follows the estimate

$$\left\| A_{r0}\mathbb{L}^{-1}(\lambda) \right\|_{B(H \overset{\circ}{\underset{p=1}{\oplus}} H^p \overset{m}{\underset{s+1}{\oplus}} \{0\}, H^r)} \leq C|\lambda|^q, \qquad \lambda \in \ell_k, \ |\lambda| \to \infty,$$

for $r = 1, \ldots, s$. Hence

$$\|R(\lambda, \underline{A})\| \leq C|\lambda|^q, \qquad \lambda \in \ell_k, \ |\lambda| \to \infty.$$

It remain to be shown that $R(\lambda, \underline{A}) \in \sigma_p(\mathcal{H})$. This follows from (4.5) and conditions 3-4. Indeed, from (4.5) and condition 4 it follows that $R(\lambda, \underline{A})$ acts boundedly from $\mathcal{H}$ into $H \oplus H_0^1 \oplus \cdots \oplus H_0^s$. On the other hand, we have

$$s_j(R(\lambda, \underline{A}); \mathcal{H}, \mathcal{H}) \leq s_j(\underline{J}R(\lambda, \underline{A}); \mathcal{H}, \mathcal{H})$$

$$\leq \|R(\lambda, \underline{A})\|_{B(\mathcal{H}, H_1 \oplus H_0^1 \oplus \cdots \oplus H_0^s)} s_j(\underline{J}; H_1 \oplus H_0^1 \oplus \cdots \oplus H_0^s, \mathcal{H}),$$

where $\underline{J} = \mathrm{diag}(J, J_1, \ldots, J_s)$. Hence, by virtue of condition 3, we have $R(\lambda, \underline{A}) \in \sigma_p(\mathcal{H})$. ∎

Chapter 3

Principally boundary value problems for ordinary differential equations with a polynomial parameter

3.0. Introduction

Many papers and books have dealt with the analysis of regular ordinary differential operators. It is sufficient to mention here such well-known books as those by J. D. Tamarkin [69], M. A. Naimark [53], M. L. Rasulov [57], N. M. Dunford and J. T. Schwartz [13]. As a rule the asymptotic methods were used, and, as a result, the investigation schemes and obtained results were close enough. We apply the modified method of Schauder, developed in details for the elliptic partial regular differential operators. Our class of regular differential operators is wider than that described by Birkhoff-Tamarkin. In this chapter principally boundary value problems for ordinary differential equations with variable coefficients are analyzed. The spectral parameter appears polynomially in both the equation and the functional conditions, and the weight of the problem $d = m/n$ is an arbitrary number. More precisely, we consider the equation

$$L(\lambda)u = \lambda^n u(x) + \sum_{k=1}^{n} \lambda^{n-k}(a_k(x)u^{(dk)}(x) + B_k u|_x) = f(x), \qquad (0.1)$$

with the functional conditions

$$L_s(\lambda)u = \sum_{k=0}^{n_s} \lambda^k [\alpha_{sk} u^{(m_s-dk)}(0) + \beta_{sk} u^{(m_s-dk)}(1)$$

$$+ \sum_{i=1}^{N_{sk}} \delta_{ski} u^{(m_s-dk)}(x_{ski}) + T_{sk} u] = f_s, \quad s = 1, \ldots, m, \qquad (0.2)$$

where $n \geq 1$, $m \geq 1$, $m_s \geq dn_s$; α_{sk}, β_{sk}, δ_{ski}, f_s are complex numbers; $x_{ski} \in (0,1)$; $a_k(x)$ are numerical functions, defined on $[0,1]$; $a_k(x) = \alpha_{sk} = \beta_{sk} = \delta_{ski} = 0$ if dk is not an integer; B_k are operators in $L_q(0,1)$ and T_{sk} are functionals in $L_q(0,1)$, $q \in [1,\infty)$.

Here both the operators B_k and the functionals T_{sk}, generally speaking, are unbounded.

The operator $L(\lambda)$ has on λ a power n and an order m. The operator $L_s(\lambda)$ has on λ a power $\leq n_s$ and an order $\leq m_s$. It is important to correctly define the numbers m_s. This is done in the following way: first the forces of all terms of the operator $L_s(\lambda)$ are defined, then the greatest of them is taken. In addition, the force of the member $\lambda^k u^{(j)}$ is equal to $mk/n + j$. For example, for $n = 1$, $m = 2$ and for the boundary value condition $L_1(\lambda)u = u'(0) + \lambda u(1) = 0$ we have $m_1 = 2$. With such a definition m_s becomes, generally speaking, the force of the operator $L_s(\lambda)$, but not its order.

3.0.1. p-regular functional conditions. The functions $\omega_j(x)$, $j = 1, \ldots, m$ are called p-separated, if there exists a straight line P passing through 0, such that no value of the functions $\omega_j(x)$ lies on it and $\omega_1(x), \ldots, \omega_p(x)$ are on one side of P while $\omega_{p+1}(x), \ldots, \omega_m(x)$ – on the other.

The conditions (0.2) are called p-regular with respect to some functions $\omega_j(x)$, $j = 1, \ldots, m$, if:

a) the functions $\omega_j(x)$ are p-separated and $\theta(0) \neq 0$, $\theta(1) \neq 0$, where

$$\theta(x) = \begin{vmatrix} \sum_{k=0}^{n_1} \alpha_{1k}\omega_1^{m_1-dk}(x) & \cdots & \sum_{k=0}^{n_1} \beta_{1k}\omega_{p+1}^{m_1-dk}(x) & \cdots \\ \vdots & \cdots & \vdots & \cdots \\ \sum_{k=0}^{n_m} \alpha_{mk}\omega_1^{m_m-dk}(x) & \cdots & \sum_{k=0}^{n_m} \beta_{mk}\omega_{p+1}^{m_m-dk}(x) & \cdots \end{vmatrix}; \quad (0.3)$$

b) $x_{ski} \in (0,1)$; for some $r \in [1,\infty)$ the functionals T_{sk} are continuous in $W_r^{m_s-dk}(0,1)$.

The problem (0.1)-(0.2) is called p-regular, if:

(1) the conditions (0.2) are p-regular with respect to the roots of the characteristic equation

$$a_n(x)\omega^m + a_{n-1}(x)\omega^{d(n-1)} + \cdots + 1 = 0, \quad x \in [0,1];$$

(2) operators B_k from $W_q^{dk}(0,1)$ into $L_q(0,1)$ are compact.

Further, for expliciteness, we will write out conditions b) and 2 every time they are used, in spite of it being in the definition of p-regularity.

Here the case $p = 0$ or $p = m$ is also admitted.

If the boundary value conditions (0.2) are local, i.e., they are given only in 0 or in 1, then it follows from the p-regularity of boundary value conditions that the number of them in 0 is equal to p and is equal to $m - p$ in 1.

3.1. Coerciveness of principally boundary value problems for ordinary differential equations with a polynomial parameter

In this section it is shown that the p-regularity condition of the problem (0.1)–(0.2) ensures the coerciveness of the problem (0.1)–(0.2) both in the space variable and in the spectral parameter. In contrast to the coerciveness in the space variable, the coerciveness in the spectral parameter may not exist in ordinary differential equations.

3.1.1. Coerciveness of the principal part of a boundary value problem for equations with constant coefficients.
Consider an ordinary differential equation with constant coefficients and with weight 1 on the whole axis

$$L_0(\lambda)u = \lambda^m u(x) + \lambda^{m-1} a_1 u'(x) + \cdots + a_m u^{(m)}(x) = f(x), \qquad (1.1)$$

where $m \geq 1$, a_k are complex numbers.

Let us enumerate the roots of the characteristic equation

$$a_m \omega^m + a_{m-1} \omega^{m-1} + \cdots + 1 = 0 \qquad (1.2)$$

ω_j, $j = 1, \ldots, m$. Let numbers ω_j be p-separated.

Let us denote

$$\begin{aligned}
\underline{\omega} &= \min\{\arg \omega_1, \ldots, \arg \omega_p, \arg \omega_{p+1} + \pi, \ldots, \arg \omega_m + \pi\}, \\
\overline{\omega} &= \max\{\arg \omega_1, \ldots, \arg \omega_p, \arg \omega_{p+1} + \pi, \ldots, \arg \omega_m + \pi\},
\end{aligned} \qquad (1.3)$$

and the value $\arg \omega_j$ is chosen up to a multiple of 2π, so that $\overline{\omega} - \underline{\omega} < \pi$.

Theorem 1.1. Let $a_m \neq 0$ and the roots of the equation (1.2) be p-separated. Then for any $\varepsilon > 0$ and for all complex numbers λ satisfying

$$\pi/2 - \underline{\omega} + \varepsilon < \arg \lambda < 3\pi/2 - \overline{\omega} - \varepsilon,$$

the operator $L_0(\lambda) : u \to L_0(\lambda)u$ from $W_q^z(\mathbb{R})$ onto $W_q^{z-m}(\mathbb{R})$, where an integer $z \geq m$, $q \in (1, \infty)$, is an isomorphism and for these λ the following estimates hold for a solution of (1.1)

$$\sum_{k=0}^{z} |\lambda|^{z-k} \|u\|_{W_q^k(\mathbb{R})} \leq C(\varepsilon)(\|f\|_{W_q^{z-m}(\mathbb{R})} + |\lambda|^{z-m} \|f\|_{L_q(\mathbb{R})}) \qquad (1.4)$$

and

$$\sum_{k=0}^{m} |\lambda|^{m-k} \|u^{(k+s)}\|_{L_q(\mathbb{R})} \leq C(\varepsilon)\|f^{(s)}\|_{L_q(\mathbb{R})}, \qquad s \geq 0. \tag{1.4'}$$

Proof. The operator $\mathbb{L}_0(\lambda)$ acts from $W_q^z(\mathbb{R})$ into $W_q^{z-m}(\mathbb{R})$ linearly and continuously. Let us prove that if $f \in W_q^{z-m}(\mathbb{R})$ then equation (1.1) has a solution $u(x)$, belonging to $W_q^z(\mathbb{R})$.

From (1.1) we obtain

$$(\lambda^m + \lambda^{m-1}a_1(i\sigma) + \cdots + a_m(i\sigma)^m)Fu = Ff,$$

where $F\varphi = (F\varphi)(\sigma)$ is the Fourier transform. It is obvious that

$$\lambda^m + \lambda^{m-1}a_1(i\sigma) + \cdots + a_m(i\sigma)^m = a_m \prod_{j=1}^{m}(i\sigma - \omega_j\lambda). \tag{1.5}$$

Since for $\pi/2 - \underline{\omega} + \varepsilon < \arg\lambda < 3\pi/2 - \overline{\omega} - \varepsilon$, $\sigma \in \mathbb{R}$, we have

$$|i\sigma - \omega_j\lambda| \geq C(\varepsilon)(|\sigma| + |\lambda|), \qquad j = 1, \ldots, m, \tag{1.6}$$

then

$$Fu = (\lambda^m + \lambda^{m-1}a_1(i\sigma) + \cdots + a_m(i\sigma)^m)^{-1}Ff.$$

Hence

$$u^{(k)}(x) = F^{-1}(i\sigma)^k Fu$$
$$= F^{-1}(i\sigma)^k(\lambda^m + \lambda^{m-1}a_1(i\sigma) + \cdots + a_m(i\sigma)^m)^{-1}Ff. \tag{1.7}$$

From (1.5) and (1.6) it follows that functions

$$T_k(\sigma, \lambda) = \lambda^{m-k}(i\sigma)^k(\lambda^m + \lambda^{m-1}a_1(i\sigma) + \cdots + a_m(i\sigma)^m)^{-1}, \qquad k = 0, \ldots, m,$$

for $\pi/2 - \underline{\omega} + \varepsilon < \arg\lambda < 3\pi/2 - \overline{\omega} - \varepsilon$ are continuously differentiable in σ on $\mathbb{R}$ and

$$|T_k(\sigma, \lambda)| < C(\varepsilon), \qquad |\frac{\partial}{\partial\sigma}T_k(\sigma, \lambda)| < \frac{C(\varepsilon)}{|\sigma|}.$$

Then, by virtue of the Mikhlin theorem [13, p.1181], the functions $T_k(\sigma, \lambda)$ are the Fourier miltipliers of the type (q, q). Hence, if $f \in L_q(\mathbb{R})$ then a function $u(x)$, obtained from (1.7) for $k = 0$, is a solutions to equation (1.1) and belongs to $W_q^m(\mathbb{R})$ and

$$\sum_{k=0}^{m} |\lambda|^{m-k} \|u^{(k+s)}\|_{L_q(\mathbb{R})} = \sum_{k=0}^{m} \|F^{-1}T_k(\cdot, \lambda)(i\sigma)^s Ff\|_{L_q(\mathbb{R})}$$
$$\leq C(\varepsilon)\|f^{(s)}\|_{L_q(\mathbb{R})}, \qquad s \geq 0. \tag{1.8}$$

So, (1.4') has been proved.

Let us show now that if $f \in W_q^{z-m}(\mathbb{R})$ then $u \in W_q^z(\mathbb{R})$ and inequality (1.4) holds. This is established by induction on z. For $z = m$ from $f \in L_q(\mathbb{R})$ it follows that $u \in W_q^m(\mathbb{R})$ and inequality (1.4) holds (if $z = m$, (1.4) is transformed into (1.8) for $s = 0$). Our statement is valid if we substitute z with $z - 1$, i.e., from $f \in W_q^{z-1-m}(\mathbb{R})$ it follows that $u \in W_q^{z-1}(\mathbb{R})$ and

$$\sum_{k=0}^{z-1} |\lambda|^{z-1-k} \|u\|_{W_q^k(\mathbb{R})} \leq C(\varepsilon) \left(\|f\|_{W_q^{z-1-m}(\mathbb{R})} + |\lambda|^{z-1-m} \|f\|_{L_q(\mathbb{R})} \right).$$

Multiplying this inequality by $|\lambda|$ and applying Lemma 1.2.4 we obtain

$$\sum_{k=0}^{z-1} |\lambda|^{z-k} \|u\|_{W_q^k(\mathbb{R})} \leq C(\varepsilon) \left(\|f\|_{W_q^{z-m}(\mathbb{R})} + |\lambda|^{z-m} \|f\|_{L_q(\mathbb{R})} \right). \qquad (1.9)$$

From (1.1) follows

$$u^{(m)}(x) = a_m^{-1} \left(f(x) - \lambda^m u(x) - \cdots - \lambda a_{m-1} u^{(m-1)}(x) \right).$$

Let $f \in W_q^{z-m}(\mathbb{R})$. Then from the last equality it follows that $u \in W_q^z(\mathbb{R})$ and

$$\|u^{(z)}\|_{L_q(\mathbb{R})} \leq C \left(\|f^{(z-m)}\|_{L_q(\mathbb{R})} + \sum_{k=z-m}^{z-1} |\lambda|^{z-k} \|u^{(k)}\|_{L_q(\mathbb{R})} \right).$$

Hence, by virtue of (1.9), we have

$$\|u^{(z)}\|_{L_q(\mathbb{R})} \leq C(\varepsilon) \left(\|f\|_{W_q^{z-m}(\mathbb{R})} + |\lambda|^{z-m} \|f\|_{L_q(\mathbb{R})} \right).$$

From this inequality and inequality (1.9) we obtain (1.4). ∎

Now, consider a boundary value problem for ordinary differential equations with constant coefficients and the weight $d = 1$ on $[0, 1]$

$$L_0(\lambda)u = \lambda^m u(x) + \lambda^{m-1} a_1 u'(x) + \cdots + a_m u^{(m)}(x) = f(x),$$

$$(1.10)$$

$$L_{s0}(\lambda)u = \sum_{k=0}^{n_s} \lambda^k \left[\alpha_{sk} u^{(m_s-k)}(0) + \beta_{sk} u^{(m_s-k)}(1) \right] = f_s, \qquad s = 1, \ldots, m,$$

where $m \geq 1$, $m_s \geq n_s$; a_k, α_{sk}, β_{sk}, f_s are complex numbers; $a_m \neq 0$.

Let us denote by m_{sk} the order of the functional

$$\alpha_{sk} u^{(m_s-k)}(0) + \beta_{sk} u^{(m_s-k)}(1).$$

Obviously, $m_{sk} = m_s - k$ if at least one of the numbers α_{sk} and β_{sk} is not equal to 0. Set $m_{sk} = 0$ if $\alpha_{sk} = \beta_{sk} = 0$. If conditions (0.2) are p-regular relative to some numbers then, even if under one k, we have $m_{sk} = m_s - k$. It is also obvious that the order m'_s of the functional $L_{s0}(\lambda)$ satisfies the relation

$$m'_s = \max\{m_{sk} : k = 0, \ldots, n_s\} \leq m_s, \quad s = 1, \ldots, m.$$

Theorem 1.2. *Let $m \geq 1$, $n_s \leq m_s$, $a_m \neq 0$ and problem (1.10) be p-regular. Then for any $\varepsilon > 0$ there exists $R_\varepsilon > 0$ such that for all complex numbers λ satisfying*

$$\pi/2 - \underline{\omega} + \varepsilon < \arg \lambda < 3\pi/2 - \overline{\omega} - \varepsilon, \qquad |\lambda| > R_\varepsilon,$$

where numbers $\underline{\omega}$, $\overline{\omega}$ are defined by equalities (1.3), the operator

$$\mathbb{L}_0(\lambda): \quad u \to (L_0(\lambda)u, L_{10}(\lambda)u, \ldots, L_{m0}(\lambda)u)$$

from $W_q^z(0,1)$ onto $W_q^{z-m}(0,1) \dotplus \mathbb{C}^m$, where an integer $z \geq \max\{m, m'_s + 1\}$, $q \in (1, \infty)$, is an isomorphism and for these λ the following estimates hold for a solution to problem (1.10)

$$\sum_{k=0}^{z} |\lambda|^{z-k} \|u\|_{W_q^k(0,1)} \leq C(\varepsilon)(\|f\|_{W_q^{z-m}(0,1)} + |\lambda|^{z-m}\|f\|_{L_q(0,1)}$$

$$+ \sum_{s=1}^{m} |\lambda|^{z-m_s-1/q}|f_s|) \tag{1.11}$$

and

$$\sum_{k=0}^{m} |\lambda|^{m-k} \|u\|_{W_q^{k+p}(0,1)} \leq C(\varepsilon)(\|f\|_{W_q^p(0,1)}$$

$$+ \sum_{s=1}^{m} |\lambda|^{m+p-m_s-1/q}|f_s|), \quad 0 \leq p \leq \min\{m_s - n_s\}. \tag{1.11'}$$

Proof. The operator $\mathbb{L}_0(\lambda)$ acts from $W_q^z(0,1)$ into $W_q^{z-m}(0,1) \dotplus \mathbb{C}^m$ linearly and continuously. Let us prove that for any $f \in W_q^{z-m}(0,1)$ and any numbers f_s, $s = 1, \ldots, m$, problem (1.10) has a unique solution belonging to $W_q^z(0,1)$. Let $\hat{f} \in W_q^z(\mathbb{R})$ be an extension of $f \in W_q^s(0,1)$ such that the extension operator $Sf = \hat{f}$ from $W_q^s(0,1)$ into $W_q^s(\mathbb{R})$ is bounded [72, p.314]. Let us show that the function $u(x)$ of the form $u(x) = u_1(x) + u_2(x)$, where $u_1(x)$ is a restriction on $[0, 1]$ of the solution $\widehat{u}_1(x)$ of the equation

$$\lambda^m \widehat{u}_1(x) + \lambda^{m-1} a_1 \widehat{u}_1'(x) + \cdots + a_m \widehat{u}_1^{(m)}(x) = \widehat{f}(x), \quad x \in \mathbb{R}, \tag{1.12}$$

and $u_2(x)$ is the solution of the problem

$$\lambda^m u_2(x) + \lambda^{m-1} a_1 u_2'(x) + \cdots + a_m u_2^{(m)}(x) = 0, \qquad x \in (0, 1), \tag{1.13}$$

$$L_{s0}(\lambda)u_2 = -L_{s0}(\lambda)u_1 + f_s, \qquad s = 1, \ldots, m, \tag{1.14}$$

is a solution to problem (1.10) that satisfies estimates (1.11) and (1.11').

By virtue of Theorem 1.1 equation (1.12) has a unique solution $\widehat{u}_1(x)$, belonging to $W_q^z(\mathbb{R})$, that satisfies estimate (1.4). Hence, for $u_1(x)$ (restriction of $\widehat{u}_1(x)$ on $[0, 1]$) the estimates[6]

$$\sum_{k=0}^{z} |\lambda|^{z-k} \|u_1\|_{k,q} \leq C(\varepsilon) \left(\|f\|_{z-m,q} + |\lambda|^{z-m} \|f\|_{0,q} \right), \tag{1.15}$$

$$\sum_{k=0}^{m} |\lambda|^{m-k} \|u_1^{(k+s)}\|_{0,q} \leq C(\varepsilon) \|f\|_{s,q}, \qquad s \geq 0 \tag{1.15'}$$

are valid for all complex numbers λ satisfying

$$\pi/2 - \underline{\omega} + \varepsilon < \arg \lambda < 3\pi/2 - \overline{\omega} - \varepsilon, \qquad |\lambda| > R_\varepsilon,$$

where numbers $\underline{\omega}, \overline{\omega}$ are defined by equalities (1.3).

Now, let us prove that for any complex f_s, $s = 1, \ldots, m$, the problem (1.13)–(1.14) has a unique solution $u_2(x)$, belonging to $W_q^z(0, 1)$ and let us estimate this solution. Since from condition (0.3) it follows that $\omega_j \neq \omega_s$ for $j \neq s$, then the general solution to equation (1.13) has the form

$$u_2(x) = \sum_{i=1}^{p} C_i e^{\omega_i \lambda x} + \sum_{i=p+1}^{m} C_i e^{\omega_i \lambda(x-1)}. \tag{1.16}$$

Substituting (1.16) into (1.14), we obtain a system for finding C_i, $i = 1, \ldots, m$,

$$\sum_{k=0}^{n_s} \lambda^k \sum_{i=1}^{p} (\omega_i \lambda)^{m_s - k} [\alpha_{sk} + \beta_{sk} e^{\omega_i \lambda}] C_i + \sum_{k=0}^{n_s} \lambda^k \sum_{i=p+1}^{m} (\omega_i \lambda)^{m_s - k} [\alpha_{sk} e^{-\omega_i \lambda} + \beta_{sk}] C_i$$
$$= -L_{s0}(\lambda)u_1 + f_s, \qquad s = 1, \ldots, m. \tag{1.17}$$

---

[6] Sometimes, for brevity, the norm in $W_q^k(0, 1)$ is denoted by $\| \cdot \|_{k,q}$.

Since for $\pi/2 - \underline{\omega} + \varepsilon < \arg \lambda < 3\pi/2 - \overline{\omega} - \varepsilon$ we have $\pi/2 + \varepsilon < \arg \omega_i \lambda < 3\pi/2 - \varepsilon$, if $i = 1, \ldots, p$ and $|\arg \omega_i \lambda| < \pi/2 - \varepsilon$, if $i = p+1, \ldots, m$, then $\operatorname{Re} \omega_i \lambda < -\delta(\varepsilon)|\lambda|$ for $i = 1, \ldots, p$ and $-\operatorname{Re} \omega_i \lambda < -\delta(\varepsilon)|\lambda|$ for $i = p+1, \ldots, m$, where $\delta(\varepsilon) = C \sin \varepsilon$. Hence, the determinant of system (1.17) has the form

$$
D(\lambda) = \lambda^{\sum\limits_{s=1}^{m} m_s} \begin{vmatrix} \sum_{k=0}^{n_1} \alpha_{1k}\omega_1^{m_1-k} & \cdots & \sum_{k=0}^{n_1} \beta_{1k}\omega_{p+1}^{m_1-k} & \cdots \\ \vdots & \cdots & \vdots & \cdots \\ \sum_{k=0}^{n_m} \alpha_{mk}\omega_1^{m_m-k} & \cdots & \sum_{k=0}^{n_m} \beta_{mk}\omega_{p+1}^{m_m-k} & \cdots \end{vmatrix}
$$

$$
+ \lambda^{\sum\limits_{s=1}^{m} m_s} R(\lambda) = [\theta + R(\lambda)]\lambda^{\sum\limits_{s=1}^{m} m_s},
$$

where $R(\lambda) \to 0$ if $\pi/2 - \underline{\omega} + \varepsilon < \arg \lambda < 3\pi/2 - \overline{\omega} - \varepsilon$ and $|\lambda| \to \infty$. Since, according to the theorem condition, $\theta \neq 0$, then for a large enough $|\lambda|$ in the angle $\pi/2 - \underline{\omega} + \varepsilon < \arg \lambda < 3\pi/2 - \overline{\omega} - \varepsilon$ we have $D(\lambda) \neq 0$. Hence, for these λ system (1.17) has a unique solution

$$
C_i = D^{-1}(\lambda) \sum_{s=1}^{m} a_{is}(\lambda)(-L_{s0}(\lambda)u_1 + f_s), \qquad i = 1, \ldots, m.
$$

where $a_{is}(\lambda)$ are determinants equal to

$$
a_{is}(\lambda) = [\theta_{is} + R_{is}(\lambda)]\lambda^{\sum\limits_{k \neq s} m_k}.
$$

Moreover $R_{is}(\lambda) \to 0$ if $\pi/2 - \underline{\omega} + \varepsilon < \arg \lambda < 3\pi/2 - \overline{\omega} - \varepsilon$ and $|\lambda| \to \infty$. Hence,

$$
C_i = \sum_{s=1}^{m} \lambda^{-m_s} \frac{\theta_{is} + R_{is}(\lambda)}{\theta + R(\lambda)}(-L_{s0}(\lambda)u_1 + f_s), \qquad i = 1, \ldots, m,
$$

Substituting these values into (1.16), we find that the problem (1.13)–(1.14) has a unique solution and this solution is given by the formula

$$
u_2(x) = \sum_{i=1}^{p} \sum_{s=1}^{m} \lambda^{-m_s} \frac{\theta_{is} + R_{is}(\lambda)}{\theta + R(\lambda)}(-L_{s0}(\lambda)u_1 + f_s)e^{\omega_i \lambda x}
$$

$$
+ \sum_{i=p+1}^{m} \sum_{s=1}^{m} \lambda^{-m_s} \frac{\theta_{is} + R_{is}(\lambda)}{\theta + R(\lambda)}(-L_{s0}(\lambda)u_1 + f_s)e^{\omega_i \lambda(x-1)}.
$$

Hence, for $n \geq 0$ and $\pi/2 - \underline{\omega} + \varepsilon < \arg \lambda < 3\pi/2 - \overline{\omega} - \varepsilon$, $|\lambda| \to \infty$ we have the estimate

$$
\begin{aligned}
\|u_2^{(n)}\|_{0,q} \leq &C \sum_{i=1}^{p} \sum_{s=1}^{m} |\lambda|^{-m_s+n}(|L_{s0}(\lambda)u_1| + |f_s|)\|e^{\omega_i \lambda \cdot}\|_{0,q} \\
&+ C \sum_{i=p+1}^{m} \sum_{s=1}^{m} |\lambda|^{-m_s+n}(|L_{s0}(\lambda)u_1| + |f_s|)\|e^{\omega_i \lambda(\cdot-1)}\|_{0,q} \\
\leq &C(\varepsilon) \sum_{s=1}^{m} |\lambda|^{-m_s-1/q+n}(|L_{s0}(\lambda)u_1| + |f_s|).
\end{aligned} \tag{1.18}
$$

To estimate $|L_{s0}(\lambda)u_1|$ we use the inequality [6, Ch.3, §10, Th.10.4]

$$
\|u^{(j)}\|_{C[0,1]} \leq C(h^{1-\gamma}\|u^{(z)}\|_{0,q} + h^{-\gamma}\|u\|_{0,q}),
$$

where $p \leq j < z$, $0 < h < h_0$, $\gamma = (j - p + 1/q)/(z - p)$. Set $p = 0$ and $h = |\lambda|^{-z}$. Then, taking inequality (1.15) into account, we obtain

$$
\begin{aligned}
|L_{s0}(\lambda)u_1| \leq &C \sum_{k=0}^{n_s} |\lambda|^k \|u_1\|_{m_s-k,\infty} \\
\leq &C \sum_{k=0}^{n_s} |\lambda|^k (|\lambda|^{-z[1-(m_s-k+1/q)/z]}\|u_1^{(z)}\|_{0,q} \\
&+ |\lambda|^{z[(m_s-k+1/q)/z]}\|u_1\|_{0,q}) \\
\leq &C(\varepsilon)|\lambda|^{-z+m_s+1/q}(\|f\|_{z-m,q} + |\lambda|^{z-m}\|f\|_{0,q}),
\end{aligned}
$$

for $\pi/2 - \underline{\omega} + \varepsilon < \arg \lambda < 3\pi/2 - \overline{\omega} - \varepsilon$, $|\lambda| \to \infty$. Substituting these estimates into (1.18), for $k = 0, \ldots, z$, the estimate

$$
\begin{aligned}
|\lambda|^{z-k}\|u_2^{(k)}\|_{0,q} \leq &C(\varepsilon)(\|f\|_{z-m,q} + |\lambda|^{z-m}\|f\|_{0,q} \\
&+ \sum_{s=1}^{m} |\lambda|^{z-m_s-1/q}|f_s|),
\end{aligned}
$$

holds as $\pi/2 - \underline{\omega} + \varepsilon < \arg \lambda < 3\pi/2 - \overline{\omega} - \varepsilon$, $|\lambda| \to \infty$. From this and from (1.15) we obtain estimate (1.11) for the solution to problem (1.10). So, for $\pi/2 - \underline{\omega} + \varepsilon < \arg \lambda < 3\pi/2 - \overline{\omega} - \varepsilon$, $|\lambda| \to \infty$, there exists a solution to problem (1.10) belonging to $W_q^z(0,1)$ and for this solution estimate (1.11) holds.

Let $0 < p \leq \min\{m_s - n_s\}$ and $m_s < m + tp$ for some $t \in \mathbb{R}$. Set $h = |\lambda|^{-m-(t-1)p}$.

Then, taking inequality $(1.15')$ into account, we obtain

$$|L_{s0}(\lambda)u_1| \leq C \sum_{k=0}^{n_s} |\lambda|^k \|u_1\|_{m_s-k,\infty}$$

$$\leq C \sum_{k=0}^{n_s} |\lambda|^k (|\lambda|^{-[m+(t-1)p][1-(m_s-k-p+1/q)]/[m+(t-1)p]} \|u_1^{(m+tp)}\|_{0,q}$$

$$+ |\lambda|^{-[m+(t-1)p](m_s-k-p+1/q)/(m+(t-1)p)} \|u_1^{(p)}\|_{0,q}$$

$$\leq C \sum_{k=0}^{n_s} \left(|\lambda|^{-m-tp+m_s+1/q} \|u_1^{(m+tp)}\|_{0,q} + |\lambda|^{m_s-p+1/q} \|u_1^{(p)}\|_{0,q} \right)$$

$$\leq C(\varepsilon) |\lambda|^{-m+m_s-p+1/q} \|f\|_{p,q},$$

for $\pi/2 - \underline{\omega} + \varepsilon < \arg \lambda < 3\pi/2 - \overline{\omega} - \varepsilon$, $|\lambda| \to \infty$. Substituting these estimates into (1.18), we find that

$$\|u_2^{(k+p)}\|_{0,q} \leq C(\varepsilon) \sum_{s=1}^{m} |\lambda|^{-m_s-1/q+k+p} (|L_{s0}(\lambda)u_1| + |f_s|)$$

$$\leq C(\varepsilon) \left(|\lambda|^{-m+k} \|f\|_{p,q} + \sum_{s=1}^{m} |\lambda|^{-m_s-1/q+k+p} |f_s| \right),$$

holds, for $\pi/2 - \underline{\omega} + \varepsilon < \arg \lambda < 3\pi/2 - \overline{\omega} - \varepsilon$, $|\lambda| \to \infty$. From this and from $(1.15')$ we prove estimate $(1.11')$ for the solution to problem (1.10).

The uniqueness of the solution to problem (1.10) follows from estimate (1.11).

∎

Let us define the analog of Theorem 1.1 for the equation

$$L_0(\lambda)u = \lambda^n u(x) + \sum_{k=1}^{n} \lambda^{n-k} a_k u^{(dk)}(x) = f(x), \qquad x \in \mathbb{R},$$

where $a_k = 0$ if dk is a non-integer, and $d = m/n$.

Theorem 1.3. Let $a_n \neq 0$ and the roots of the characteristic equation

$$a_n \omega^m + a_{n-1} \omega^{d(n-1)} + \cdots + 1 = 0 \tag{1.19}$$

be p-separated. Then for any $\varepsilon > 0$ and for all complex numbers λ satisfying

$$(\pi/2 - \underline{\omega} - 2\pi\gamma)d + \varepsilon < \arg \lambda < (3\pi/2 - \overline{\omega} - 2\pi\gamma)d - \varepsilon \tag{1.20}$$

for some $\gamma = 0, \ldots, n-1$, where $\underline{\omega}$ and $\overline{\omega}$ as in (1.3) and ω_j being roots of equation (1.19), the oeprator $\mathbb{L}_0(\lambda) : \quad u \to L_0(\lambda)u$ from $W_q^z(\mathbb{R})$ onto $W_q^{z-m}(\mathbb{R})$, where $z \geq m$, $q \in (1, \infty)$ is an isomorphism and for these λ the following estimates

$$\sum_{k=0}^{z} |\lambda|^{d^{-1}(z-k)} \|u\|_{W_q^k(\mathbb{R})} \leq C(\varepsilon) \left(\|f\|_{W_q^{z-m}(\mathbb{R})} + |\lambda|^{d^{-1}(z-m)} \|f\|_{L_q(\mathbb{R})} \right)$$

and

$$\sum_{k=0}^{m} |\lambda|^{d^{-1}(m-k)} \|u^{(k+s)}\|_{L_q(\mathbb{R})} \le C(\varepsilon)\|f^{(s)}\|_{L_q(\mathbb{R})}, \qquad s \ge 0$$

hold.

Proof. After substituting $\lambda = \mu^d$ into the equation $L_0(\lambda)u = f(x)$, it is transformed into the equation

$$\tilde{L}_0(\mu)u = \mu^m u(x) + \sum_{k=1}^{n} \mu^{m-dk} a_k u^{(dk)}(x) = f(x), \qquad x \in \mathbb{R},$$

to which we apply Theorem 1.1. ∎

3.1.2. Fredholm property of the problem with general functional conditions. Let an operator A, from a Banach space E into a Banach space F, act boundedly. The set of solutions of the homogeneous equation

$$Au = 0$$

is called the *kernel* of the operator A and is denoted by $\ker A$. So,

$$\ker A = \{u \mid u \in E, \ Au = 0\}.$$

The set of functionals from F' equal to 0 on $R(A)$ is called the *cokernel* of the operator A and is denoted by $\operatorname{coker} A$. So,

$$\operatorname{coker} A = \{v' \mid v' \in F', \ < Au, v' > \ = 0, \ u \in E\}.$$

The bounded operator A from E into F is *fredholm*, if
a) $R(A)$ is closed in F;
b) $\ker A$ and $\operatorname{coker} A$ are finitely dimensional subspaces in E and F' respectively;
c) $\dim \ker A = \dim \operatorname{coker} A$.

Theorem 1.4. *Let the following conditions be satisfied:*

(1) $a \in C^{z-m}[0,1]$, *where an integer* $z \ge m$; $a(x) \ne 0$;
(2) *operator* B *from* $W_q^z(0,1)$ *into* $W_q^{z-m}(0,1)$ *is compact, where* $q \in (1,\infty)$;
(3) *functionals* L_s *in* $W_q^z(0,1)$ *are continuous.*

Then the operator $\mathbb{L}: \ u \to (a(x)u^{(m)}(x) + Bu, L_1 u, \ldots, L_m u)$ *from* $W_q^z(0,1)$ *into* $W_q^{z-m}(0,1) \dotplus \mathbb{C}^m$ *is fredholm.*

Proof. The operator $\mathbb{L}$ can be represented in the form $\mathbb{L} = \mathbb{L}_0 + \mathbb{L}_1$, where

$$\mathbb{L}_0 u = (a(x)u^{(m)}(x), u(0), \ldots, u^{(m-1)}(0)),$$
$$\mathbb{L}_1 u = (Bu, L_1 u - u(0), \ldots, L_m u - u^{(m-1)}(0)).$$

The operator $\mathbb{L}_0$ from $W_q^z(0,1)$ onto $W_q^{z-m}(0,1) \dotplus \mathbb{C}^m$ is an isomorphism. By virtue of conditions 2 and 3 the operator $\mathbb{L}_1$ from $W_q^z(0,1)$ into $W_q^{z-m}(0,1) \dotplus \mathbb{C}^m$ is compact. Then it is possible to apply the theorem of fredholm operator perturbation [28, p.238] to the operator $\mathbb{L} = \mathbb{L}_0 + \mathbb{L}_1$, from which the theorem statement follows. ∎

Theorem 1.4 seems rather strange, but it is true. For example, if $a(x) = 1$, $B = 0$, $L_s = 0$, then

$$\dim \ker \ \mathbb{L} = m, \qquad \dim \operatorname{coker} \mathbb{L} = m.$$

3.1.3. Coerciveness of the problem for equations with constant coefficients. Consider a boundary value problem for ordinary differential equations with constant coefficients, when the spectral parameter appears polynomially in both the equation and the boundary conditions and the weight $d = m/n$ of the problem is an arbitrary number. More precisely, we deal with the problem

$$L_0(\lambda)u = \lambda^n u(x) + \sum_{k=1}^{n} \lambda^{n-k} a_k u^{(dk)}(x) = f(x),$$

$$(1.21)$$

$$L_{s0}(\lambda)u = \sum_{k=0}^{n_s} \lambda^k \left(\alpha_{sk} u^{(m_s - dk)}(0) + \beta_{sk} u^{(m_s - dk)}(1) \right) = f_s, \qquad s = 1, \ldots, m,$$

where $n \geq 1$, $m \geq 1$, $m_s \geq dn_s$; a_k, α_{sk}, β_{sk}, f_s are complex numbers; $a_k = \alpha_{sk} = \beta_{sk} = 0$ if dk is non-integer.

Theorem 1.5. *Let $a_n \neq 0$ and problem (1.21) is p-regular. Then for any $\varepsilon > 0$ there exists $R_\varepsilon > 0$ such that for all complex numbers λ, satisfying $|\lambda| > R_\varepsilon$ and for some $\gamma = 0, \ldots, n - 1$ lying inside angle (1.20) where $\underline{\omega}$ and $\overline{\omega}$ are as in (1.3) and ω_j are roots of equation (1.19), the operator*

$$\mathbb{L}_0(\lambda): \ u \to (L_0(\lambda)u, L_{10}(\lambda)u, \ldots, L_{m0}(\lambda)u)$$

from $W_q^z(0,1)$ into $W_q^{z-m}(0,1) \dotplus \mathbb{C}^m$, where the integer $z \geq \max\{m, m_s + 1\}$, $q \in (1, \infty)$, is an isomorphism and for these λ the following estimates hold for a solution to problem (1.21)

$$\sum_{k=0}^{z} |\lambda|^{d^{-1}(z-k)} \|u\|_{W_q^k(0,1)} \leq C(\varepsilon)(\|f\|_{W_q^{z-m}(0,1)} + |\lambda|^{d^{-1}(z-m)} \|f\|_{L_q(0,1)}$$

$$+ \sum_{s=1}^{m} |\lambda|^{d^{-1}(z-m_s-1/q)} |f_s|) \qquad (1.22)$$

and

$$\sum_{k=0}^{m}|\lambda|^{d^{-1}(m-k)}\|u\|_{W_q^{k+p}(0,1)} \le C(\varepsilon)(\|f\|_{W_q^p(0,1)} + \sum_{s=1}^{m}|\lambda|^{d^{-1}(m-m_s-1/q+p)}|f_s|),$$

$$0 \le p \le \min\{m_s - n_s\}. \tag{1.22'}$$

Proof. Substituting $\lambda = \mu^d$ into problem (1.21), it is reduced to

$$\mu^m u(x) + \sum_{k=1}^{n} \mu^{m-dk} a_k u^{(dk)}(x) = f(x),$$

$$\sum_{k=0}^{n_s} \mu^{dk} \left[\alpha_{sk} u^{(m_s-dk)}(0) + \beta_{sk} u^{(m_s-dk)}(1) \right] = f_s, \qquad s = 1, \dots, m. \tag{1.23}$$

Now we apply Theorem 1.2 to problem (1.23).

Let $u \in W_q^z(0,1)$ be a solution to problem (1.21), i.e., (1.23). Then, by virtue of Theorem 1.2, for any $\varepsilon > 0$, $\pi/2 - \underline{\omega} + \varepsilon < \arg\mu < 3\pi/2 - \overline{\omega} - \varepsilon$, $|\mu| > R_\varepsilon$, we have

$$\sum_{k=0}^{z}|\mu|^{z-k}\|u\|_{k,q} \le C(\varepsilon)(\|f\|_{z-m,q} + |\mu|^{z-m}\|f\|_{0,q}$$

$$+ \sum_{s=1}^{m}|\mu|^{z-m_s-1/q}|f_s|).$$

This implies that for λ from the angle (1.20) with sufficiently large $|\lambda|$, estimate (1.22) holds for a solution to problem (1.21). Analogously, (1.11') implies (1.22'). Hence, for the indicated λ, a solution to problem (1.21) in $W_q^z(0,1)$ is unique.

By virtue of Theorem 1.4, the operator $\mathbb{L}_0(\lambda)$ from $W_q^z(0,1)$ into $W_q^{z-m}(0,1) \dot{+} \mathbb{C}^m$ is fredholm. From this and the fact that the operator $\mathbb{L}_0(\lambda)$ is one-to-one, the theorem statement follows. ∎

3.1.4. Coerciveness of the problem for equations with variable coefficients. Remember that, according to (1.2.10), for $s \ge 0$ and non-integers s

$$W_q^s(G) = (W_q^{s_0}(G), W_q^{s_1}(G))_{\theta,q},$$

where $0 \le s_0, s_1$ are integers, $0 < \theta < 1$, $1 < q < \infty$, $s = (1-\theta)s_0 + \theta s_1$ and G is a bounded domain of the class C^∞ in $\mathbb{R}^n$.

Lemma 1.6. An operator B from $W_q^z(G)$ into $W_q^k(G)$ $\big(\text{resp. } W_q^{k-1/q}(\Gamma)\big)$ is compact, where $z > k \ge 0$, $1 < q < \infty$, if and only if for any $\varepsilon > 0$, $u \in W_q^z(G)$

$$\|Bu\|_{W_q^k(G)} \le \varepsilon\|u\|_{W_q^z(G)} + C(\varepsilon)\|u\|_{W_q^k(G)}, \tag{1.24}$$

$$\left(resp. \ \|Bu\|_{W_q^{k-1/q}(\Gamma)} \leq \varepsilon \|u\|_{W_q^z(G)} + C(\varepsilon)\|u\|_{W_q^k(G)}\right).$$

Proof. Let us show that Lemmas 2.2.6 and 2.2.7 can be applied to the operator B. By virtue of [72, p.350/14], the space $E = W_q^z(G)$ is compactly embedded into $F = G = W_q^k(G)$. Then, by virtue of Lemma 2.2.6, from (1.24) it follows that the operator B from $W_q^z(G)$ into $W_q^k(G)$ is compact.

By virtue of [72, p.338] the space $W_q^s(G)$, $s \geq 0$, $1 < q < \infty$, has a basis. By virtue of [72, p.337] $W_q^s(G)$ is isomorphic to $L_q(0,1)$ for integer s, and is isomorphic to ℓ_q for non-integer s. Since, for $1 < q < \infty$, spaces $L_q(0,1)$ and ℓ_q are reflexive, then the space $W_q^s(G)$ for $s > 0$ and $1 < q < \infty$ is reflexive. Let us now show that the embedding $W_q^z(G) \subset W_q^k(G)$, $z > k$, is dense. If k is an integer, then it follows from the embeddings $W_q^{[z]+1}(G) \subset W_q^z(G) \subset W_q^k(G)$. If k is non-integer, it follows from the formula

$$W_q^k(G) = (L_q(G), W_q^z(G))_{k/z,q}$$

and the theorem [72, p.39] about the density $E_0 \cap E_1$ in $(E_0, E_1)_{\theta,p}$. Hence, by virtue of Lemma 2.2.7, from the operator B from $W_q^z(G)$ into $W_q^k(G)$ being compact follows the estimate (1.24).

In the second case we have to use a continuation operator from $W_q^{k-1/q}(\Gamma)$ onto $W_q^k(G)$ [72, p.330]. ∎

Consider a principally boundary value problem for ordinary differential equations with variable coefficients, when the spectral parameter appears polynomially in both the equation and the functional conditions, and the weight $d = m/n$

$$L(\lambda)u = \lambda^n u(x) + \sum_{k=1}^{n} \lambda^{n-k} \left(a_k(x)u^{(dk)}(x) + B_k u|_x \right) = f(x), \tag{1.25}$$

$$L_s(\lambda)u = \sum_{k=0}^{n_s} \lambda^k [\alpha_{sk} u^{(m_s - dk)}(0) + \beta_{sk} u^{(m_s - dk)}(1)$$

$$+ \sum_{i=1}^{N_{sk}} \delta_{ski} u^{(m_s - dk)}(x_{ski}) + T_{sk} u] = f_s, \quad s = 1, \ldots, m, \tag{1.26}$$

where $n \geq 1$, $m \geq 1$, $m_s \geq dn_s$; $\alpha_{sk}, \beta_{sk}, \delta_{ski}, f_s$ are complex numbers; $x_{ski} \in (0,1)$; $a_k(x)$ are scalar functions defined on $[0,1]$; $a_k(x) = \alpha_{sk} = \beta_{sk} = \delta_{ski} = 0$ if dk is a non-integer; B_k are operators in $L_q(0,1)$; T_{sk} are functionals in $L_q(0,1)$, $q \in [1,\infty)$.

Here both the operators B_k and the functionals T_{sk}, generally speaking, are unbounded.

Consider the characteristic equation

$$a_n(x)\omega^m + a_{n-1}(x)\omega^{d(n-1)} + \cdots + 1 = 0, \quad x \in [0,1], \tag{1.27}$$

and denote its roots by $\omega_j(x)$, $j = 1, \ldots, m$. If the problem (1.25)–(1.26) is p-regular, then $\omega_j(0) \neq \omega_k(0)$ and $\omega_j(1) \neq \omega_k(1)$ for $j \neq k$. At other points of the segment $[0, 1]$ the equality $\omega_j(x) = \omega_k(x)$ for $j \neq k$ is admitted.

Theorem 1.7. *Let the following conditions be satisfied:*

(1) $a_k \in C^{z-m}[0, 1]$, *where an integer* $z \geq \max\{m, m_s + 1\}$; $a_n(x) \neq 0$; $a_k(x) = \alpha_{sk} = \beta_{sk} = \delta_{ski} = 0$, *if* dk *is not an integer;* $a_j(0) = a_j(1)^7$;

(2) *the problem* (1.25)–(1.26) *is p-regular;*

(3) *operators* B_k *from* $W_q^{dk}(0, 1)$ *into* $L_q(0, 1)$ *and from* $W_q^{z-m+dk}(0, 1)$ *into* $W_q^{z-m}(0, 1)$ *are compact, where* $q \in (1, \infty)$;

(4) *for some* $r \in [1, \infty)$ *the functionals* T_{sk} *in* $W_r^{m_s - dk}(0, 1)$ *are continuous.*

Then for any $\varepsilon > 0$ *there exists* $R_\varepsilon > 0$ *such that for all complex numbers* λ *that satisfy* $|\lambda| > R_\varepsilon$ *and for some* $\gamma = 0, \ldots, n - 1$ *lying inside the angle*

$$(\pi/2 - \underline{\omega} - 2\pi\gamma)d + \varepsilon < \arg \lambda < (3\pi/2 - \overline{\omega} - 2\pi\gamma)d - \varepsilon, \qquad (1.28)$$

where

$$\underline{\omega} = \inf_{x \in [0,1]} \min\{\arg \omega_j(x) : j = 1, \ldots, p; \ \arg \omega_s(x) + \pi : s = p+1, \ldots, m\},$$

$$\overline{\omega} = \sup_{x \in [0,1]} \max\{\arg \omega_j(x) : j = 1, \ldots, p; \ \arg \omega_s(x) + \pi : s = p+1, \ldots, m\},$$

and $\omega_j(x)$ *are roots of equation* (1.27) *(the value* $\arg \omega_j(x)$ *is chosen up to a multiple of* 2π*, so that* $\overline{\omega} - \underline{\omega} < \pi$*), the operator*

$$\mathbb{L}(\lambda) : \ u \to (L(\lambda)u, L_1(\lambda)u, \ldots, L_m(\lambda)u)$$

from $W_q^z(0, 1)$ *onto* $W_q^{z-m}(0, 1) \dotplus \mathbb{C}^m$ *is an isomorphism and for these* λ *the following estimates hold for a solution to the problem* (1.25)–(1.26)

$$\sum_{k=0}^{z} |\lambda|^{d^{-1}(z-k)} \|u\|_{W_q^k(0,1)} \leq C(\varepsilon)(\|f\|_{W_q^{z-m}(0,1)} + |\lambda|^{d^{-1}(z-m)} \|f\|_{L_q(0,1)}$$

$$+ \sum_{s=1}^{m} |\lambda|^{d^{-1}(z-m_s-1/q)} |f_s|) \qquad (1.29)$$

and

$$\sum_{k=0}^{m} |\lambda|^{d^{-1}(m-k)} \|u\|_{W_q^{k+p}(0,1)} \leq C(\varepsilon)(\|f\|_{W_q^p(0,1)} + \sum_{s=1}^{m} |\lambda|^{d^{-1}(m-m_s-1/q+p)} |f_s|),$$

$$0 \leq p \leq \min\{m_s - n_s\}. \qquad (1.29')$$

---

^{7}From $a_j(0) = a_j(1)$ it follows that $\theta(0) = \theta(1)$. If conditions (1.26) principally are local, then the condition $a_j(0) = a_j(1)$ should be omitted.

Proof. First we obtain an a priori estimate (1.29) for the principal part of the problem (1.25)–(1.26), i.e., for the problem

$$L_0(\lambda)u = \lambda^n u(x) + \sum_{k=1}^{n} \lambda^{n-k} a_k(x) u^{(dk)}(x) = f(x), \tag{1.30}$$

$$L_{s0}(\lambda)u = \sum_{k=0}^{n_s} \lambda^k [\alpha_{sk} u^{(m_s - dk)}(0) + \beta_{sk} u^{(m_s - dk)}(1)] = f_s, \qquad s = 1, \dots, m, \tag{1.31}$$

Consider the domain G_1 and the intervals G_i, $i = 2, \dots, N$, with measure less than ρ, which cover the segment $[0, 1]$, such that points 0, 1 belong only to G_1 and every point of the segment $[0, 1]$ is situated in no more than two G_i, $i = 1, \dots, N$.

Let $\{\varphi_i(x)\}$ be a partition of unity, subordinate to a cover of $[0, 1]$ by $\{G_i\}$ [6, Ch.2, §8.3]. The functions $\varphi_i(x)$ have the following properties:

(1) the support of the function $\varphi_i(x)$ belongs to the set G_i, i.e., $\varphi_i(x) = 0$ outside of G_i;

(2) functions $\varphi_i(x)$ are infinitely differentiable on the segment $[0, 1]$;

(3) $0 \le \varphi_i(x) \le 1$; $\sum_{i=1}^{N} \varphi_i(x) = 1$.

An a priori estimate (1.29) for a solution of the problem (1.30)–(1.31) is established in two stages. First, (1.29) is established for such solutions from $W_q^z(0, 1)$, to the problem (1.30)–(1.31), the support of which belongs to G_i.

Consider the fixed points $x_i \in G_i$, where $x_1 = 0$. Equation (1.30) can be rewritten in the form

$$\lambda^n u(x) + \sum_{k=1}^{n} \lambda^{n-k} a_k(x_i) u^{(dk)}(x) = f(x) + \sum_{k=1}^{n} \lambda^{n-k} (a_k(x_i) - a_k(x)) u^{(dk)}(x). \tag{1.32}$$

Let $i = 1$ and the support of a solution $u(x)$ of the problem (1.30)–(1.31) belong to G_1. Since $x_1 = 0$, then, by virtue of Theorem 1.5, for all λ satisfying the inequalities (1.28) and with large enough moduli, for the solution to the problem (1.32)–(1.31) the estimate

$$\sum_{k=0}^{z} |\lambda|^{d^{-1}(z-k)} \|u\|_{k,q} \le C(\varepsilon) \{ \|f\|_{z-m,q} + |\lambda|^{d^{-1}(z-m)} \|f\|_{0,q}$$

$$+ \sum_{k=1}^{n} |\lambda|^{n-k} [\|(a_k(0) - a_k(\cdot)) u^{(dk)}(\cdot)\|_{z-m,q}$$

$$+ |\lambda|^{d^{-1}(z-m)} \|(a_k(0) - a_k(\cdot)) u^{(dk)}(\cdot)\|_{0,q}]$$

$$+ \sum_{s=1}^{m} |\lambda|^{d^{-1}(z-m_s - 1/q)} |f_s| \} \tag{1.33}$$

is true.

Since, by virtue of condition 1 $a_j(0) = a_j(1)$ then for considered solutions with small ρ the inequality with sufficiently small δ

$$\sum_{k=1}^{n} |\lambda|^{n-k}[\|(a_k(0) - a_k(\cdot))u^{(dk)}(\cdot)\|_{z-m,q} + |\lambda|^{d^{-1}(z-m)}\|(a_k(0) - a_k(\cdot))u^{(dk)}(\cdot)\|_{0,q}]$$

$$\leq \delta \sum_{k=0}^{z} |\lambda|^{d^{-1}(z-k)}\|u\|_{k,q},$$

is true. Hence, from (1.33) follows (1.29).

Let $i = 2, \ldots, N$ and the support of a solution $u(x)$ to problem (1.30) belongs to G_i. Without loss of generality one can assume that $G_i \subset (0,1)$. Then the function $u(x)$ extended as 0 outside $[0,1]$ is a solution of (1.32) on the whole axis, where the function $f(x)$ is also extended as 0 outside $[0,1]$. Hence, by virtue of Theorem 1.3 the estimate

$$\sum_{k=0}^{z} |\lambda|^{d^{-1}(z-k)}\|u\|_{W_q^k(\mathbb{R})} \leq C(\varepsilon)\{\|f\|_{W_q^{z-m}(\mathbb{R})} + |\lambda|^{d^{-1}(z-m)}\|f\|_{L_q(\mathbb{R})}$$

$$+ \sum_{k=1}^{n} |\lambda|^{n-k}[\|(a_k(x_i) - a_k(\cdot))u^{(dk)}(\cdot)\|_{W_q^{z-m}(\mathbb{R})}$$

$$+ |\lambda|^{d^{-1}(z-m)}\|(a_k(x_i) - a_k(\cdot))u^{(dk)}(\cdot)\|_{L_q(\mathbb{R})}]\}$$

is true.

Since functions $u(x)$ and $f(x)$ are equal to 0 outside $[0,1]$ and $L_{s0}(\lambda)u = 0$, $s = 1, \ldots, m$, this estimate coincides with the estimate (1.33). So, the estimate (1.29) for solutions to problem (1.30) from $W_q^z(0,1)$, the support of which belongs to G_i, has also been established for $i = 2, \ldots, N$.

Now, let $u(x)$ be any solution of the problem (1.30)–(1.31) belonging to $W_q^z(0,1)$. Then from $u = \sum_{i=1}^{N} \varphi_i u$ and that the support of the function $\varphi_i(x)u(x)$ belongs to G_i, the estimate

$$\sum_{k=0}^{z} |\lambda|^{d^{-1}(z-k)}\|u\|_{k,q} \leq \sum_{k=0}^{z} |\lambda|^{d^{-1}(z-k)} \sum_{i=1}^{N} \|\varphi_i u\|_{k,q} \leq C(\varepsilon) \sum_{i=1}^{N}\{\|L_0(\lambda)(\varphi_i u)\|_{z-m,q}$$

$$+ |\lambda|^{d^{-1}(z-m)}\|L_0(\lambda)(\varphi_i u)\|_{0,q}$$

$$+ \sum_{s=1}^{m} |\lambda|^{d^{-1}(z-m_s-1/q)}|L_{s0}(\lambda)(\varphi_i u)|\} \tag{1.34}$$

follows.

Using the Leibniz formula it can be written as

$$L_0(\lambda)(\varphi_i u) = \varphi_i L_0(\lambda)u + \cdots,$$

where in the terms substituted by dots φ_i is differentiated at least one time. On the other hand, since $L_{s0}(\lambda)u = f_s$, $s = 1, \ldots, m$, by virtue of the fact that the functions $\varphi_i(x)$, $i = 2, \ldots, N$ are equal to 0 in the neighborhood of points 0, 1, we have

$$L_{s0}(\lambda)(\varphi_i u) = 0, \qquad s = 1, \ldots, m, \ i = 2, \ldots, N, \tag{1.35}$$

and taking into account properties $\varphi_1(0) = \varphi_1(1) = 1$, $\varphi_1^{(n)}(0) = \varphi_1^{(n)}(1) = 0$, $n \geq 1$, we have

$$\begin{aligned}
L_{s0}(\lambda)(\varphi_1 u) &= \sum_{k=0}^{n_s} \lambda^k \{\alpha_{sk}[\varphi_1(x)u(x)]^{(m_s-dk)}|_{x=0} \\
&\quad + \beta_{sk}[\varphi_1(x)u(x)]^{(m_s-dk)}|_{x=1}\} = \varphi_1(0)L_{s0}(\lambda)u \\
&= f_s, \qquad s = 1, \ldots, m.
\end{aligned} \tag{1.36}$$

Hence, from (1.34), by virtue of (1.35)–(1.36) we obtain

$$\begin{aligned}
\sum_{k=0}^{z} |\lambda|^{d^{-1}(z-k)} \|u\|_{k,q} &\leq C(\varepsilon)[\sum_{i=1}^{N}(\|\varphi_i f\|_{z-m,q} + |\lambda|^{d^{-1}(z-m)}\|\varphi_i f\|_{0,q}) \\
&\quad + \sum_{k=1}^{n} |\lambda|^{n-k}(\|u\|_{z-m+dk-1,q} + |\lambda|^{d^{-1}(z-m)}\|u\|_{dk-1,q}) \\
&\quad + \sum_{s=1}^{m} |\lambda|^{d^{-1}(z-m_s-1/q)}|f_s|] \leq C(\varepsilon)(\|f\|_{z-m,q} \\
&\quad + |\lambda|^{d^{-1}(z-m)}\|f\|_{0,q} + \sum_{s=1}^{m} |\lambda|^{d^{-1}(z-m_s-1/q)}|f_s| \\
&\quad + |\lambda|^{-d^{-1}} \sum_{k=0}^{z} |\lambda|^{d^{-1}(z-k)}\|u\|_{k,q}).
\end{aligned}$$

From this for all λ satisfying inequalities (1.28) and with large enough moduli we establish estimate (1.29) for a solution to the problem (1.30)–(1.31).

Let us now prove estimate (1.29) for a solution to the complete problem (1.25)–(1.26). Let $u \in W_q^z(0,1)$ be a solution to the problem (1.25)–(1.26). Denote

$$T_{sk}^0 u = \sum_{i=1}^{N_{sk}} \delta_{ski} u^{(m_s-dk)}(x_{ski}).$$

Then

$$
\sum_{k=0}^{z} |\lambda|^{d^{-1}(z-k)} \|u\|_{k,q} \leq C(\varepsilon)(\|f - \sum_{k=1}^{n} \lambda^{n-k} B_k u\|_{z-m,q}
$$

$$
+ |\lambda|^{d^{-1}(z-m)} \|f - \sum_{k=1}^{n} \lambda^{n-k} B_k u\|_{0,q}
$$

$$
+ \sum_{s=1}^{m} |\lambda|^{d^{-1}(z-m_s-1/q)} |f_s - \sum_{k=0}^{n_s} \lambda_k (T_{sk}^0 u + T_{sk} u)|)
$$

$$
\leq C(\varepsilon)[\|f\|_{z-m,q} + |\lambda|^{d^{-1}(z-m)} \|f\|_{0,q}
$$

$$
+ \sum_{s=1}^{m} |\lambda|^{d^{-1}(z-m_s-1/q)} |f_s|
$$

$$
+ \sum_{k=1}^{n} |\lambda|^{n-k}(\|B_k u\|_{z-m,q} + |\lambda|^{d^{-1}(z-m)} \|B_k u\|_{0,q})
$$

$$
+ \sum_{s=1}^{m} \sum_{k=0}^{n_s} |\lambda|^{d^{-1}(z-m_s-1/q)+k}(|T_{sk}^0 u| + |T_{sk} u|)].
\tag{1.37}
$$

Let us introduce

$$
\eta \in C_0^\infty(\mathbb{R}), \quad \eta(x) = \begin{cases} 1, & x \in [\delta, 1-\delta], \\ 0, & x \in [0, \delta/2] \cup [1 - \delta/2, 1] \end{cases}, \quad 0 \leq \eta(x) \leq 1,
$$

where $\delta = \inf_{s,k,i} \{|x_{ski}|, |x_{ski} - 1|\}$. Then

$$
|T_{sk}^0 u| \leq C \|u\|_{m_s - dk, \infty, [\delta, 1-\delta]} \leq C \|\eta u\|_{m_s - dk, \infty, [0,1]}.
$$

We now use the inequality [6, Ch.3, §10, th.10.4]

$$
\|u^{(j)}\|_{0,\infty} \leq C(h^{1-\gamma} \|u^{(z)}\|_{0,q} + h^{-\gamma} \|u\|_{0,q}),
$$

where $0 \leq j < z$, $0 < h < h_0$, $\gamma = (j + 1/q)/z$, for $h = |\lambda|^{-d^{-1}z}$, and Theorem 1.5. Then for all λ satisfying inequalities (1.28) and with large enough moduli we find

$$
|T_{sk}^0 u| \leq C(|\lambda|^{-d^{-1}z[1-(m_s-dk+1/q)/z]} \|\eta u\|_{z,q}
$$

$$
+ |\lambda|^{d^{-1}z[(m_s-dk+1/q)/z]} \|\eta u\|_{0,q}
$$

$$
\leq C(\varepsilon)(|\lambda|^{-d^{-1}(1-m_s+dk-1/q)} \|L_0(\lambda)(\eta u)\|_{z-m,q}
$$

$$+ |\lambda|^{d^{-1}(z-m)} \|L_0(\lambda)(\eta u)\|_{0,q}$$

$$\leq C(\varepsilon)[(|\lambda|^{-d^{-1}(1-m_s+dk-1/q)} \|L(\lambda)(\eta u)\|_{z-m,q}$$

$$+ |\lambda|^{d^{-1}(z-m)} \|L(\lambda)(\eta u)\|_{0,q}$$

$$+ \sum_{k=1}^{n} |\lambda|^{n-k} (\|B_k(\eta u)\|_{z-m,q} + |\lambda|^{d^{-1}(z-m)} \|B_k(\eta u)\|_{0,q})]$$

$$\leq C(\varepsilon) |\lambda|^{-d^{-1}(1-m_s+dk-1/q)} [\|f\|_{z-m,q} + |\lambda|^{d^{-i}(z-m)} \|f\|_{0,q}$$

$$+ \sum_{k=1}^{n} |\lambda|^{n-k} (\|u\|_{z-m+dk-1,q} + |\lambda|^{d^{-1}(z-m)} \|u\|_{dk-1,q})$$

$$+ \sum_{k=1}^{n} |\lambda|^{n-k} (\|B_k(\eta u)\|_{z-m,q} + |\lambda|^{d^{-1}(z-m)} \|B_k(\eta u)\|_{0,q})].$$

Hence, by virtue of conditions 3 and 4 of Lemma 1.6, for any $\delta > 0$ we have

$$\sum_{k=1}^{n} |\lambda|^{n-k} (\|B_k u\|_{z-m,q} + |\lambda|^{d^{-1}(z-m)} \|B_k u\|_{0,q})$$

$$+ \sum_{s=1}^{m} \sum_{k=0}^{n_s} |\lambda|^{d^{-1}(z-m_s-1/q)+k} (|T_{sk}^0 u| + |T_{sk} u|)$$

$$\leq C(\varepsilon) \{ \|f\|_{z-m,q} + |\lambda|^{d^{-1}(z-m)} \|f\|_{0,q}$$

$$+ \sum_{k=1}^{n} |\lambda|^{n-k} [\delta(\|u\|_{z-m+dk,q} + |\lambda|^{d^{-1}(z-m)} \|u\|_{dk,q})$$

$$+ C(\delta)(\|u\|_{z-m,q} + |\lambda|^{d^{-1}(z-m)} \|u\|_{0,q})]$$

$$+ \sum_{s=1}^{m} \sum_{k=0}^{n_s} |\lambda|^{d^{-1}(z-m_s-1/q)+k} \|u\|_{m_s-dk,r} \}.$$

Setting $\lambda = \mu^d$, then applying Lemma 1.2.5 and the interpolation Lemma 1.2.4, we obtain

$$\sum_{k=1}^{n} |\lambda|^{n-k} (\|B_k u\|_{z-m,q} + |\lambda|^{d^{-1}(z-m)} \|B_k u\|_{0,q})$$

$$+ \sum_{s=1}^{m} \sum_{k=0}^{n_s} |\lambda|^{d^{-1}(z-m_s-1/q)+k} (|T_{sk}^0 u| + |T_{sk} u|)$$

$$\leq C(\varepsilon) \{ \|f\|_{z-m,q} + |\lambda|^{d^{-1}(z-m)} \|f\|_{0,q}$$

$$+ \sum_{k=1}^{n} [\delta(|\mu|^{m-dk} \|u\|_{z-m+dk,q} + |\mu|^{z-dk} \|u\|_{dk,q})$$

$$+ C(\delta) |\lambda|^{-k} (|\mu|^{m} \|u\|_{z-m,q} + |\mu|^{z} \|u\|_{0,q})]$$

$$+ \sum_{s=1}^{m} \sum_{k=0}^{n_s} |\mu|^{z-m_s-1/q+dk}(\delta \|u\|_{m_s-dk+1/q,q} + C(\delta)\|u\|_{0,q})\}$$

$$\le C(\varepsilon)(\|f\|_{z-m,q} + |\lambda|^{d^{-1}(z-m)}\|f\|_{0,q})$$

$$+ (C(\varepsilon)\delta + C(\varepsilon,\delta)|\lambda|^{-\gamma})(\|u\|_{z,q} + |\mu|^{z}\|u\|_{0,q})$$

$$\le C(\varepsilon)(\|f\|_{z-m,q} + |\lambda|^{d^{-1}(z-m)}\|f\|_{0,q})$$

$$+ (C(\varepsilon)\delta + C(\varepsilon,\delta)|\lambda|^{-\gamma})(\|u\|_{z,q} + |\lambda|^{d^{-1}z}\|u\|_{0,q}), \qquad (1.38)$$

where $\gamma = \min\{1, (dq)^{-1}\}$. Substituting (1.38) into (1.37) we have

$$\sum_{k=0}^{z} |\lambda|^{d^{-1}(z-k)}\|u\|_{k,q} \le C(\varepsilon)[\|f\|_{z-m,q} + |\lambda|^{d^{-1}(z-m)}\|f\|_{0,q}$$

$$+ \sum_{s=1}^{m} |\lambda|^{d^{-1}(z-m_s-1/q)}|f_s|]$$

$$+ (C(\varepsilon)\delta + C(\varepsilon,\delta)|\lambda|^{-\gamma})(\|u\|_{z,q} + |\lambda|^{d^{-1}z}\|u\|_{0,q}).$$

It is clear that for a fixed $\varepsilon > 0$ it is possible to choose $\delta > 0$ so small and $|\lambda|$ so large that $C(\varepsilon)\delta + C(\varepsilon,\delta)|\lambda|^{-\gamma} < 1$. Hence, for such λ from angle (1.28) we obtain an a priori estimate (1.29). Analogously (1.22') implies (1.29').

From (1.29) it follows that for a large enough $|\lambda|$, where λ from angle (1.28), a solution to the problem (1.25)–(1.26) in $W_q^z(0,1)$ is unique.

For $k = 1, \ldots, n$ the embedding $W_q^z(0,1) \subset W_q^{z-m+dk}(0,1)$ is continuous. Then from condition 3 it follows that operators B_k from $W_q^z(0,1)$ into $W_q^{z-m}(0,1)$ are compact. Hence, by virtue of Theorem 1.4 the operator $\mathbb{L}(\lambda)$ from $W_q^z(0,1)$ into $W_q^{z-m}(0,1) \dotplus \mathbb{C}^m$ is fredholm. From this and the operator $\mathbb{L}(\lambda)$ is one-to-one the theorem statement follows. ■

Lemma 1.8. Let m_p, $p = 1, \ldots, m$ be non-negative integers, $m_i \ne m_k$ for $i \ne k$. Then for any system of mutually distinct numbers $\omega_1, \ldots, \omega_M$, where $M = \max\{m_p\} - \min\{m_p\} + 1$, such that $\omega_i \ne \omega_k$, for $i \ne k$ and $\omega_i \ne 0$, it is possible to choose a subsystem $\omega_{k_1}, \ldots, \omega_{k_m}$ such that the determinant

$$\begin{vmatrix} \omega_{k_1}^{m_1} & \cdots & \omega_{k_m}^{m_1} \\ \vdots & \vdots & \vdots \\ \omega_{k_1}^{m_m} & \cdots & \omega_{k_m}^{m_m} \end{vmatrix} \ne 0.$$

Proof. Without loss of generality we can assume that $0 \le m_1 < \cdots < m_m$. Let us expand the Vandermond determinant of order M

$$\begin{vmatrix} 1 & \cdots & 1 \\ \omega_1 & \cdots & \omega_M \\ \vdots & \vdots & \vdots \\ \omega_1^{M-1} & \cdots & \omega_M^{M-1} \end{vmatrix}$$

by rows $1, m_2 - m_1 + 1, \ldots, m_m - m_1 + 1$. Since, under conditions of the lemma, the Vandermond determinant is not equal to 0, at least one of the obtained minors of order m (in decomposition) is not equal to 0. This minor differs from the determinant pointed out in the lemma by the multiplier $\omega_{k_1}^{m_1}, \cdots, \omega_{k_m}^{m_1}$. ∎

Denote by s the number of functional conditions with the same order.

A system

$$L_k u = \alpha_k u^{(m_k)}(0) + \beta_k u^{(m_k)}(1) + \sum_{i=0}^{N_k} \delta_{ki} u^{(m_k)}(x_{ki}) + T_k u, \qquad k = 1, \ldots, m,$$

is called *normal*, if

(1) $s \leq 2$;

(2) for $s = 1$ at least one of the numbers α_k, β_k is not equal to 0;

(3) for $s = 2$ in one functional condition $\alpha_k \neq 0$, $\beta_k = 0$ and in another one with the same order $\alpha_k = 0$, $\beta_k \neq 0$;

(4) $x_{ki} \in (0,1)$; for some $r \in [1, \infty)$ the functionals T_k in $W_r^{m_k}(0,1)$ are continuous.

By p we denote a number of functional conditions with $\alpha_k \neq 0$. Then $\alpha_k = 0$ and $\beta_k \neq 0$ for the rest of the functional conditions of a normal system.

Lemma 1.9. *A normal system is p-regular with respect to some numbers ω_k,* $k = 1, \ldots, m$.

Proof. After the enumeration we obtain

$$\alpha_k \neq 0, \qquad k = 1, \ldots, p$$
$$\alpha_k = 0 \text{ and } \beta_k \neq 0, \qquad k = p+1, \ldots, m.$$

By virtue of Lemma 1.8 there exist numbers $\omega_1, \ldots, \omega_m$ with properties

a) $0 < \omega_1 < \omega_2 < \cdots < \omega_p$ such that the determinant

$$\begin{vmatrix} \omega_1^{m_1} & \cdots & \omega_p^{m_1} \\ \vdots & \vdots & \vdots \\ \omega_1^{m_p} & \cdots & \omega_p^{m_p} \end{vmatrix} \neq 0$$

b) $\omega_{p+1} < \omega_{p+2} < \cdots < \omega_m < 0$ such that the determinant

$$\begin{vmatrix} \omega_{p+1}^{m_{p+1}} & \cdots & \omega_m^{m_{p+1}} \\ \vdots & \vdots & \vdots \\ \omega_{p+1}^{m_m} & \cdots & \omega_m^{m_m} \end{vmatrix} \neq 0.$$

Then

$$\begin{vmatrix} \alpha_1\omega_1^{m_1} & \cdots & \alpha_1\omega_p^{m_1} & \beta_1\omega_{p+1}^{m_1} & \cdots & \beta_1\omega_m^{m_1} \\ \vdots & \vdots & \vdots & \vdots & \vdots & \vdots \\ \alpha_p\omega_1^{m_p} & \cdots & \alpha_p\omega_p^{m_p} & \beta_p\omega_{p+1}^{m_p} & \cdots & \beta_p\omega_m^{m_p} \\ 0 & \cdots & 0 & \beta_{p+1}\omega_{p+1}^{m_{p+1}} & \cdots & \beta_{p+1}\omega_m^{m_{p+1}} \\ \vdots & \vdots & \vdots & \vdots & \vdots & \vdots \\ 0 & \cdots & 0 & \beta_m\omega_{p+1}^{m_m} & \cdots & \beta_m\omega_m^{m_m} \end{vmatrix} \neq 0. \ \blacksquare$$

Corollary 1.10. *Let the system*

$$L_k u = \alpha_k u^{(m_k)}(0) + \beta_k u^{(m_k)}(1) + \sum_{i=0}^{N_k} \delta_{ki} u^{(m_k)}(x_{ki}) + T_k u, \qquad k = 1, \ldots, m,$$

be normal. Then the operator $u \to (L_1 u, \ldots, L_m u)$ *from* $W_q^\ell(0,1)$ *onto* $\mathbb{C}^m$, *where* $\ell \geq \max\{m_k + 1\}$, $q \in (1, \infty)$, *has a continuous right-inverse. In other words, there exists such an operator* $R(f_1, \ldots, f_m) = u$ *continuous from* $\mathbb{C}^m$ *into* $W_q^\ell(0,1)$, *where* u *is a solution to the system*

$$L_k u = f_k, \qquad k = 1, \ldots, m.$$

Moreover, the inverse operator does not depend on $\ell \leq \ell_0$. *Here the fixed number* $\ell_0 \geq \max\{m, m_k + 1\}$.

Proof. By virtue of Lemma 1.9 the system L_k is p-regular with respect to some numbers ω_k, $k = 1, \ldots, m$. Let us find such complex numbers a_k, $k = 1, \ldots, m$ that the roots of the equations $a_m \omega^m + a_{m-1}\omega^{m-1} + \cdots + 1 = 0$ coincide with the given numbers ω_k, $k = 1, \ldots, m$. It is enough to apply Theorem 1.7 to the problem

$$\lambda^m u(x) + \lambda^{m-1} a_1 u'(x) + \cdots + a_m u^{(m)}(x) = 0,$$
$$L_k u = f_k, \qquad k = 1, \ldots, m,$$

for some λ and $\ell_0 \geq \max\{m, m_k + 1\}$. $\blacksquare$

Let the set $\{1, \ldots, m\}$ be divided into two nonintersecting subsets, i.e.,

$$\{1, \ldots, m\} = \{1, \ldots, m\}' \cup \{1, \ldots, m\}'', \quad \{1, \ldots, m\}' \cap \{1, \ldots, m\}'' = \emptyset.$$

Corollary 1.11. *Let the system*

$$L_k u = \alpha_k u^{(m_k)}(0) + \beta_k u^{(m_k)}(1) + \sum_{i=0}^{N_k} \delta_{ki} u^{(m_k)}(x_{ki}) + T_k u, \qquad k = 1, \ldots, m,$$

be normal. Then the operator $u \rightarrow (L_k u, \ k \in \{1, \ldots, m\}')$, *from* $W_q^\ell(0,1)$
onto $\underset{k \in \{1, \ldots, m\}'}{+} \mathbb{C}$, *where* $\ell \geq \max\{m_k + 1 \ : \ k \in \{1, \ldots, m\}'\}, q \in (1, \infty)$,
has a continuous right-inverse. In other words, there exists such an operator
$R(f_k, \ k \in \{1, \ldots, m\}') = u$ *continuous from* $\underset{k \in \{1, \ldots, m\}'}{+} \mathbb{C}$ *into* $W_q^\ell(0,1)$, *where*
u is a solution to the system

$$L_k u = f_k, \quad k \in \{1, \ldots, m\}'.$$

Moreover, the inverse operator does not depend on $\ell \leq \ell_0$. *Here the fixed number*
$\ell_0 \geq \max\{m, m_k + 1\}$.

Proof. It follows from Corollary 1.10 since (it is easy to see) every subsystem
of a normal system is also clearly normal. ∎

3.2. Fold completeness of root functions of
principally boundary value problems for ordinary
differential equations with a polynomial parameter.

Consider a principally boundary value problem for homogeneous ordinary differential equation with variable coefficients, when the spectral parameter appears
polynomially in both the equation and the functional conditions and the weight of
the problem $d = m/n$ is an integer, i.e.,

$$L(\lambda)u = \lambda^n u(x) + \sum_{k=1}^{n} \lambda^{n-k}(a_k(x)u^{(dk)}(x) + B_k u|_x) = 0, \tag{2.1}$$

$$L_p(\lambda)u = \sum_{k=1}^{n_p} \lambda^k [\alpha_{pk} u^{(m_p - dk)}(0) + \beta_{pk} u^{(m_p - dk)}(1)$$

$$+ \sum_{i=1}^{N_{pk}} \delta_{pki} u^{(m_p - dk)}(x_{pki}) + T_{pk} u] = 0, \qquad p = 1 \ldots, m. \tag{2.2}$$

where $x_{pki} \in (0,1)$, $n \geq 1$, $m \geq 1$, $m_p \geq dn_p$, $n_p \leq n - 1$.

A number λ_0 is called an *eigenvalue* of the problem (2.1)–(2.2) if the problem

$$L(\lambda_0)u = 0, \qquad L_p(\lambda_0)u = 0, \quad p = 1, \ldots, m$$

has a nontrivial solution that belongs to $W_2^m(0,1)$. The nontrivial solution $u(x)$ of this problem that belongs to $W_2^m(0,1)$ is called an *eigenfunction* of the problem (2.1)–(2.2), corresponding to the eigenvalue of λ_0. A solution to the problem

$$L(\lambda_0)u_k + \frac{1}{1!}L'(\lambda_0)u_{k-1} + \cdots + \frac{1}{k!}L^{(k)}(\lambda_0)u_0 = 0,$$

$$L_p(\lambda_0)u_k + \frac{1}{1!}L'_p(\lambda_0)u_{k-1} + \cdots + \frac{1}{k!}L_p^{(k)}(\lambda_0)u_0 = 0, \quad p = 1,\ldots,m, \tag{2.3}$$

$u_k(x)$ belonging to $W_2^m(0,1)$ is called an *associated function of the k-rank* of the eigenfunction $u_0(x)$ of the problem (2.1)–(2.2).

The eigenfunction and associated functions of the problem (2.1)–(2.2) are combined under the general name *root functions* of the problem (2.1)–(2.2).

A complex number λ is called a *regular point* of the problem (2.1)–(2.2), if the problem

$$L(\lambda)u = f, \qquad L_p(\lambda)u = f_p, \quad p = 1.\ldots,m,$$

for any $f \in L_2(0,1)$ and $f_p \in \mathbb{C}$, has a unique solution that belongs to $W_2^m(0,1)$, and in addition, the estimate

$$\|u\|_{m,2} \leq C(\lambda)(\|f\|_{0,2} + \sum_{p=1}^{m} |f_p|)$$

is satisfied.

The complement of the set of regular point in the complex plane is called the *spectrum* of the problem (2.1)–(2.2).

The definition of the discreteness of the spectrum of the problem (2.1)–(2.2) is analogous to the definition of the discreteness of spectrum of the operator pencil system from 2.3.1.

Consider a system of differential equations

$$L(D_t)u(t,x) = 0, \quad t > 0, \; x \in (0,1),$$

$$L_p(D_t)u(t,x) = 0, \quad t > 0, \; p = 1,\ldots,m.$$

By virtue of Lemma 2.0.1 a function of the form

$$u(t,x) = e^{\lambda_0 t}\left(\frac{t^k}{k!}u_0(x) + \frac{t^{k-1}}{(k-1)!}u_1(x) + \cdots + u_k(x)\right) \tag{2.4}$$

is a solution to this system if and only if a system of functions $u_0(x), u_1(x), \ldots, u_k(x)$ is a chain of root functions of the problem (2.1)–(2.2), corresponding to the eigenvalue λ_0, i.e., satisfying (2.3). A solution of the form (2.4) is called an *elementary solution*.

Let $\mathcal{H}$ be a Hilbert space, continuously embedded into $[L_2(0,1)]^n = \bigoplus\limits^{n} L_2(0,1)$.

A system of root functions of the problem (2.1)–(2.2) is called *n-fold complete* in $\mathcal{H}$ if a system of functions

$$(u(0,x), u_t'(0,x), \ldots, u_t^{(n-1)}(0,x)),$$

where $u(t,x)$ is an elementary solution of the form (2.4), is complete in the space $\mathcal{H}$.

From Theorem 1.4 it follows that under conditions of Theorem 1.7 every complex number is an eigenvalue of the problem (2.1)–(2.2) or its regular point.

3.2.1. Dense sets in spaces of Sobolev type.
Let us prove two theorems, which are not only repeatedly applied later on, but also are of independent interest.

Theorem 2.1. *Let m functionals of the form*

$$L_p u = \alpha_p u^{(m_p)}(0) + \beta_p u^{(m_p)}(1) + \sum_{i=1}^{N_p} \delta_{pi} u^{(m_p)}(x_{pi}) + T_p u, \qquad p = 1, \ldots, m,$$

be given, where $x_{pi} \in (0,1)$; α_p, β_p, δ_{pi} are complex numbers and

(1) *the system L_p, $p = 1, \ldots, m$ is normal;*
(2) *for some $r \in [1, \infty)$ the functionals T_p in $W_r^{m_p}(0,1)$ are continuous.*

Then for integers $\ell \geq k \geq 0$, $\ell \geq \max\{m_p + 1\}$ and arbitrary $q \in (1, \infty)$

$$\overline{W_q^\ell((0,1), L_p u = 0, p = 1, \ldots, m)}|_{W_q^k(0,1)} = W_q^k((0,1), L_p u = 0, m_p \leq k - 1).$$

Proof. Let $u \in W_q^k((0,1), L_p u = 0, m_p \leq k - 1)$. Then there exists a sequence of infinitely differentiable functions $\varphi_n(x)$, $n = 1, \ldots, \infty$, such that

$$\lim_{n \to \infty} \|\varphi_n - u\|_{k,q} = 0. \tag{2.5}$$

By virtue of the embedding theorem [72, p.328/6] and condition 2 we have

$$\lim_{n \to \infty} |L_p \varphi_n - L_p u| \leq C \lim_{n \to \infty} \|\varphi_n - u\|_{m_p, \infty} \leq C \lim_{n \to \infty} \|\varphi_n - u\|_{k,q}.$$

for $m_p \leq k - 1$. Thus

$$\lim_{n \to \infty} L_p \varphi_n = L_p u = 0, \qquad m_p \leq k - 1.$$

By virtue of Corollary 1.10 there exists a sequence of functions $g_n \in W_q^\ell(0,1)$, where $\ell \geq \max\{m_p + 1\}$, which satisfy the following relations

$$L_p g_n = -L_p \varphi_n, \qquad p = 1, \ldots, m, \tag{2.6}$$

$$\lim_{n \to \infty} \|g_n\|_{k,q} = 0. \tag{2.7}$$

Now it is easy to note that for a sequence of functions

$$u_n = \varphi_n + g_n \in W_q^\ell(0,1)$$

the following relations

$$L_p u_n = 0, \quad p = 1, \dots, m, \tag{2.8}$$

$$\lim_{n \to \infty} \|u_n - u\|_{k,q} = 0 \tag{2.9}$$

hold. Relations (2.8) follow from (2.6). Relation (2.9) follows from (2.5) and (2.7). ■

Corollary 2.2. *Under the conditions of Theorem 2.1 and for any integers $\ell \geq k \geq 0$*

$$\overline{W_q^\ell((0,1), L_p u = 0, m_p \leq \ell - 1)}|_{W_q^k(0,1)} = W_q^k((0,1), L_p u = 0, m_p \leq k - 1).$$

Let us introduce notations

$$L_{p,n_p - k} u = \alpha_{pk} u^{(m_p - dk)}(0) + \beta_{pk} u^{(m_p - dk)}(1)$$

$$+ \sum_{i=1}^{N_{pk}} \delta_{pki} u^{(m_p - dk)}(x_{pki}) + T_{pk} u,$$

$$k = 0, \dots, n_p, \ p = 1, \dots, m. \tag{2.10}$$

Then the functional conditions (2.2) are rewritten in the form

$$L_p(\lambda) u = \lambda^{n_p} L_{p0} u + \lambda^{n_p - 1} L_{p1} u + \dots + L_{pn_p} u = 0, \qquad p = 1, \dots, m.$$

Theorem 2.3. *Let the following conditions be satisfied:*

(1) $n \geq 1, \ m \geq 1, \ n_p \leq n - 1, \ m_p \geq dn_p$; *the weight $d = m/n$ is an integer;* $\max\{m_p\} - \min\{m_p - dn_p\} \leq m - 1$;

(2) $|\alpha_{p0}| + |\beta_{p0}| \neq 0$; *the system $L_{pn_p}, \ p = 1, \dots, m$ is normal;*

(3) *the functionals T_{pk} are continuous in $W_r^{m_p - dk}(0,1)$ for a number $r \in [1, \infty)$.*

Then the set

$$\mathcal{H}_d = \{v \mid v = (v_1, \dots, v_n) \in \overset{n-1}{\underset{k=0}{+}} W_q^{z + d(n-k)}(0,1), \ \sum_{k=0}^{n_p} L_{p,n_p - k} v_{k+s} = 0,$$

$$m_p \leq z + d(n - s + 1) - 1, \ s = 1, \dots, n - n_p, \ p = 1, \dots, m\}$$

is dense in the space

$$\mathcal{H} = \{v \mid v = (v_1,\dots,v_n) \in \overset{n-1}{\underset{k=0}{+}} W_q^{z+d(n-k-1)}(0,1), \ \sum_{k=0}^{n_p} L_{p,n_p-k} v_{k+s} = 0,$$

$$m_p \le z + d(n-s) - 1, \ s = 1,\dots,n-n_p-1, \ p = 1,\dots,m\}$$

for an integer $z \in [max\{0, m_p - (m-1)\}, \min\{m_p - dn_p\}], \ q \in (1,\infty)$.

Proof. Let $\varepsilon > 0$ and $v = (v_1,\dots,v_n) \in \mathcal{H}$. Set $t = \min\{n_p\}$. Construct functions $\varphi_s \in C^\infty[0,1], \ s = n-t+1,\dots,n$ such that

$$\|\varphi_s - v_s\|_{z+d(n-s),q} \le \varepsilon, \qquad s = n-t+1,\dots,n. \tag{2.11}$$

From the conditions of the connections in $\mathcal{H}_d$, for $n_p = t$ and $s = n-t$, we obtain

$$L_{pn_p}\varphi_{n-t} = -\sum_{k=1}^{n_p} L_{p,n_p-k}\varphi_{k+n-t}, \qquad m_p \le z + d(t+1) - 1, \ n_p = t. \tag{2.12}$$

Let us prove that there exist a function $\varphi_{n-t} \in W_q^{z+d(t+1)}(0,1)$ that satisfies (2.12) and

$$\|\varphi_{n-t} - v_{n-t}\|_{z+dt,q} \le \varepsilon. \tag{2.13}$$

By virtue of Corollary 1.11 there exist functions $\varphi_{n-t} \in W_q^{z+d(t+1)}(0,1)$ that satisfy the relations

$$L_{pn_p}\varphi_{n-t}^0 = -\sum_{k=1}^{n_p} L_{p,n_p-k}\varphi_{k+n-t}, \qquad m_p \le z + d(t+1) - 1, \ n_p = t. \tag{2.14}$$

Since $z \le m_p - dn_p$, then for $n_p = t$ one has $m_p \ge z + dt$. Then by virtue of Corollary 2.2 there exist functions $g_{n-t} \in W_q^{z+d(t+1)}(0,1)$ such that

$$L_{pn_p} g_{n-t} = 0, \qquad m_p \le z + d(t+1) - 1, \ n_p = t, \tag{2.15}$$

$$\|g_{n-t} - (v_{n-t} - \varphi_{n-t}^0)\|_{z+dt,q} \le \varepsilon. \tag{2.16}$$

Now it is enough to note that the function

$$\varphi_{n-t} = g_{n-t} + \varphi_{n-t}^0 \in W_q^{z+d(t+1)}(0,1)$$

satisfies both (2.12) and (2.13). Relation (2.12) follows from (2.15) and (2.14). Relation (2.13) follows from (2.16). From the conditions of the connections in $\mathcal{H}_d$, for $n_p \le t + i$, $s = n - t - i$, $m_p \le z + d(t + i + 1) - 1$, $i = 0, \ldots, n - t - 1$, one obtains

$$L_{pn_p} \varphi_{n-t-i} = -\sum_{k=1}^{n_p} L_{p,n_p-k} \varphi_{k+n-t-i}. \tag{2.17}$$

By induction on i we now show, that there exist functions

$$\varphi_{n-t-i} \in W_q^{z+d(t+i+1)}(0,1), \qquad i = 0, \ldots, n - t - 1,$$

that satisfy relations (2.17) and

$$\|\varphi_{n-t-i} - v_{n-t-i}\|_{z+d(t+i),q} \le C\varepsilon. \tag{2.18}$$

Assume that there exist functions $\varphi_{n-t-i} \in W_q^{z+d(t+i+1)}(0,1)$, $i = 0, \ldots, s - 1$, where $s = 1, \ldots, n - t - 1$, that satisfy relations (2.17) and (2.18). Next we show that there exists a function $\varphi_{n-t-s} \in W_q^{z+d(t+s+1)}(0,1)$ that satisfies relations (2.17) and (2.18) for $i = s$.

By virtue of (2.11), (2.13) and (2.18) for $m_p \le z + d(t + s)$, $n_p \le t + s$ we have

$$\left| \sum_{k=1}^{n_p} L_{p,n_p-k}(\varphi_{k+n-t-s} - v_{k+n-t-s}) \right| \le C \sum_{k=1}^{n_p} \|\varphi_{k+n-t-s} - v_{k+n-t-s}\|_{m_p-dk,q}$$

$$\le C \sum_{k=1}^{n_p} \|\varphi_{k+n-t-s} - v_{k+n-t-s}\|_{z+d(t+s-k),q}$$

$$\le C\varepsilon.$$

By virtue of Corollary 1.11 there exist functions

$$\varphi_{n-t-s}^0 \in W_q^{z+d(t+s+1)}(0,1), \qquad \varphi_{n-t-s}^{00} \in W_q^{z+d(t+s)}(0,1)$$

that satisfy the relations

$$L_{pn_p} \varphi_{n-t-s}^0 = -\sum_{k=1}^{n_p} L_{p,n_p-k} \varphi_{k+n-t-s} \tag{2.19}$$

for $m_p \le z + d(t + s + 1) - 1$, $n_p \le t + s$, and

$$L_{pn_p} \varphi_{n-t-s}^{00} = -\sum_{k=1}^{n_p} L_{p,n_p-k} v_{k+n-t-s} \tag{2.20}$$

for $m_p \le z + d(t + s) - 1$, $n_p \le t + s - 1$, and also

$$\|\varphi_{n-t-s}^0 - \varphi_{n-t-s}^{00}\|_{z+d(t+s),q} \le C\varepsilon. \tag{2.21}$$

Since $v \in \mathcal{H}$ then by virtue of (2.20)

$$L_{pn_p}(v_{n-t-s} - \varphi^{00}_{n-t-s}) = 0, \qquad m_p \leq z + d(t+s) - 1, \ n_p \leq t + s - 1.$$

Since $z \leq m_p - dn_p$, then for $n_p = t + s$ we have $m_p \geq z + d(t+s)$. Then, by virtue of Corollary 2.2 there exist functions $g_{n-t-s} \in W_q^{z+d(t+s+1)}(0,1)$, such that

$$L_{pn_p} g_{n-t-s} = 0, \qquad m_p \leq z + d(t+s+1) - 1, \ n_p = t + s, \tag{2.22}$$

$$\|g_{n-t-s} - (v_{n-t-s} - \varphi^{00}_{n-t-s})\|_{z+d(t+s),q} \leq \varepsilon. \tag{2.23}$$

Now it is enough to note that the functions

$$\varphi_{n-t-s} = g_{n-t-s} + \varphi^0_{n-t-s} \in W_q^{z+d(t+s+1)}(0,1)$$

satisfy both (2.17) and (2.18) for $i = s$. Relation (2.17) for $i = s$ follows from (2.22) and (2.19). Relation (2.18) for $i = s$ follows from (2.23) and (2.21). ■

Corollary 2.4. *Under the conditions of Theorem 2.3 the set*

$$\{v \mid v = (v_1, \ldots, v_n), v_k \in W_q^\ell(0,1), \ \sum_{k=0}^{n_p} L_{p,n_p-k} v_{k+s} = 0,$$

$$s = 1, \ldots, n - n_p, \ p = 1, \ldots, m\}$$

is dense everywhere in the space

$$\mathcal{H} = \{v \mid v = (v_1, \ldots, v_n) \in \overset{n-1}{\underset{k=0}{+}} W_q^{z+d(n-k-1)}(0,1), \ \sum_{k=0}^{n_p} L_{p,n_p-k} v_{k+s} = 0,$$

$$m_p \leq z + d(n - s) - 1, \ s = 1, \ldots, n - n_p - 1, \ p = 1, \ldots, m\}$$

for integers $z \in [max\{0, m_p - (m-1)\}, \min\{m_p - dn_p\}]$ *and* $\ell \geq \max\{m_p + 1\}$, *with an arbitrary* $q \in (1, \infty)$.

3.2.2. n-fold completeness of root functions. Consider the principally boundary value problems (2.1)–(2.2) with integer weight $d = m/n$

$$\lambda^n u(x) + \sum_{k=1}^{n} \lambda^{n-k}(a_k(x) u^{(dk)}(x) + B_k u|_x) = 0, \tag{2.24}$$

$$L_p(\lambda)u = \sum_{k=1}^{n_p} \lambda^k [\alpha_{pk} u^{(m_p-dk)}(0) + \beta_{pk} u^{(m_p-dk)}(1)$$

$$+ \sum_{i=1}^{N_{pk}} \delta_{pki} u^{(m_p-dk)}(x_{pki}) + T_{pk} u] = 0, \quad p = 1, \ldots, m, \tag{2.25}$$

where $x_{pki} \in (0,1)$, $n \geq 1$, $m \geq 1$, $m_p \geq dn_p$, $n_p \leq n - 1$, $d = m/n$.

Theorem 2.5. *Let the following conditions be satisfied:*

(1) $n \geq 1$, $m \geq 1$, $n_p \leq n-1$, $m_p \geq dn_p$; *the weight $d = m/n \geq 2$ is an integer;* $\max\{m_p\} - \min\{m_p - dn_p\} \leq m - 1$;

(2) $a_j \in C^s[0,1]$ *for an integer $s \geq \max\{m_p\}$; $a_j(0) = a_j(1)^8$; $a_n(x) \neq 0$;*

(3) *operators B_k from $W_2^{dk}(0,1)$ into $L_2(0,1)$ and from $W_2^{s+dk}(0,1)$ into $W_2^s(0,1)$ are compact;*

(4) *the problem (2.24)–(2.25) is p-regular;*

(5) $|\alpha_{p0}| + |\beta_{p0}| \neq 0$; *the system L_{pn_p}, $p = 1, \ldots, m$ is normal;*

(6) *functionals T_{pk} are continuous in $W_r^{m_p - dk}(0,1)$ for a number $r \in [1, \infty)$.*

Then the spectrum of the problem (2.24)–(2.25) is discrete and the system of root functions of the problem (2.24)–(2.25) is n-fold complete in the space

$$\mathcal{H} = \{v \mid v = (v_1, \ldots, v_n) \in \bigoplus_{k=0}^{n-1} W_2^{z+d(n-k-1)}(0,1), \sum_{k=0}^{n_p} L_{p,n_p-k} v_{k+s} = 0,$$

$$m_p \leq z + d(n - s) - 1, \ s = 1, \ldots, n - n_p - 1, \ p = 1, \ldots, m\}$$

for an integer $z \in [max\{0, m_p - (m - 1)\}, \min\{m_p - dn_p\}]$.[9]

Proof. Consider, in the space $H = W_2^z(0,1)$, operators A_k, $k = 1, \ldots, n$, defined by the equalities

$$D(A_k) = W_2^{z+dk}(0,1),$$

$$A_k u = a_k(x) u^{(dk)}(x) + B_k u|_x.$$

Let us denote

$$H_k = W_2^{z+dk}(0,1), \quad k = 0, \ldots, n,$$

$$H^p = H_0^p = \mathbb{C}, \quad p = 1, \ldots, m.$$

Consider operators A_{pk}, $p = 1, \ldots, m$, $k = 0, \ldots, n_p$, defined by the equalities

$$A_{p,n_p-k} u = L_{p,n_p-k} u = \alpha_{pk} u^{(m_p-dk)}(0) + \beta_{pk} u^{(m_p-dk)}(1)$$

$$+ \sum_{i=1}^{N_{pk}} \delta_{pki} u^{(m_p-dk)}(x_{pki}) + T_{pk} u, \quad k = 0, \ldots, n_p, \ p = 1, \ldots, m.$$

Since $m_p \leq z + m - 1$ then $m_p - d(n_p - k) \leq z + d(n - n_p + k) - 1$, which implies that the operators A_{pk} from $H_{n-n_p+k} = W_2^{z+d(n-n_p+k)}(0,1)$ into $H^p = \mathbb{C}$ are bounded, i.e., condition 4 of Theorem 2.3.4 is satisfied.

---

[8]See the footnote on p.100.

[9]The system of root functions of the problem (2.24)–(2.25) is also n-fold complete in the space $\mathcal{H}_d$ (see Theorem 2.3).

The problem (2.24)–(2.25) is rewritten in the operator form as the following system of pencil equations

$$L(\lambda)u = \lambda^n u + \lambda^{n-1} A_1 u + \cdots + A_n u = 0,$$
$$L_p(\lambda)u = \lambda^{n_p} A_{p0} u + \lambda^{n_p-1} A_{p1} u + \cdots + A_{pn_p} u = 0, \qquad p = 1, \ldots, m, \tag{2.26}$$

where $u \in D(A_n) = W_2^{z+m}(0,1)$.

Let us apply Theorem 2.3.4 to (2.26). By virtue of [72, p.350/14] the compact embeddings

$$W_2^{z+m}(0,1) \subset W_2^{z+d(n-1)}(0,1) \subset \cdots \subset W_2^{z}(0,1)$$

hold, and

$$s_j(J, W_2^{z+dk}(0,1), W_2^{z+d(k-1)}(0,1)) \sim j^{-d}, \qquad k = 1, \ldots, n.$$

Hence, for $q > d^{-1}$

$$J \in \sigma_q(W_2^{z+dk}(0,1), W_2^{z+d(k-1)}(0,1)),$$

i.e.,

$$J \in \sigma_q(H_k, H_{k-1}), \qquad q > d^{-1}, \ k = 1, \ldots, n. \tag{2.27}$$

So, conditions 1 and 2 of Theorem 2.3.4 have been checked. Condition 3 of Theorem 2.3.4 is obvious, and condition 4 has also been checked. Condition 5 of the theorem follows from Theorem 2.3, while condition 6 – from Theorem 1.7. Indeed, Theorem 1.7 can be used for the problem

$$L(\lambda)u = f, \qquad L_p(\lambda)u = f_p, \quad p = 1, \ldots, m. \tag{2.28}$$

Then all complex numbers λ with large enough moduli, lying inside the angle

$$(\pi/2 - \underline{\omega})d + \varepsilon < \arg \lambda < (3\pi/2 - \overline{\omega})d - \varepsilon, \tag{2.29}$$

where

$$\underline{\omega} = \inf_{x \in [0,1]} \min\{\arg\omega_j(x) : j = 1, \ldots, p; \ \arg\omega_s(x) + \pi : s = p+1, \ldots, m\},$$

$$\tag{2.30}$$

$$\overline{\omega} = \sup_{x \in [0,1]} \max\{\arg\omega_j(x) : j = 1, \ldots, p; \ \arg\omega_s(x) + \pi : s = p+1, \ldots, m\},$$

are the regular points of the system of pencils (2.26) (here $\omega_j(x)$ are roots of equation (1.27)) and for a solution to problem (2.28) the estimate

$$\sum_{k=0}^{z+m} |\lambda|^{d^{-1}(z+m-k)} \|u\|_{k,2} \le C(\varepsilon)\{\|f\|_{z,2} + |\lambda|^{d^{-1}z} \|f\|_{0,2}$$

$$+ \sum_{s=1}^{m} |\lambda|^{d^{-1}(z+m-m_s-1/2)} |f_s|\} \tag{2.31}$$

holds.

Since $d \ge 2$ and relation (2.27) is fulfilled for any $q > d^{-1}$ then it is possible to choose π/q close enough to 2π. Hence, Theorem 2.3.4 is applicable to the system of pencils (2.26) if the estimate

$$\|L^{-1}(\lambda)\|_{B(W_2^z(0,1)\oplus\mathbb{C}^m, W_2^{z+m}(0,1))} \le C|\lambda|^h, \qquad |\lambda| \to \infty \tag{2.32}$$

is fulfilled in some angle of the complex plane for some $h \in \mathbb{R}$. In turn, from the estimate (2.31) it follows that estimate (2.32) is fulfilled in angle (2.29). ∎

Let us show that for $m = n$, i.e., $d = 1$, it is necessary to require strengthened regularity of condition (2.25).

So, consider the problem

$$\lambda^m u(x) + \sum_{k=1}^{m} \lambda^{m-k}(a_k(x)u^{(k)}(x) + B_k u|_x) = 0, \tag{2.33}$$

$$L_p(\lambda)u = \sum_{k=1}^{n_p} \lambda^k [\alpha_{pk} u^{(m_p-k)}(0) + \beta_{pk} u^{(m_p-k)}(1)$$

$$+ \sum_{i=1}^{N_{pk}} \delta_{pki} u^{(m_p-k)}(x_{pki}) + T_{pk}u] = 0, \quad p = 1,\ldots,m, \tag{2.34}$$

where $x_{pki} \in (0,1)$, $m \ge 1$, $m_p \ge n_p$, $n_p \le m - 1$, and the corresponding characteristic equation

$$a_m(x)\omega^m + a_{m-1}(x)\omega^{m-1} + \cdots + 1 = 0. \tag{2.35}$$

Theorem 2.6. Let $m \ge 1$, $0 \le n_p \le m - 1$, $m_p \ge n_p$, $\max\{m_p\} - \min\{m_p - n_p\} \le m - 1$ and

(1) $a_j \in C^s[0,1]$, $j = 1,\ldots,m$, for an integer $s \ge \max\{m_p\}$, $a_j(0) = a_j(1)$[10], $a_m(x) \ne 0$;

---

[10]See the footnote on p.100.

(2) conditions (2.34) are p-regular with respect to functions $\omega_j(x)$, $j = 1, \ldots, m$, and $(m-p)$-regular with respect to functions $\omega_{p+1}(x), \ldots, \omega_m(x)$, $\omega_1(x), \ldots, \omega_p(x)$, where $\omega_j(x)$ are roots of the equation (2.35);

(3) $|\alpha_{p0}| + |\beta_{p0}| \neq 0$; the system L_{pn_p}, $p = 1, \ldots, m$, is normal;

(4) operators B_k from $W_2^k(0,1)$ into $L_2(0,1)$ and from $W_2^{s+k}(0,1)$ into $W_2^s(0,1)$ are compact;

(5) functionals T_{pk} are continuous in $W_r^p(0,1)$ for a number $r \in [1, \infty)$.

Then the spectrum of the problem (2.33)–(2.34) is discrete and the system of root functions of the problem (2.33)–(2.34) is m-fold complete in the space

$$\mathcal{H} = \{v \mid v = (v_1, \ldots, v_m) \in \bigoplus_{k=0}^{m-1} W_2^{z+m-k-1}(0,1), \ \sum_{k=0}^{n_p} L_{p,n_p-k} v_{k+s} = 0,$$
$$m_p \leq z + m - s - 1, \ s = 1, \ldots, m - n_p - 1, \ p = 1, \ldots, m\}$$

for an integer $z \in [max\{0, m_p - (m-1)\}, \min\{m_p - n_p\}]$.

Proof. We show the slight changes that must be made in the last part of the proof of Theorem 2.5 in order to prove Theorem 2.6. From Theorem 1.7 it follows that for a solution to problem (2.28) estimate (2.31) with $d = 1$ is fulfilled in the angles

$$\pi/2 - \underline{\omega} + \varepsilon < \arg \lambda < 3\pi/2 - \overline{\omega} - \varepsilon, \tag{2.36}$$

$$\pi/2 - \underline{\omega_1} + \varepsilon < \arg \lambda < 3\pi/2 - \overline{\omega_1} - \varepsilon, \tag{2.36'}$$

where numbers $\underline{\omega}$, $\overline{\omega}$ are defined by formulas (2.30) (here $\omega_j(x)$ are roots of equation (2.35)), and

$$\underline{\omega_1} = \inf_{x \in [0,1]} \min\{\arg \omega_j(x) : j = p+1, \ldots, m; \ \arg \omega_s(x) + \pi : s = 1, \ldots, p\}$$
$$= \underline{\omega} + \pi,$$

$$\overline{\omega_1} = \sup_{x \in [0,1]} \max\{\arg \omega_j(x) : j = p+1, \ldots, m; \ \arg \omega_s(x) + \pi : s = 1, \ldots, p\}$$
$$= \overline{\omega} + \pi.$$

Hence, angle (2.36') has the form

$$-\pi/2 - \underline{\omega} + \varepsilon < \arg \lambda < \pi/2 - \overline{\omega} - \varepsilon. \tag{2.37}$$

Let us calculate four angles, formed between neighboring sides of angles (2.36) and (2.37). The magnitudes of angles (2.36) and (2.37) are equal to $\pi - (\overline{\omega} - \underline{\omega}) - 2\varepsilon$.

The magnitudes of the other two angles are equal to $\overline{\omega} - \underline{\omega} + 2\varepsilon$. Consider a number q, that satisfies that inequality

$$1 < q < \min\{\pi/(\overline{\omega} - \underline{\omega}), \pi/[\pi - (\overline{\omega} - \underline{\omega})]\}.$$

Then all above mentioned angles are less than π/q for a small enough $\varepsilon > 0$. Hence, Theorem 2.3.4 can be applied in this case to the system of pencils (2.26) as well. ∎

The following example shows that in Theorem 2.6 the *strengthened regularity* (condition 2 of Theorem 2.6) cannot be substituted with the condition of *p-regularity*. Let us choose a_1 and a_2 so that the roots of the equation

$$a_2\omega^2 + a_1\omega + 1 = 0 \tag{2.38}$$

have properties: $\arg\omega_1 = \arg\omega_2$, $\omega_1 \neq \omega_2$. Then the Cauchy problem

$$\lambda^2 u(x) + \lambda a_1 u'(x) + a_2 u''(x) = 0,$$
$$u(0) = 0, \ u'(0) = 0$$

is 2-regular with respect to the roots ω_1, ω_2. Indeed, taking for P an arbitrary straight line that passes through the origin and does not contain ω_1, ω_2 we obtain

$$\theta = \begin{vmatrix} 1 & 1 \\ \omega_1 & \omega_2 \end{vmatrix} \neq 0.$$

But, on the other hand, since the Cauchy problem has no eigenvalues, the statement of Theorem 2.6 does not hold.

It is interesting to note that Theorem 2.6 in the complicated situation mentioned above, i.e., when $\arg\omega_1 = \arg\omega_2$, $\omega_1 \neq \omega_2$, allows those problems for which the 2-fold completeness holds to be separated. From Theorem 2.6 for the problem

$$\lambda^2 u(x) + \lambda(a_1 u'(x) + B_1 u|_x) + a_2 u''(x) + B_2 u|_x = 0,$$

$$\tag{2.39}$$

$$L_p u = \alpha_p u^{(m_p)}(0) + \beta_p u^{(m_p)}(1) + \sum_{i=1}^{N_p} \delta_{pi} u^{(m_p)}(x_{pi}) + T_p u = 0, \qquad p = 1, \ldots, m,$$

where $x_{pi} \in (0, 1)$, follows:

Theorem 2.7. *Let the following conditions be satisfied:*

(1) $\arg \omega_1 = \arg \omega_2$, $\omega_1 \neq \omega_2$, where ω_1, ω_2 are roots of (2.38);

(2) $\alpha_p \neq 0$, $\beta_p \neq 0$, $\max\{m_1, m_2\} - \min\{m_1, m_2\} \leq 1$, $m_1 \neq m_2$;

(3) *operators* B_k *from* $W_2^k(0,1)$ *into* $L_2(0,1)$ *and from* $W_2^{s+k}(0,1)$ *into* $W_2^s(0,1)$ *are compact, for* $s \geq \max\{m_p\}$;

(4) *functionals* T_p *are continuous in* $W_r^{m_p}(0,1)$ *for a number* $r \in [1,\infty)$.

Then the spectrum of problem (2.39) is discrete and a system of root functions of problem (2.39) is 2-fold complete in the space

$$W_2^{z+1}((0,1), L_p u = 0, p = 1,2) \oplus W_2^z(0,1),$$

for an integer $z \in [\max\{0, m_p - 1\}, \min\{m_p\}]$.

From Theorem 2.6 for the problem

$$a(x)u'(x) + Bu|_x + \lambda u(x) = 0,$$

$$\alpha u^{(m)}(0) + \beta u^{(m)}(1) + \sum_{p=1}^{N} \delta_p u^{(m)}(x_p) + Tu = 0, \tag{2.40}$$

where $x_p \in (0,1)$, follows:

Theorem 2.8. *Let the following condition be satisfied:*

(1) $a \in C^m[0,1]$, $a(0) = a(1)$, $a(x) \neq 0$, $\arg a(x) \in [\gamma, \delta]$, where $\delta - \gamma < \pi$;

(2) *the operator* B *from* $W_2^1(0,1)$ *into* $L_2(0,1)$ *and from* $W_2^{m+1}(0,1)$ *into* $W_2^m(0,1)$ *is compact;*

(3) $\alpha \neq 0$, $\beta \neq 0$; *the functional* T *is continuous in* $W_r^m(0,1)$ *for a number* $r \in [1,\infty)$.

Then the spectrum of problem (2.40) is discrete and a system of root functions of problem (2.40) is complete in the space $W_2^m(0,1)$.

If $a(x) = e^{i2\pi x}$, $B = 0$, $\delta_p = 0$, $T = 0$ the finite spectrum of problem (2.40) is empty. In other words, for $\delta - \gamma = 2\pi$ Theorem 2.8 is not true.

Remark 2.9. *If* $a \in C[0,1]$, $a(x) \neq 0$, $\arg a(x) \in [\gamma, \delta]$, $\delta - \gamma < \pi$, *then the Ward problem [53, Ch.2, §5.4]*

$$a(x)u'''(x) = \lambda u(x),$$

$$u(0) = 0, \ u'(0) = 0, \ u(1) = 0, \tag{2.41}$$

has a discrete spectrum, and a system of root functions of problem (2.41) is complete in $W_2^3((0,1), u(0) = 0, u'(0) = 0, u(1) = 0)$ *and* $L_2(0,1)$.

Note that the boundary value conditions in (2.41) are irregular in the sense of Birkhoff-Tamarkin [53, Ch.2, §5.4].

3.3. Completeness of root functions of principally boundary value problems for ordinary differential equations with a linear parameter

Consider a principally boundary value problem for an ordinary differential equation with a variable coefficient in the case when the spectral parameter appears linearly not only in the equation, but in the boundary conditions as well

$$L(\lambda)u = \lambda u(x) + a(x)u^{(m)}(x) + Bu|_x = 0, \qquad x \in (0,1), \qquad (3.1)$$

$$
\begin{aligned}
L_n(\lambda)u &= \lambda[\alpha_{n1}u^{(m_n)}(0) + \beta_{n1}u^{(m_n)}(1) \\
&\quad + \sum_{i=1}^{N_{n1}} \delta_{n1i}u^{(m_n)}(x_{n1i}) + T_{n1}u] + T_{n0}u = 0, \qquad n = 1,\ldots,s, \\
L_n u &= \alpha_{n0}u^{(m_n)}(0) + \beta_{n0}u^{(m_n)}(1) \\
&\quad + \sum_{i=1}^{N_{n0}} \delta_{n0i}u^{(m_n)}(x_{n0i}) + T_{n0}u = 0, \qquad n = s+1,\ldots,m,
\end{aligned}
$$

$$(3.2)$$

where $x_{nki} \in (0,1)$, $m \geq 2$, $m_n \geq 0$, B is an operator in $L_2(0,1)$, T_{nk} is a functional in $L_2(0,1)$.

Conditions (3.2) will be p-regular with respect to the numbers $\omega_j = e^{i2\pi(j-1)/m}$, $j = 1,\ldots,m$, if

(1)

$$
\begin{vmatrix}
\alpha_{11}\omega_1^{m_1} & \cdots & \alpha_{11}\omega_p^{m_1} & \beta_{11}\omega_{p+1}^{m_1} & \cdots & \beta_{11}\omega_m^{m_1} \\
\vdots & \vdots & \vdots & \vdots & \vdots & \vdots \\
\alpha_{s1}\omega_1^{m_s} & \cdots & \alpha_{s1}\omega_p^{m_s} & \beta_{s1}\omega_{p+1}^{m_s} & \cdots & \beta_{s1}\omega_m^{m_s} \\
\alpha_{s+1,0}\omega_1^{m_s+1} & \cdots & \alpha_{s+1,0}\omega_p^{m_s+1} & \beta_{s+1,0}\omega_{p+1}^{m_s+1} & \cdots & \beta_{s+1,0}\omega_m^{m_s+1} \\
\vdots & \vdots & \vdots & \vdots & \vdots & \vdots \\
\alpha_{m0}\omega_1^{m_m} & \cdots & \alpha_{m0}\omega_p^{m_m} & \beta_{m0}\omega_{p+1}^{m_m} & \cdots & \beta_{m0}\omega_m^{m_m}
\end{vmatrix} \neq 0,
$$

where $p = m/2$, if m is even; $p = [m/2]$ or $p = [m/2] + 1$ if m is odd;

(2) $x_{nki}(0,1)$; for some $r \in [1,\infty)$ functionals T_{nk}, $n = s+1,\ldots,m$, $k = 0,1$ in $W_r^{m_n+m-km}(0,1)$ and functionals T_{n0}, $n = s+1,\ldots,m$ in $W_r^{m_n}(0,1)$ are continuous.

Under the order m_{nk} of the functional T_{nk} we mean infimum of those q for which T_{nk} is continuous in $C^q[0,1]$.

Let us denote

$$A_{n0}u = \alpha_{n1}u^{(m_n)}(0) + \beta_{n1}u^{(m_n)}(1) + \sum_{i=1}^{N_{n1}} \delta_{n1i}u^{(m_n)}(x_{n1i})$$

$$+ T_{n1}u, \qquad n = 1, \dots, s, \tag{3.3}$$

$$A_{n1}u = T_{n0}u = 0, \qquad n = 1, \dots, s, \tag{3.4}$$

$$A_{n1}u = \alpha_{n0}u^{(m_n)}(0) + \beta_{n0}u^{(m_n)}(1) + \sum_{i=1}^{N_{n0}} \delta_{n0i}u^{(m_n)}(x_{n0i})$$

$$+ T_{n0}u, \qquad n = s+1, \dots, m, \tag{3.5}$$

From the p-regularity of conditions (3.2) with respect to the numbers ω_j, $j = 1, \dots, m$, it follows that for all $n = 1, \dots, s$, we have $A_{n0} \neq 0$.

Theorem 3.1. *Let the following conditions be satisfied:*

(1) $m \geq 1$, $m_n \geq 0$, $n = 1, \dots, m$;
(2) *conditions (3.2) are p-regular with respect to the numbers* $\omega_j = e^{i2\pi(j-1)/m}$, $j = 1, \dots, m$.

Then the linear manifold $\{v \mid v = (u, A_{10}u, \dots, A_{s0}u),\ u \in W_q^\ell(0,1),\ A_{n1}u = 0,\ n = s+1, \dots, m\}$, where $\ell \geq \max\{m_n + 1\}$ and functionals A_{n0}, A_{n1} are defined by the equalities (3.3)–(3.5), is dense in the space $W_q^z(0,1) \dotplus \mathbb{C}^s$, $z \leq \min\{m_n\}$, $q \in (1, \infty)$.

Proof. Let $(u, v_1, \dots, v_s) \in W_q^z(0,1) \dotplus \mathbb{C}^s$. By virtue of Corollary 1.10 there exists a function $u^0 \in W_q^\ell(0,1)$ that satisfies the relations

$$\begin{aligned} A_{n0}u^0 &= v_n, & n &= 1, \dots, s, \\ A_{n1}u^0 &= 0, & n &= s+1, \dots, m. \end{aligned} \tag{3.6}$$

Since $z \leq \min\{m_n\}$ then by virtue of Theorem 2.1 the set

$$\{u \mid u \in W_q^\ell(0,1),\ A_{n0}u = 0,\ n = 1, \dots, s,\ A_{n1}u = 0,\ n = s+1, \dots, m\}$$

is dense in the space $W_q^z(0,1)$. Hence, there exist functions $g \in W_q^\ell(0,1)$ such that

$$A_{n0}g = 0, \quad n = 1, \dots, s, \tag{3.7}$$

$$A_{n1}g = 0, \quad n = s+1, \dots, m, \tag{3.8}$$

$$\|g - (u - u^0)\|_{z,q} \leq \varepsilon.$$

From (3.6)–(3.8) it follows that for functions

$$\varphi = u^0 + g \in W_q^\ell(0,1)$$

the relations

$$A_{n0}\varphi = v_n, \quad n = 1, \ldots, s,$$
$$A_{n1}\varphi = 0, \quad n = s+1, \ldots, m, \tag{3.9}$$
$$\|\varphi - u\|_{z,q} \le \varepsilon,$$

are satisfied. On the other hand, from (3.9) it follows that

$$\|(\varphi, A_{10}\varphi, \ldots, A_{s0}\varphi) - (u, v_1, \ldots, v_s)\|_{W_q^z(0,1)+\mathbb{C}^s} < \varepsilon. \quad \blacksquare$$

Theorem 3.2. *Let the following conditions be satisfied:*

(1) $m \ge 2$; $m_n \ge 0$; $\max\{m_n\} - \min\{m_n\} \le m - 1$; $\max\{m_n : n = 1, \ldots, s\} - \min\{m_n\} \le m - 1$;

(2) $a \in C^{\ell}[0,1]$ *for an integer* $\ell \in [\max\{m_n, m_{n0} : n = 1, \ldots, s\} - (m - 1), \min\{m_n\}]$; $a(x) \ne 0$; $a(0) = a(1)^{11}$;

(3) *the operator* B *from* $W_2^m(0,1)$ *into* $L_2(0,1)$ *and from* $W_2^{\ell+m}(0,1)$ *into* $W_2^{\ell}(0,1)$ *is compact;*

(4) *conditions (3.2) are p-regular with respect to the numbers* $\omega_j = e^{i2\pi(j-1)/m}$, $j = 1, \ldots, m$;

(5) *for some* $r \in [1,\infty)$ *functionals* T_{nk}, $n = 1, \ldots, s$, $k = 0,1$ *in* $W_r^{m_n+m-km}(0,1)$ *and functionals* T_{n0}, $n = s+1, \ldots, m$ *in* $W_r^{m_n}(0,1)$ *are continuous.*

Then the spectrum of the problem (3.1)–(3.2) is discrete and a system of the vectors $\{(u_k, A_{10}u_k, \ldots, A_{s0}u_k)\}$, where u_k are root functions of the problem (3.1)–(3.2) and functionals A_{n0} are defined by equalities (3.3), is complete in the space $W_2^{\ell}(0,1) \oplus \mathbb{C}^s$.

Proof. Consider the operator A, defined by the equality

$$Au = a(x)u^{(m)}(x) + Bu|_x,$$

and functionals A_{n0}, $n = 1, \ldots, s$, A_{n1}, $n = 1, \ldots, m$, defined by the equalities (3.3)–(3.5). From conditions 2 and 3 it follows that the operator A from $H_1 = W_2^{\ell+m}(0,1)$ into $H = W_2^{\ell}(0,1)$ is bounded. Since for $n = 1, \ldots, m$ we have $m_n \le \ell + m - 1$ and for $n = 1, \ldots, s$ we have $m_{n0} \le \ell + m - 1$, then from condition 5 it follows that functionals A_{n0}, $n = 1, \ldots, s$, and A_{n1}, $n = 1, \ldots, m$, in $H_1 = W_2^{\ell+m}(0,1)$ are continuous. So, conditions 4 and 5 of Theorem 2.4.1 are satisfied

---

[11] See the footnote on p.100.

with $H^n = H_0^n = \mathbb{C}$. We now rewrite the problem (3.1)–(3.2) in the form of a system of operator pencil equations

$$
\begin{aligned}
&L(\lambda)u = \lambda u + Au = 0, \\
&L_n(\lambda)u = \lambda A_{n0}u + A_{n1}u = 0, \quad n = 1, \ldots, s, \\
&L_n u = A_{n1}u, \quad n = s+1, \ldots, m,
\end{aligned}
\tag{3.10}
$$

where $u \in H_1$. Apply Theorem 2.4.1 to (3.10).

By virtue of [72, p.350/14] the compact embedding

$$
W_2^{\ell+m}(0,1) \subset W_2^{\ell}(0,1)
$$

and the equivalence relation

$$
s_j(J, W_2^{\ell+m}(0,1), W_2^{\ell}(0,1)) \sim j^{-m},
$$

hold. Hence, for $q > m^{-1}$

$$
J \in \sigma_p(W_2^{\ell+m}(0,1), W_2^{\ell}(0,1)),
$$

i.e.,

$$
J \in \sigma_p(H_1, H), \quad q > m^{-1}.
\tag{3.11}
$$

So, conditions 1–3 of Theorem 2.4.1 are also satisfied. By virtue of conditions 4 and 5 and Theorem 3.1 it follows that the linear manifold

$$
\{v \mid v = (u, A_{10}u, \ldots, A_{s0}u),\ u \in W_2^{\ell+m}(0,1),\ A_{n1}u = 0,\ n = s+1, \ldots, m\}
$$

is dense in the Hilbert space $H \overset{s}{\underset{n=1}{\oplus}} H^n = W_2^{\ell}(0,1) \oplus \mathbb{C}^s$, i.e., condition 6 of Theorem 2.4.1 is satisfied.

Let us now establish condition 7 of Theorem 2.4.1. Since the roots of the characteristic equation

$$
a(x)\omega^m + 1 = 0
\tag{3.12}
$$

have the form

$$
\omega_j(x) = |a(x)|^{-1/m}\, e^{i(\pi - \arg a(x))\omega_j/m}, \quad j = 1, \ldots, m,
$$

then conditions (3.2) are p-regular with respect to the roots of the equation (3.12) too. In other words, the problem (3.1)–(3.2) is p-regular. Hence, by virtue of

Theorem 1.7, for $\ell + m \geq \max\{m_n : n = 1, \ldots, m; \ m_{n0} : n = 1, \ldots, s\} + 1$ and for all complex numbers λ lying inside the angle

$$(\pi/2 - \underline{\omega})m + \varepsilon < \arg \lambda < (3\pi/2 - \overline{\omega})m - \varepsilon, \tag{3.13}$$

where

$$\underline{\omega} = \inf_{x \in [0,1]} \min\{\arg \omega_j(x) : j = 1, \ldots, p; \ \arg \omega_s(x) - \pi : s = p+1, \ldots, m\},$$

$$\overline{\omega} = \sup_{x \in [0,1]} \max\{\arg \omega_j(x) : j = 1, \ldots, p; \ \arg \omega_s(x) - \pi : s = p+1, \ldots, m\},$$

and with large enough moduli, the problem

$$L(\lambda)u = f,$$
$$L_n(\lambda)u = f_n, \qquad n = 1, \ldots, s,$$
$$L_n u = f_n, \qquad n = s+1, \ldots, m,$$

for $f \in W_2^\ell(0,1)$, and $f_n \in \mathbb{C}$, has a unique solution $u \in W_2^{\ell+m}(0,1)$ and the estimate

$$\sum_{k=0}^{\ell+m} |\lambda|^{m^{-1}(\ell+m-k)} \|u\|_{k,2} \leq C(\varepsilon)\{\|f\|_{\ell,2} + |\lambda|^{m^{-1}\ell}\|f\|_{0,2}$$

$$+ \sum_{s=1}^{m} |\lambda|^{m^{-1}(\ell+m-m_s-1/2)}|f_s|\} \tag{3.14}$$

is true.

Since $m \geq 2$ and relation (3.11) is fulfilled for any $q > m^{-1}$ then it is possible to choose q to make π/q very close to 2π. Hence, Theorem 2.4.1 is aplicable to the system of pencils (3.10), if the estimate

$$\|\mathbf{L}^{-1}(\lambda)\|_{B(W_2^\ell(0,1)\oplus\mathbb{C}^m, W_2^{\ell+m}(0,1))} \leq C|\lambda|^h, \quad |\lambda| \to \infty, \tag{3.15}$$

is fulfilled in some angle of the complex plane for some $h \in \mathbb{R}$. In turn, from the estimate (3.14) it follows that the estimate (3.15) is fulfilled in the angle (3.13). ∎

Examples. Let us give examples of operators B_k and functionals T_n that satisfy the conditions of the theorems in this chapter. For simplicity we suppose that the weight $d = m/n$ is an integer.

1. If $b_{kj} \in W_q^{z-m}(0,1)$, the operator

$$B_k u = \sum_{j \leq dk-1} b_{kj}(x)u^{(j)}(x)$$

from $W_q^{dk}(0,1)$ into $L_q(0,1)$ and from $W_q^{z-m+dk}(0,1)$ into $W_q^{z-m}(0,1)$ is compact.

2. If $b_{kji} \in W_q^{z-m}(0,1)$, then the operator

$$B_k u = \sum_{j \leq dk-1} \sum_{i=1}^{N_{kj}} b_{kji}(x) u^{(j)}(x_{kji}),$$

where $x_{kji} \in [0,1]$, from $W_q^{dk}(0,1)$ into $L_q(0,1)$ and from $W_q^{z-m+dk}(0,1)$ into $W_q^{z-m}(0,1)$ is compact.

3. If $b_{kji} \in W_q^{z-m}(0,1)$ and these functions $\varphi_{kji}(x)$ map the segment [0,1] into itself and belong to $W_q^{z-m}(0,1)$, then the operator

$$B_k u = \sum_{j \leq dk-1} \sum_{i=1}^{N_{kj}} b_{kji}(x) u^{(j)}(\varphi_{kji}(x)),$$

from $W_q^{dk}(0,1)$ into $L_q(0,1)$ and from $W_q^{z-m+dk}(0,1)$ into $W_q^{z-m}(0,1)$ is compact.

Examples 1 and 2 are special cases of example 3, because they are obtained from example 3 under $\varphi_{kji}(x) = x$ and $\varphi_{kji}(x) = x_{kji}$ respectively. Let us prove example 3.

Proof. By virtue of [72, p.350/15] the embedding $W_q^{s+dk}(0,1) \subset C^{s+dk-1}[0,1]$, $s = 0, \ldots, z-m$ is compact. On the other hand, from

$$\|B_k u\|_{W_q^s(0,1)} \leq C \sum_{j \leq dk-1} \sum_{i=1}^{N_{kj}} \|u^{(j)}(\varphi_{kji}(\cdot))\|_{C^s[0,1]}$$
$$\leq C \|u\|_{C^{s+dk-1}[0,1]}$$

it follows that the operator B_k from $C^{s+dk-1}[0,1]$ into $W_q^s(0,1)$ is bounded. Hence, the operator B_k from $W_q^{s+dk}(0,1)$ into $W_q^s(0,1)$ is compact. ∎

4. If kernels $B_{kj}(x,y)$ are such that under some $\sigma > 1$

$$\int_0^1 |B_{kj}(x,y)|^\sigma dy + \int_0^1 |B_{kj}(x,y)|^\sigma dx \leq C$$

then the operator

$$B_k u = \sum_{j \leq dk} \int_0^1 B_{kj}(x,y) u^{(j)}(y) \, dy$$

from $W_q^{dk}(0,1)$ into $L_q(0,1)$ is compact [27, Ch.10, §2.2].

5. If γ_{sjp} are complex numbers then the functional

$$T_s u = \sum_{j=0}^{m_s-1} \sum_{p=1}^{N_{sj}} \gamma_{sjp} u^{(j)}(x_{sjp}),$$

where $x_{sjp} \in [0,1]$, in $W_q^{m_s}(0,1)$ is continuous. Here $N_{sj} = \infty$ is admitted. In this case it is required that $\sum_{p=1}^{\infty} |\gamma_{sjp}| < \infty$.

6. If $g_j \in L_1(0,1)$, $j = 0, \ldots, m_s - 1$, $g_{m_s} \in L_{q'}(0,1)$, then the functional

$$T_s u = \sum_{j=0}^{m_s} \int_0^1 g_j(x) u^{(j)}(x) \, dx$$

in $W_q^{m_s}(0,1)$ is continuous, where $1/q + 1/q' = 1$.

Chapter 4

Principally elliptic boundary value problems with a polynomial parameter

4.0. Introduction

In the papers by M. V. Keldysh [30] and F. E. Browder [9] the completeness of root functions of elliptic boundary value problems with a self-adjoint principal part is proved. S. Agmon [1] proved the completeness of root functions of elliptic boundary value problems, if the principal part is not selfadjoint.

In this chapter the completeness of root functions of a problem, with the principal part being an elliptic boundary value problem, is proved. In this case some difficulties appear, one of which is the following: given, in the principal part, differential operators

$$L_p u = \sum_{|\alpha|=m_p} b_{p\alpha}(x') D^\alpha u(x') + T_p u|_{x'}, \qquad x' \in \Gamma, \ p = 1, \dots, m,$$

where Γ is the $(r-1)$-dimensional infinitely smooth boundary of a bounded domain $G \subset \mathbb{R}^r$, and the operators T_p from $W_q^{m_p}(G)$ into $L_q(\Gamma)$ are bounded. It is necessary to prove that the set

$$H_0 = \{u \mid u \in W_q^{2m}(G), \ L_p u = 0, \ x' \in \Gamma, \ p = 1, \dots, m\}, \quad m_p \le 2m - 1$$

is dense in $L_q(G)$. When T_p are differential operators, i.e.,

$$T_p u = \sum_{|\alpha| \le m_p - 1} b_{p\alpha}(x') D^\alpha u(x'),$$

this is known, since $C_0^\infty(G) \subset H_0 \subset L_q(G)$ and $\overline{C_0^\infty(G)} = L_q(G)$. As we will see below, the general case is not so simple.

Consider a spectral problem in G for which the prncipal part is a regular elliptic problem with a polynomial parameter

$$L(x, D, \lambda)u = \lambda^n u(x) + \sum_{k=1}^{n} \lambda^{n-k} \left(\sum_{|\alpha|=dk} a_{k\alpha}(x)D^\alpha u(x) + B_k u|_x \right) = 0,$$

$$x \in G, \tag{0.1}$$

$$L_p(x', D, \lambda)u = \sum_{k=0}^{n_p} \lambda^k \left(\sum_{|\alpha|=m_p-dk} b_{pk\alpha}(x')D^\alpha u(x') + T_{pk}u|_{x'} \right) = 0,$$

$$x' \in \Gamma, \quad p = 1, \ldots, m, \tag{0.2}$$

where the problem weight $d = 2m/n$ is, in general, a non-integer, and $a_{k\alpha}(x) = b_{pk\alpha}(x) = 0$ if dk is a non-integer, $D^\alpha = D_1^{\alpha_1} \cdots D_r^{\alpha_r}$, $D_j = -i\frac{\partial}{\partial x_j}$, $j = 1, \ldots, r$, $\alpha = (\alpha_1, \ldots, \alpha_r)$ is a multi-index, $|\alpha| = \sum_{j=1}^{r} \alpha_j$, $\lambda \in \mathbb{C}$, $x = (x_1, \ldots, x_r)$, $a_{k\alpha} \in C^{\ell-2m}(\overline{G})$, $b_{pk\alpha} \in C^{\ell-m_p}(\overline{G})$, $\Gamma \in C^\ell$, $\ell \geq \max\{2m, m_p + 1\}$. Here operators B_k from $W_q^{dk}(G)$ into $L_q(G)$ are compact, operators T_{pk} from $W_q^{m_p-dk+1/q}(G)$ into $L_q(\Gamma)$ are compact, $q \in (1, \infty)$.

A number λ_0 is called an *eigenvalue* of the problem (0.1)–(0.2) if the problem

$$L(x, D, \lambda_0)u = 0, \quad x \in \Gamma, \tag{0.3}$$

$$L_p(x', D, \lambda_0)u = 0, \quad x' \in \Gamma, \ p = 1, \ldots, m \tag{0.4}$$

has a nontrivial solution that belongs to $W_2^{2m}(G)$. The nontrivial solution $u_0(x)$ of the problem (0.3)–(0.4) that belongs to $W_2^{2m}(G)$ is called an *eigenfunction* of the problem (0.1)–(0.2), and corresponds to the eigenvalue λ_0. A solution $u_p(x)$ to the problem

$$L(\lambda_0)u_p + \frac{1}{1!}L'(\lambda_0)u_{p-1} + \cdots + \frac{1}{p!}L^{(p)}(\lambda_0)u_0 = 0,$$

$$L_k(\lambda_0)u_p + \frac{1}{1!}L_k'(\lambda_0)u_{p-1} + \cdots + \frac{1}{p!}L_k^{(p)}(\lambda_0)u_0 = 0, \quad k = 1, \ldots, m,$$

belongs to $W_2^{2m}(G)$, and is called an *associated function* of the p-th rank to the eigenfunction $u_0(x)$ of the problem (0.1)–(0.2).

Eigenfunctions and associated functions of the problem (0.1)–(0.2) are combined under the general name *root functions* of the problem (0.1)–(0.2).

A complex number λ is called a *regular point* of the problem (0.1)–(0.2), if the problem

$$L(x, D, \lambda)u = f(x), \quad x \in G, \tag{0.5}$$

$$L_p(x', D, \lambda)u = f_p(x'), \quad x' \in \Gamma, \ p = 1, \ldots, m, \tag{0.6}$$

for any $f \in L_2(G)$, $f_p \in W_2^{2m-m_p-1/2}(\Gamma)$, has a unique solution belonging to $W_2^{2m}(G)$ and the estimate

$$\|u\|_{2m,2,G} \le C(\lambda) \left(\|f\|_{0,2,G} + \sum_{p=1}^{m} \|f_p\|_{2m-m_p-1/2,2,\Gamma} \right)$$

is satisfied.[12]

The complement of the set of regular points in the complex plane is called the *spectrum* of the problem (0.1)–(0.2).

The definition of the discreteness of the problem (0.1)–(0.2) spectrum is analogous to that of the discreteness of the operator pencil system spectrum from 2.3.1.

The spectrum of the problem (0.1)–(0.2) is called *discrete*, if

a) all points λ, not coinciding with eigenvalues of the problem (0.1)–(0.2), are regular points of the problem (0.1)–(0.2);

b) the eigenvalues are isolated and have finite algebraic multiplicities;

c) infinity is the only limit point of th set of eigenvalues of the problem (0.1)–(0.2).

Consider a system of differential equations

$$L(x, D, D_t)u(t, x) = 0, \quad t > 0, \ x \in G, \tag{0.7}$$

$$L_p(x', D, D_t)u(t, x') = 0, \quad t > 0, \ x' \in \Gamma, \ p = 1, \dots, m, \tag{0.8}$$

where $D_t = \frac{\partial}{\partial t}$.

By virtue of Lemma 2.0.1, a function of the form

$$u(t, x) = e^{\lambda_0 t} \left(\frac{t^k}{k!} u_0(x) + \frac{t^{k-1}}{(k-1)!} u_1(x) + \cdots + u_k(x) \right) \tag{0.9}$$

is a solution to the system (0.7)–(0.8) if and only if the system of functions $u_0(x), u_1(x), \dots, u_k(x)$ is a chain of root functions of the problem (0.1)–(0.2), that corresponds to the eigenvalue λ_0.

A solution of the form (0.9) is called an *elementary solution* of the system (0.7)–(0.8).

Let $\mathcal{H}$ be a Hilbert space, continuously embedded into $[L_2(G)]^n = \bigoplus^n L_2(G)$.

A system of root functions of the problem (0.1)–(0.2) is called *n-fold complete* in $\mathcal{H}$ if the system of functions

$$(u(0, x)u_t'(0, x), \dots, u_t^{(n-1)}(0, x))$$

---

[12]In the general case, i.e., when a boundary value problem is an irregular, it should be required that a solution belongs to $L_2(G)$ and the estimate $\|u\|_{0,2,G} \le C(\lambda)(\|f\|_{0,2,G} + \sum_{p=1}^{m} \|f_p\|_{2m-m_p-1/2,2,\Gamma})$ holds.

is complete in the space $\mathcal{H}$.

If (0.1)–(0.2) is a regular elliptic problem, i.e.,

$$B_k u = \sum_{|\alpha| \leq dk-1} a_{k\alpha}(x) D^\alpha u(x), \tag{0.10}$$

$$T_{pk} u = \sum_{|\alpha| \leq m_p - dk - 1} b_{pk\alpha}(x') D^\alpha u(x'), \tag{0.11}$$

then conditions of coercive solvability of the problem (0.5)–(0.6) in the space $W_q^\ell(G)$ have been found in the works of S. Agmon and L. Nirenberg [2], M. S. Agranovich and M. I. Vishik [3], and in the space $C(\overline{G})$, in the paper by H. B. Stewart [68], for $n = 1$.

The completeness of root functions for (0.1)–(0.2) even with operators (0.10)–(0.11), i.e., in the regular elliptic problem case, was shown by S. Ya. Yakubov [88], [91].

In papers by S. Agmon [1], M. S. Agranovich [4], G. Geymonat and P. Grisvard [17], Z. A. Kotko and S. G. Krein [35], A. N. Kozhevnikov [36], [37], and S. Ya. Yakubov [88], [90] and [91] completeness of root functions is proved when the principal part of the problem is non-selfadjoint. In the paper by S. Agmon [1] a spectral parameter enters linearly into the equation and does not appear in the boundary value conditions. Except in the paper by S. Ya. Yakubov [91] in other works it is assumed that $m_p \leq 2m - 1$. A completeness condition of the form

$$\max\{m_p\} - \min\{m_p - dn_p\} \leq 2m - 1$$

appears for the first time in [91]. In Agmon's case $(n = 1, n_p = 0)$ [1] this condition is transformed into the following

$$\max\{m_p\} - \min\{m_p\} \leq 2m - 1,$$

i.e., the case $m_p \leq 2m - 1$, as it relates to the work of S. Agmon and others, is covered in [91].

4.1. Principally regular elliptic boundary value problems with a polynomial parameter

Let G be a bounded domain in the Euclidean space $\mathbb{R}^r$ with an $(r-1)$-dimensional boundary Γ. Consider the spectral problem in G

$$L(x, D, \lambda)u = \lambda^n u(x) + \sum_{k=1}^{n} \lambda^{n-k} \left(\sum_{|\alpha|=dk} a_{k\alpha}(x)D^\alpha u(x) + B_k u|_x \right) = 0,$$

$$x \in G, \tag{1.1}$$

$$L_p(x', D, \lambda)u = \sum_{k=0}^{n_p} \lambda^k \left(\sum_{|\alpha|=m_p-dk} b_{pk\alpha}(x')D^\alpha u(x') + T_{pk}u|_{x'} \right) = 0,$$

$$x' \in \Gamma, \ p = 1, \ldots, m, \tag{1.2}$$

where the problem weigh $d = 2m/n$ is, in general, a non-integer, and $a_{k\alpha}(x) = b_{pk\alpha}(x) = 0$ if dk is a non-integer, $D^\alpha = D_1^{\alpha_1} \cdots D_r^{\alpha_r}$, $D_j = -i\frac{\partial}{\partial x_j}$, $j = 1, \ldots, r$, $\alpha = (\alpha_1, \ldots, \alpha_r)$ is a multi-index, $|\alpha| = \sum_{j=1}^{r} \alpha_j$, $\lambda \in \mathbb{C}$, $x = (x_1, \ldots, x_r)$, $a_{k\alpha} \in C^{\ell-2m}(\overline{G})$, $b_{pk\alpha} \in C^{\ell-m_p}(\overline{G})$, $\Gamma \in C^\ell$, $\ell \geq \max\{2m, m_p + 1\}$. Here operators B_k from $W_q^{dk}(G)$ into $L_q(G)$ are compact, operators T_{pk} from $W_q^{m_p-dk+1/q}(G)$ into $L_q(\Gamma)$ are compact, $q \in (1, \infty)$.

Let us denote

$$L_0(x, \sigma, \lambda) = \lambda^n + \sum_{k=1}^{n} \lambda^{n-k} \sum_{|\alpha|=dk} a_{k\alpha}(x)\sigma^\alpha,$$

$$L_{p0}(x', \sigma, \lambda) = \sum_{k=0}^{n_p} \lambda^k \sum_{|\alpha|=m_p-dk} b_{pk\alpha}(x')\sigma^\alpha, \quad p = 1, \ldots, m,$$

where $\sigma^\alpha = \sigma_1^{\alpha_1} \cdots \sigma_r^{\alpha_r}$, $\sigma \in \mathbb{R}^r$.

Let S be a set in the complex plane $\mathbb{C}$.

CONDITION I. Let for $x \in \overline{G}$, $\sigma \in \mathbb{R}^r$, $\lambda \in S$, $|\sigma| + |\lambda| \neq 0$

$$L_0(x, \sigma, \lambda) \neq 0.$$

If $r = 1$, in addition, suppose that roots of the equation $L_0(x, \sigma, \lambda) = 0$, with respect to σ for $x \in \overline{G}$, $\lambda \in S$, are equally distributed between upper and lower half-planes.

CONDITION II. Let x' be any point on Γ. Let the vector σ' be tangent and σ be a normal vector to Γ at the point $x' \in \Gamma$. Consider the following ordinary differential problem on a half line $y > 0$

$$L_0\left(x', \sigma' - i\sigma\frac{d}{dy}, \lambda\right)u(y) = 0, \quad y > 0,$$

$$L_{p0}\left(x', \sigma' - i\sigma\frac{d}{dy}, \lambda\right)u(y)|_{y=0} = h_p, \quad p = 1, \ldots, m.$$

It is required that for $\lambda \in S$ this problem should have one and only one solution including all its derivatives, tending to 0 as $y \to \infty$ for any numbers $h_p \in \mathbb{C}$.

In addition condition I or II is fulfilled on the set S.

Condition II has many equivalent formulations [44, Ch.2, 4.1]. Let us show one of them. Let x', σ' and σ be the same as in condition II. Consider τ-roots of the polynomial

$$L_0(x', \sigma' + \tau\sigma, \lambda) = \lambda^n + \sum_{k=1}^{n} \lambda^{n-k} \sum_{|\alpha|=dk} a_{k\alpha}(x')(\sigma' + \tau\sigma)^\alpha = 0.$$

By virtue of condition I τ-roots are non-real and

$$\operatorname{Im}\tau_k^+(x', \sigma', \sigma, \lambda) > 0, \quad k = 1, \ldots, m,$$
$$\operatorname{Im}\tau_k^-(x', \sigma', \sigma, \lambda) < 0, \quad k = 1, \ldots, m,$$

CONDITION II'. *For any tangent vector σ' and normal vector σ to Γ at the point $x' \in \Gamma$, $\lambda \in S$ polynomials on τ*

$$L_{p0}(x', \sigma' + \tau\sigma, \lambda) = \sum_{k=1}^{n_p} \lambda^k \sum_{|\alpha|=m_p-dk} b_{pk\alpha}(x')(\sigma' + \tau\sigma)^\alpha, \quad p = 1, \ldots, m,$$

are linearly independent on modulo a polynomial

$$M^+(x', \sigma', \sigma, \lambda, \tau) = \prod_{k=1}^{m}(\tau - \tau_k^+(x', \sigma', \sigma, \lambda)).$$

In other words, if the polynomial

$$L'_{p0}(x', \sigma', \sigma, \lambda, \tau) = \sum_{k=0}^{m-1} b'_{pk}(x', \sigma', \sigma, \lambda)\tau^k$$

is the residue of the division of the polynomial L_{p0} on M^+, then

$$\det |b'_{pk}(x', \sigma', \sigma, \lambda)| \underset{\substack{p=1,\ldots,m \\ k=0,\ldots,m-1}}{} \neq 0.$$

4.1.1. Coerciveness of principally regular elliptic boundary value problems.

Consider a principally boundary value problem

$$L(x, D, \lambda)u = \lambda^n u(x) + \sum_{k=1}^{n} \lambda^{n-k} \left(\sum_{|\alpha|=dk} a_{k\alpha}(x)D^\alpha u(x) + B_k u|_x \right) = f(x),$$

$$x \in G, \tag{1.3}$$

$$L_p(x', D, \lambda)u = \sum_{k=0}^{n_p} \lambda^k \left(\sum_{|\alpha|=m_p-dk} b_{pk\alpha}(x')D^\alpha u(x') + T_{pk}u|_{x'} \right) = f_p(x'),$$

$$x' \in \Gamma, \ p = 1, \ldots, m, \tag{1.4}$$

where the weight $d = 2m/n$, in general, is a non-integer, and $a_{k\alpha}(x) = b_{pk\alpha}(x) = 0$ if dk is a non-integer.

A system

$$L_p(x, D) = \sum_{|\alpha|=m_p} b_{p\alpha}(x)D^\alpha + T_p, \quad p = 1, \ldots, m,$$

is called normal, if $m_j \neq m_k$ for $j \neq k$ and for any vector σ normal to the boundary Γ at the point $x' \in \Gamma$ the following condition is fulfilled

$$L_p(x', \sigma) = \sum_{|\alpha|=m_p} b_{p\alpha}(x')\sigma^\alpha \neq 0, \quad p = 1, \ldots, m,$$

and T_p from $W_q^{m_p+1/q}(G)$ into $L_q(\Gamma)$ are compact.

Let us denote

$$L_{p,n_p-k}(x', D)u = \sum_{|\alpha|=m_p-dk} b_{pk\alpha}(x')D^\alpha u(x') + T_{pk}u|_{x'}, \tag{1.5}$$

$k = 0, \ldots, n_p, \ p = 1, \ldots, m.$

Theorem 1.1. *Let the following conditions be satisfied:*

(1) $n \geq 1, \ m \geq 1, \ d = 2m/n, \ m_p \geq dn_p$;

(2) $a_{k\alpha} \in C^{\ell-2m}(\overline{G}), \ b_{pk\alpha} \in C^{\ell-m_p}(\overline{G}), \ \Gamma \in C^\ell$, where $\ell \geq \max\{2m, m_p + 1\}$;

(3) operators B_k from $W_q^{dk}(G)$ into $L_q(G)$ and from $W_q^{\ell-2m+dk}(G)$ into $W_q^{\ell-2m}(G)$ are compact, where $q \in (1, \infty)$;

(4) operators T_{pk} from $W_q^{m_p-dk+1/q}(G)$ into $L_q(\Gamma)$ and from $W_q^{\ell-dk}(G)$ into $W_2^{\ell-m_p-1/q}(\Gamma)$ are compact; $T_{pn_p} = 0$, if $m_p - dn_p = 0$;

(5) the system $L_{pn_p} = \sum_{|\alpha|=m_p} b_{p0\alpha}(x')D^\alpha + T_{p0}, \ p = 1, \ldots, m$, is normal;

(6) conditions I and II are fulfilled on the rays $\ell(\varphi)$.

Then the operator

$$\mathbb{L}(\lambda): \quad u \rightarrow (L(x, D, \lambda)u, L_1(x', D, \lambda)u, \ldots, L_m(x', D, \lambda)u)$$

from $W_q^\ell(G)$ into $W_q^{\ell-2m}(G) \overset{m}{\underset{p=1}{\dotplus}} W_q^{\ell-m_p-1/q}(\Gamma)$ for $\lambda \in \ell(\varphi)$, $|\lambda| \rightarrow \infty$, is an isomorphism, and for a solution of the problem (1.3)–(1.4) the estimate

$$\sum_{k=0}^{\ell} |\lambda|^{d^{-1}(\ell-k)} \|u\|_{k,q,G} \leq C[\|f\|_{\ell-2m,q,G} + |\lambda|^{d^{-1}(\ell-2m)} \|f\|_{0,q,G}$$

$$+ \sum_{s=1}^{m} (|\lambda|^{d^{-1}(\ell-m_s-1/q)} \|f_s\|_{0,q,\Gamma} + \|f_s\|_{\ell-m_s,-1/q,q,\Gamma})] \tag{1.6}$$

is valid.

Proof. Using substitution $\lambda = \mu^d$ the problem (1.3)–(1.4) is reduced to

$$\mu^{2m} u(x) + \sum_{k=1}^{n} \mu^{2m-dk} \left(\sum_{|\alpha|=dk} a_{k\alpha}(x) D^\alpha u(x) + B_k u|_x \right)$$
$$= f(x), x \in G, \tag{1.7}$$

$$\sum_{k=0}^{n_s} \mu^{dk} \left(\sum_{|\alpha|=m_s-dk} b_{sk\alpha}(x') D^\alpha u(x') + T_{sk} u|_{x'} \right) = f_s(x'),$$
$$x' \in \Gamma, \ s = 1, \ldots, m. \tag{1.8}$$

Let $u \in W_q^\ell(G)$ be a solution to the problem (1.3)–(1.4), i.e., (1.7)–(1.8). Then by virtue of [2] and [3], for any $\mu \in \ell(d^{-1}\varphi)$, $|\mu| \rightarrow \infty$, we have

$$\sum_{k=0}^{\ell} |\mu|^{\ell-k} \|u\|_{k,q,G} \leq C[\|f - \sum_{k=1}^{n} \mu^{2m-dk} B_k u\|_{\ell-2m,q,G}$$

$$+ |\mu|^{\ell-2m} \|f - \sum_{k=1}^{n} \mu^{2m-dk} B_k u\|_{0,q,G}$$

$$+ \sum_{s=1}^{m} (|\mu|^{\ell-m_s-1/q} \|f_s - \sum_{k=0}^{n_s} \mu^{dk} T_{sk} u\|_{0,q,\Gamma}$$

$$+ \|f_s - \sum_{k=0}^{n_s} \mu^{dk} T_{sk} u\|_{\ell-m_s,-1/q,q,\Gamma})]$$

$$\leq C[\|f\|_{\ell-2m,q,G} + |\mu|^{\ell-2m} \|f\|_{0,q,G}$$

$$+ \sum_{s=1}^{m}(|\mu|^{\ell-m_s-1/q}\|f_s\|_{0,q,\Gamma} + \|f_s\|_{\ell-m_s-1/q,q,\Gamma})$$

$$+ \sum_{k=1}^{n}|\mu|^{2m-dk}(\|B_k u\|_{\ell-2m,q,G} + |\mu|^{\ell-2m}\|B_k u\|_{0,q,G})$$

$$+ \sum_{s=1}^{m}\sum_{k=0}^{n_s}|\mu|^{dk}(|\mu|^{\ell-m_s-1/q}\|T_{sk}u\|_{0,q,\Gamma}$$

$$+ \|T_{sk}u\|_{\ell-m_s-1/q,q,\Gamma})]. \tag{1.9}$$

From (1.9), by virtue of conditions 3–4 and Lemma 3.1.6, we obtain

$$\sum_{k=0}^{\ell}|\mu|^{\ell-k}\|u\|_{k,q,G} \leq C\{\|f\|_{\ell-2m,q,G} + |\mu|^{\ell-2m}\|f\|_{0,q,G}$$

$$+ \sum_{s=1}^{m}(|\mu|^{\ell-m_s-1/q}\|f_s\|_{0,q,\Gamma} + \|f_s\|_{\ell-m_s-1/q,q,\Gamma})$$

$$+ \delta\sum_{k=1}^{n}(|\mu|^{2m-dk}\|u\|_{\ell-2m+dk,q,G} + |\mu|^{\ell-dk}\|u\|_{dk,q,G})$$

$$+ C(\delta)\sum_{k=1}^{n}(|\mu|^{2m-dk}\|u\|_{\ell-2m,q,G} + |\mu|^{\ell-dk}\|u\|_{0,q,G})$$

$$+ \sum_{s=1}^{m}\sum_{k=0}^{n_s}[|\mu|^{\ell-(m_s-dk)-1/q}(\delta\|u\|_{m_s-dk+1/q,q,G} + C(\delta)\|u\|_{1/q,q,G})$$

$$+ |\mu|^{dk}(\delta\|u\|_{\ell-dk,q,G} + C(\delta)\|u\|_{\ell-m_s,q,G})]\},$$

for any $\delta > 0$, besides $m_s - dn_s \neq 0$. Using Lemma 1.2.4., i.e.,

$$|\mu|^s\|u\|_{\ell-s,q,G} \leq C(\|u\|_{\ell,q,G} + |\mu|^{\ell}\|u\|_{0,q,G}),$$

we have

$$\sum_{k=0}^{\ell}|\mu|^{\ell-k}\|u\|_{k,q,G} \leq C[\|f\|_{\ell-2m,q,G} + |\mu|^{\ell-2m}\|f\|_{0,q,G}$$

$$+ \sum_{s=1}^{m}(|\mu|^{\ell-m_s-1/q}\|f_s\|_{0,q,\Gamma} + \|f_s\|_{\ell-m_s-1/q,q,\Gamma})]$$

$$+ (C\delta + C(\delta)|\mu|^{-\theta})(\|u\|_{\ell,q,G} + |\mu|^{\ell}\|u\|_{0,q,G}),$$

where $\theta = \min_{m_s-dn_s\neq 0}\{d, m_s - dn_s\}$. It is clear that one can choose $\delta > 0$ so small and $|\mu|$ so large that $C\delta + C(\delta)|\mu|^{-\theta} < 1$. Hence, for $\mu \in \ell(d^{-1}\varphi)$, $|\mu| \to \infty$ we obtain an a priori estimate

$$\sum_{k=0}^{\ell} |\mu|^{\ell-k} \|u\|_{k,q,G} \le C[\|f\|_{\ell-2m,q,G} + |\mu|^{\ell-2m}\|f\|_{0,q,G}$$

$$+ \sum_{s=1}^{m} (|\mu|^{\ell-m_s-1/q}\|f_s\|_{0,q,\Gamma} + \|f_s\|_{\ell-m_s-1/q,q,\Gamma})]. \tag{1.10}$$

From (1.10) it follows that, for $\lambda \in \ell(\varphi)$, $|\lambda| \to \infty$, the estimate (1.6) is valid. Hence, for the indicated λ, a solution of the problem (1.3)–(1.4) in $W_q^\ell(G)$ is unique.

Let us represent the operator $L(\lambda)$ in the form

$$\mathbb{L}(\lambda) = \mathbb{L}_0(\lambda) + \mathbb{L}_1(\lambda),$$

where

$$\mathbb{L}_0(\lambda) : u \to (L_0(x, D, \lambda)u, L_{10}(x', D, \lambda)u, \ldots, L_{m0}(x', D, \lambda)u)$$

$$L_0(x, D, \lambda) = \lambda^n + \sum_{k=1}^{n} \lambda^{n-k} \sum_{|\alpha|=dk} a_{k\alpha}(x)D^\alpha,$$

$$L_{p0}(x', D, \lambda) = \sum_{k=0}^{n_p} \lambda^k \sum_{|\alpha|=m_p-dk} b_{pk\alpha}(x')D^\alpha, \quad p = 1, \ldots, m,$$

$$\mathbb{L}_1(\lambda) = \left(\sum_{k=1}^{n} \lambda^{n-k} B_k u|_x, \sum_{k=0}^{n_1} \lambda^k T_{1k} u|_{x'}, \ldots, \sum_{k=0}^{n_m} \lambda^k T_{mk} u|_{x'} \right)$$

From [2] and [3] it follows that the operator $\mathbb{L}_0(\lambda)$ from $W_q^\ell(G)$ into $W_q^{\ell-2m}(G) \overset{m}{\underset{p=1}{+}}$ $W_q^{\ell-m_p-1/q}(\Gamma)$ is an isomorphism. By virtue of conditions 3 and 4 the operator $\mathbb{L}_1(\lambda)$ from $W_q^\ell(G)$ into $W_q^{\ell-2m}(G) \overset{m}{\underset{p=1}{+}} W_q^{\ell-m_p-1/q}(\Gamma)$ is compact. Then it is possible to apply the perturbation theorem of fredholm operators [28, p.238] to the operator $\mathbb{L}(\lambda) = \mathbb{L}_0(\lambda) + \mathbb{L}_1(\lambda)$, from which the Theorem 1.1 statement follows. ∎

Corollary 1.2. *If, in condition 6 of Theorem 1.1, conditions I and II are fulfilled in some angle S, then the statement of Theorem 1.1 is also true for $\lambda \in S$, $|\lambda| \to \infty$.*

CONDITION III. *Let there exist a differential operator $L(x, D, \lambda) = \lambda^{2m} + \sum_{k=1}^{2m} \lambda^{2m-k} \sum_{|\alpha|\le k} C_{k\alpha}(x)D^\alpha$ and rays $\ell(\varphi)$ such that for a system $(L(x, D, \lambda), L_1(x, D), \ldots, L_m(x, D))$, where*

$$L_p(x, D) = \sum_{|\alpha|=m_p} b_{p\alpha}(x)D^\alpha + T_p,$$

and T_p from $W_q^{m_p+1/q}(G)$ into $L_q(\Gamma)$ are compact, conditions I and II are fulfilled on the rays $\ell(\varphi)$.

If condition III is fulfilled then we say that the system $L_p(x, D)$, $p = 1, \ldots, m$ has property III.

Remark 1.3. For a one-dimensional case Lemma 3.1.9 implies that every normal system has property III.

Apparently in a multi-dimensional case it is not so. Let us represent the set $\{1, \ldots, m\}$ by a union of two disjoint subsets, i.e.,

$$\{1, \ldots, m\} = \{1, \ldots, m\}' \cup \{1, \ldots, m\}'', \qquad \{1, \ldots, m\}' \cap \{1, \ldots, m\}'' = \emptyset.$$

Theorem 1.4. Let the following conditions be satisfied:

(1) $b_{p\alpha} \in C^{s-m_p}(\overline{G})$, operators T_p from $W_q^{m_p+1/q}(G)$ into $L_q(\Gamma)$ and from $W_q^s(G)$ into $W_q^{s-m_p-1/q}(\Gamma)$ are compact, where $s \geq \max\{2m, m_p + 1\}$, $q \in (1, \infty)$; $T_p = 0$, if $m_p = 0$;

(2) the system $L_p(x', D) = \sum\limits_{|\alpha|=m_p} b_{p\alpha}(x')D^\alpha + T_p$, $p = 1, \ldots, m$, is normal and has property III.

Then the operator $u \to (L_p u, p \in \{1, \ldots, m\}')$ from $W_q^\ell(G)$ onto

$$\dot{+}_{p \in \{1, \ldots, m\}'} W_q^{\ell-m_p-1/q}(\Gamma)$$

has a continuous right-inverse, in other words, there exists such an operator $R(f_p, p \in \{1, \ldots, m\}') = u$ continuous from $\dot{+}_{p \in \{1, \ldots, m\}'} W_q^{\ell-m_p-1/q}(\Gamma)$ into $W_q^\ell(G)$, where u is a solution of the system

$$L_p(x', D)u = \sum\limits_{|\alpha|=m_p} b_{p\alpha}(x')D^\alpha u(x') + T_p u|_{x'} = f_p, \quad p \in \{1, \ldots, m\}',$$

and $\ell \geq \max\{m_p + 1 : p \in \{1, \ldots, m\}'\}$. Moreover, the inverse operator does not depend on $\ell \in [\max\{m_p + 1\} : p \in \{1, \ldots, m\}', s]$.

Proof. By virtue of Theorem 1.1 the operator

$$\mathbb{L}(\lambda) : \quad u \to (L(x, D, \lambda)u, L_1(x', D)u, \ldots, L_m(x', D)u),$$

from $W_q^\ell(G)$ into $W_q^{\ell-2m}(G) \overset{m}{\underset{p=1}{+}} W_q^{\ell-m_p-1/q}(\Gamma)$ for $\lambda \in \ell(\varphi)$, $|\lambda| \to \infty$, where $\ell \geq \max\{2m, m_p+1\}$, is an isomorphism. Then for some indicated λ, $f(x) = 0$, $f_p(x) = 0$, $p \in \{1, \ldots, m\}''$, and from (1.6) follows the estimate

$$\sum_{k=0}^{\ell} \|u\|_{k,q,G} \leq C \sum_{p \in \{1,\ldots,m\}'} \|f_p\|_{\ell-m_p-1/q,q,\Gamma}.$$

Now it is enough to close the inverse operator $L^{-1}(\lambda)$ on the subspace

$$\{0\} \underset{p\in\{1,\ldots,m\}'}{+} W_q^{\ell-m_p-1/q}(\Gamma) \underset{p\in\{1,\ldots,m\}''}{+} \{0\}. \blacksquare$$

4.1.2. Dense sets in Sobolev spaces. Before we pass to the main results of this item, we prove a theorem of dense sets in direct sums of Sobolev spaces, which is also of independent mathematical interest. In what follows this theorem is essential.

Theorem 1.5. *Let the following conditions be satisfied:*

(1) $b_{p\alpha} \in C^{s-m_p}(\overline{G})$, *operators T_p from $W_q^{m_p+1/q}(G)$ into $L_q(\Gamma)$ and from $W_q^s(G)$ into $W_q^{s-m_p-1/q}(\Gamma)$ are compact, where $s \geq \max\{2m, m_p + 1\}$, $q \in (1, \infty)$; $T_p = 0$, if $m_p = 0$;*

(2) *the system*

$$L_p = \sum_{|\alpha|=m_p} b_{p\alpha}(x')D^\alpha + T_p, \quad p = 1, \ldots, m$$

is normal and has property III.

Then for integer $z \in [\max\{m_p + 1\}, s]$ and $0 \leq k \leq z$

$$\overline{W_q^z(G, L_p u|_{x'} = 0, \ x' \in \Gamma, p = 1, \ldots, m)}|_{W_q^k(G)} = W_q^k(G, L_p u|_{x'} = 0, x' \in \Gamma, m_p \leq k - 1). \tag{1.11}$$

Proof. Let $u \in W_q^k(G, L_p u|_{x'} = 0, x' \in \Gamma, m_p \leq k - 1)$. Then there exists a sequence of functions $\varphi_n \in C^\infty(\overline{G})$ such that

$$\lim_{n\to\infty} \|\varphi_n - u\|_{k,q,G} = 0. \tag{1.12}$$

Since $W_q^s = (W_q^{s_0}, W_q^{s_1})_{\theta,q}$, $0 < s \neq$ an integer, $0 < \theta < 1$, and $s = (1-\theta)s_0 + \theta s_1$, then by virtue of interpolation theorems [72, p.117], operators T_p from $W_q^k(G)$ into

$W_q^{k-m_p-1/q}(\Gamma)$, for $m_p+1 \le k \le s$, are also compact. Then by virtue of embedding theorems we have

$$\lim_{n \to \infty} \|L_p \varphi_n - L_p u\|_{k-m_p-1/q,q,\Gamma} \le C \lim_{n \to \infty} \|\varphi_n - u\|_{k,q,G} = 0$$

for $m_p \le k-1$. Hence, for $m_p \le k-1$

$$\lim_{n \to \infty} \|L_p \varphi_n\|_{k-m_p-1/q,q,\Gamma} = \lim_{n \to \infty} \|L_p u\|_{k-m_p-1/q,q,\Gamma} = 0.$$

By virtue of Theorem 1.4 there exists a sequence of functions $g_n \in W_q^z(G)$ that satisfy the following relations

$$L_p g_n = -L_p \varphi_n, \quad p = 1, \dots, m, \tag{1.13}$$

$$\lim_{n \to \infty} \|g_n\|_{k,q,G} = 0. \tag{1.14}$$

It is easy to see now that for a sequence of functions

$$u_n = \varphi_n + g_n \in W_q^z(G)$$

the following relations

$$L_p u_n = 0, \quad p = 1, \dots, m, \tag{1.15}$$

$$\lim_{n \to \infty} \|u_n - u\|_{k,q,G} = 0 \tag{1.16}$$

exist. Relations (1.15) follow from (1.13) and relation (1.16) follows from (1.12) and (1.14). ∎

Remark 1.6. *Along by side with (1.11) the equality*

$$\overline{W_q^z(G, L_p u|_{x'} = 0, x' \in \Gamma, m_p \le z-1)}|_{W_q^k(G)}$$
$$= W_q^k(G, L_p u|_{x'} = 0, x' \in \Gamma, m_p \le k-1), \quad s \ge z \ge k \ge 0,$$

also holds.

Consider operators (1.2) on Γ. Then by virtue of (1.5), operators (1.2) have the form

$$L_p(x', D, \lambda)u = \sum_{k=0}^{n_p} \lambda^k L_{p,n_p-k}(x', D)u.$$

Theorem 1.7. *Let the following conditions be satisfied:*

(1) $n \ge 1$, $m \ge 1$, $0 \le n_p \le n-1$, $d = 2m/n$ *is an integer,* $m_p \ge dn_p$;

(2) $b_{pk\alpha} \in C^{\ell-m_p}(\overline{G})$, operators T_{pk} from $W_q^{m_p-dk+1/q}(G)$ into $L_q(\Gamma)$ and from $W_q^{\ell-dk}(G)$ into $W_q^{\ell-m_p-1/q}(\Gamma)$ are compact; $T_{pn_p} = 0$, if $m_p - dn_p = 0$;

(3) the system

$$L_{pn_p} = \sum_{|\alpha|=m_p} b_{p0\alpha}(x')D^\alpha + T_{p0}, \quad p = 1,\ldots,m$$

is normal and has property III.

Then the set

$$\mathcal{H}_d = \{v \mid v = (v_1,\ldots,v_n) \in \overset{n-1}{\underset{k=0}{+}} W_q^{z+d(n-k)}(G), \sum_{k=0}^{n_p} L_{p,n_p-k}v_{k+s}(x') = 0,$$

$$x' \in \Gamma, \; m_p \leq z + d(n-s+1) - 1, \; s = 1,\ldots,n-n_p, \; p = 1,\ldots,m\}$$

is dense in the space

$$\mathcal{H} = \{v \mid v = (v_1,\ldots,v_n) \in \overset{n-1}{\underset{k=0}{+}} W_q^{z+d(n-k-1)}(G), \sum_{k=0}^{n_p} L_{p,n_p-k}v_{k+s}|_{x'} = 0,$$

$$x' \in \Gamma, \; m_p \leq z + d(n-s) - 1, \; s = 1,\ldots,n-n_p-1, \; p = 1,\ldots,m\}$$

for an integer $z \in [max\{0, m_p - (2m-1)\}, min\{m_p - dn_p\}]$, and an arbitrary $q \in (1,\infty)$.

Proof. Let $\varepsilon > 0$ and $v = (v_1,\ldots,v_n) \in \mathcal{H}$. Set $t = min\{n_p\}$. Construct functions $\varphi_s \in C^\infty(G)$, $s = n-t+1,\ldots,n$ such that

$$\|\varphi_s - v_s\|_{z+d(n-s),q,G} < \varepsilon, \quad s = n-t+1,\ldots,n. \tag{1.17}$$

From the conditions of connection in $\mathcal{H}_d$, for $n_p = t$ and $s = n-t$, we obtain

$$L_{pn_p}\varphi_{n-t} = -\sum_{k=1}^{n_p} L_{p,n_p-k}\varphi_{k+n-t}, \quad m_p \leq z + d(t+1) - 1, \; n_p = t. \tag{1.18}$$

Since $\varphi_s \in C^\infty(G)$, $s = n-t+1,\ldots,n$, then by virtue of Theorem 1.4 there exists a function $\varphi_{n-t}^0 \in W_q^{z+d(t+1)}(G)$ such that

$$L_{pn_p}\varphi_{n-t}^0 = -\sum_{k=1}^{n_p} L_{p,n_p-k}\varphi_{k+n-t}, \quad m_p \leq z + d(t+1) - 1, \; n_p = t. \tag{1.19}$$

Since $z \leq m_p - dn_p$, then for $n_p = t$ we have $m_p \geq z + dt$. Then by virtue of Remark 1.6 there exists a function $g_{n-t} \in W_q^{z+d(t+1)}(G)$ such that

$$L_{pn_p}g_{n-t} = 0, \quad m_p \leq z + d(t+1) - 1, \; n_p = t, \tag{1.20}$$

$$\|g_{n-t} - (v_{n-t} - \varphi_{n-t}^0)\|_{z+dt,q,G} < \varepsilon.$$

It is easy to see now that the function

$$\varphi_{n-t} = g_{n-t} + \varphi_{n-t}^0 \in W_q^{z+d(t+1)}(G)$$

by virtue of (1.19)-(1.20), satisfies (1.18). Obviously,

$$\|\varphi_{n-t} - v_{n-t}\|_{z+dt,q,G} < \varepsilon.$$

From the conditions of connection in $\mathcal{H}_d$, for $n_p \leq t + i$, $s = n - t - i$, $m_p \leq z + d(t + i + 1) - 1$, $i = 0, \ldots, n - t - 1$, one obtains

$$L_{pn_p}\varphi_{n-t-i} = - \sum_{k=1}^{n_p} L_{p,n_p-k}\varphi_{k+n-t-i}. \tag{1.21}$$

Let us show by induction on i that there exist functions $\varphi_{n-t-i} \in W_q^{z+d(t+i+1)}(G)$, $i = 0, \ldots, n - t - 1$ that satisfy relations (1.21) and

$$\|\varphi_{n-t-i} - v_{n-t-i}\|_{z+d(t+i),q,G} < C\varepsilon. \tag{1.22}$$

Suppose that there exist functions $\varphi_{n-t-i} \in W_q^{z+d(t+i+1)}(G)$, that satisfy relations (1.21) and (1.22) for $i = 0, \ldots, s - 1$, where $s = 1, \ldots, n - t - 1$. We will now show that there exists a function $\varphi_{n-t-s} \in W_q^{z+d(t+s+1)}(G)$ satisfying (1.21) and (1.22) for $i = s$. Obviously, if $m_p \leq z + d(t + s + 1) - 1$, $n_p \leq t + s$, then from

$$\left\| \sum_{k=1}^{n_p} L_{p,n_p-k}\varphi_{k+n-t-s} \right\|_{z+d(t+s+1)-m_p-1/q,q,\Gamma}$$

$$\leq C \sum_{k=1}^{n_p} \|\varphi_{k+n-t-s}\|_{z+d(t+s-k+1),q,G}$$

it follows that $\sum_{k=1}^{n_p} L_{p,n_p-k}\varphi_{k+n-t-s} \in W_q^{z+d(t+s+1)-m_p-1/q}(\Gamma)$, and if $m_p \leq z + d(t + s) - 1$, $n_p \leq t + s - 1$ then

$$\left\| \sum_{k=1}^{n_p} L_{p,n_p-k}v_{k+n-t-s} \right\|_{z+d(t+s)-m_p-1/q,q,\Gamma}$$

$$\leq C \sum_{k=1}^{n_p} \|\varphi_{k+n-t-s}\|_{z+d(t+s-k),q,G}$$

implies $\sum_{k=1}^{n_p} L_{p,n_p-k}v_{k+n-t-s} \in W_q^{z+d(t+s)-m_p-1/q}(\Gamma)$. Then, by virtue of Theorem 1.4, there exist functions

$$\varphi_{n-t-s}^0 \in W_q^{z+d(t+s+1)}(G), \qquad \varphi_{n-t-s}^{00} \in W_q^{z+d(t+s)}(G)$$

such that

$$L_{pn_p}\varphi^0_{n-t-s} = -\sum_{k=1}^{n_p} L_{p,n_p-k}\varphi_{k+n-t-s} \tag{1.23}$$

for $m_p \leq z + d(t+s+1) - 1$, $n_p \leq t+s$, and

$$L_{pn_p}\varphi^{00}_{n-t-s} = -\sum_{k=1}^{n_p} L_{p,n_p-k}v_{k+n-t-s} \tag{1.24}$$

for $m_p \leq z + d(t+s) - 1$, $n_p \leq t+s-1$. By virtue of (1.17) and (1.22) we have

$$\|\sum_{k=1}^{n_p} L_{p,n_p-k}(\varphi_{k+n-t-s} - v_{k+n-t-s})\|_{z+d(t+s)-m_p-1/q,q,\Gamma}.$$

$$\leq C \sum_{k=1}^{n_p} \|\varphi_{k+n-t-s} - v_{k+n-t-s}\|_{z+d(t+s-k),q,G} < C\varepsilon,$$

under $m_p \leq z + d(t+s) - 1$, $n_p \leq t+s-1$. Then, by virtue of the same Theorem 1.4

$$\|\varphi^0_{n-t-s} - \varphi^{00}_{n-t-s}\|_{z+d(t+s),q,G} < C\varepsilon. \tag{1.25}$$

Since $v \in \mathcal{H}$, then by virtue of (1.24), $m_p \leq z + d(t+s) - 1$, and $n_p \leq t+s-1$ we have

$$L_{pn_p}(v_{n-t-s} - \varphi^{00}_{n-t-s}) = 0. \tag{1.26}$$

Since $z \leq m_p - dn_p$, then for $n_p = t+s$ we have $m_p \geq z + d(t+s)$. Then, by virtue of (1.26) and Theorem 1.5 there exist functions $g_{n-t-s} \in W_q^{z+d(t+s+1)}(G)$, such that

$$L_{pn_p}g_{n-t-s} = 0, \quad m_p \leq z + d(t+s+1) - 1, \quad n_p = t+s, \tag{1.27}$$

$$\|g_{n-t-s} - (v_{n-t-s} - \varphi^{00}_{n-t-s})\|_{z+d(t+s),q,G} < \varepsilon. \tag{1.28}$$

Now it is enough to note that the functions

$$\varphi_{n-t-s} = g_{n-t-s} + \varphi^0_{n-t-s} \in W_q^{z+d(t+s+1)}(G)$$

satisfy both (1.21) and (1.22) for $i = s$. The relation (1.21) follows from (1.27) and (1.23) for $i = s$. The relation (1.22) follows from (1.28) and (1.25) for $i = s$. ∎

4.1.3. Fold completeness of root functions of principally regular elliptic boundary value problems. Let us formulate and prove the main theorem of completeness of root functions for regular elliptic boundary value problems.

Theorem 1.8. Let the following conditions be satisfied:

(1) $n \geq 1$, $m \geq 1$, $n_p \leq n - 1$, $m_p \geq dn_p$, $\max\{m_p\} - \min\{m_p - dn_p\} \leq m - 1$;

(2) the weight $d = 2m/n$ of the problem (1.1)-(1.2) is an integer;

(3) $a_{k\alpha} \in C^{q-2m}(\overline{G})$, operators B_k from $W_2^{dk}(G)$ into $L_2(G)$ and from $W_2^{q-2m+dk}(G)$ into $W_2^{q-2m}(G)$ are compact, where $q \geq \max\{2m, m_p + 1\}$; $\Gamma \in C^q$;

(4) $b_{pk\alpha} \in C^{q-m_p}(\overline{G})$, operators T_{pk} from $W_2^{m_p - dk + 1/2}(G)$ into $L_2(\Gamma)$ and from $W_2^{q-dk}(G)$ into $W_2^{q-m_p-1/2}(\Gamma)$ are compact; $T_{pn_p} = 0$, if $m_p - dn_p = 0$;

(5) the system

$$L_{pn_p} = \sum_{|\alpha|=m_p} b_{p0\alpha}(x')D^\alpha + T_{p0}, \quad p = 1,\ldots,m,$$

is normal and has property III;

(6) there exist rays $\ell_k(\varphi_k)$ with the angles between the neighboring rays less than $2m\pi/r$, such that conditions I and II are fulfilled on the rays ℓ_k.

Then the spectrum of the problem (1.1)–(1.2) is discrete and the system of root functions of the problem (1.1)–(1.2) is n-fold complete in the space

$$\mathcal{H} = \{u \mid u = (u_1,\ldots,u_n) \in \bigoplus_{k=0}^{n-1} W_2^{\ell+d(n-k-1)}(G), \sum_{k=0}^{n_p} L_{p,n_p-k}u_{k+s}|_{x'} = 0,$$
$$x' \in \Gamma, \ m_p \leq \ell + d(n-s) - 1, \ s = 1,\ldots,n-n_p-1, \ p = 1,\ldots,m\}$$

for an integer $\ell \in [max\{0, m_p - (2m-1)\}, \min\{m_p - dn_p\}]$.

Proof. Consider operators A_k and A_{pk} which are defined by the equalities

$$D(A_k) = W_2^{\ell+dk}(G), \quad k = 1,\ldots,n,$$
$$A_k u = \sum_{|\alpha|=dk} a_{k\alpha}(x)D^\alpha u(x) + B_k u|_x, \quad k = 1,\ldots,n,$$

$$A_{p,n_p-k}u = \sum_{|\alpha|=m_p-dk} b_{pk\alpha}(x')D^\alpha u(x') + T_{pk}u|_{x'}, k = 0,\ldots,n_p, \ p = 1,\ldots,m.$$
$$\tag{1.29}$$

Let us denote

$$H_k = W_2^{\ell+dk}(G), \quad k = 0,\ldots,n,$$
$$H^p = W_2^{\ell+2m-m_p-1/2}(\Gamma), \qquad H_0^p = W_2^{1/2}(\Gamma), \quad p = 1,\ldots,m.$$

Then the problem (1.1)–(1.2) can be rewritten in the form of a system of operator pencil equations

$$L(\lambda)u = \lambda^n u + \lambda^{n-1} A_1 u + \cdots + A_n u = 0,$$
$$L_p(\lambda)u = \lambda^{n_p} A_{p0} u + \lambda^{n_p - 1} A_{p1} u + \cdots + A_{pn_p} u = 0, \quad p = 1, \ldots, m,$$

(1.30)

where $u \in D(A_n) = W_2^{\ell+2m}(G)$.

Let us apply Theorem 2.3.4 to (1.30). By virtue of [72, p.258] the compact embeddings

$$W_2^{\ell+2m}(G) \subset W_2^{\ell+d(n-1)}(G) \subset \cdots \subset W_2^\ell(G)$$

hold, and by virtue of [72, p.350]

$$s_j(J, W_2^{\ell+dk}(G), W_2^{\ell+d(k-1)}(G)) \sim j^{-d/r}, \quad k = 1, \ldots, n.$$

Hence, for $p > d^{-1}r$

$$J \in \sigma_p(W_2^{\ell+dk}(G), W_2^{\ell+d(k-1)}(G)),$$

i.e.,

$$J \in \sigma_p(H_k, H_{k-1}), \quad p > d^{-1}r, \ k = 1, \ldots, n.$$

So, conditions 1 and 2 of Theorem 2.3.4 have been checked. Condition 3 of Theorem 2.3.4 is obvious. By virtue of [72, p.330], A_{pk}, defined by the equalities (1.29), act boundedly from $H_{n-n_p+k} = W_2^{\ell+d(n-n_p+k)}(G)$ into $H^p = W_2^{\ell+2m-m_p-1/2}(\Gamma)$, for $\ell + 2m \geq m + 1$, i.e., condition 4 of Theorem 2.3.4 holds. By virtue of Theorem 1.7, condition 5 of Theorem 2.3.4 also holds. Condition 6 of Theorem 2.3.4 follows from Theorem 1.1. Indeed, from Theorem 1.1 it follows that all complex numbers $\lambda \in \ell_k$ and with large enough moduli are regular points of the pencil

$$\mathbb{L}(\lambda) = (L(\lambda), L_1(\lambda), \ldots, L_m(\lambda))$$

and for a solution to the problem

$$L(x, D, \lambda)u = f(x), \quad x \in G,$$
$$L_p(x', D, \lambda)u = f_p(x'), \quad x' \in \Gamma, \ p = 1, \ldots, m.$$

the following estimate (for the above indicated λ)

$$\sum_{k=0}^{\ell+2m} |\lambda|^{d^{-1}(\ell+2m-k)} \|u\|_{k,2,G} \leq C\{\|f\|_{\ell,2,G} + |\lambda|^{d^{-1}\ell} \|f\|_{0,2,G}$$

$$+ \sum_{s=1}^{m} |\lambda|^{d^{-1}(\ell+2m-m_s-1/2)} \|f_s\|_{0,2,\Gamma}$$

$$+ \|f_s\|_{\ell+2m-m_s-1/2,2,\Gamma}\}$$

holds.

From here, in particular, it follows that for the pencil $\mathbb{L}(\lambda)$ the estimate

$$\|\mathbb{L}^{-1}(\lambda)\|_{B(W_2^\ell(G) \oplus W_2^{\ell+2m-m_p-1/2}(\Gamma), W_2^{\ell+2m}(G))} \leq C|\lambda|^h, \quad \lambda \in \ell_k, \ |\lambda| \to \infty$$

is fulfilled for some $h \in \mathbb{R}$.

So, under $nr/2m < p < \pi/\varphi$, where $\varphi = \max(\varphi_k - \varphi_{k-1}) < 2m\pi/nr$, Theorem 2.3.4 is applicable to the system of pencils (1.30). ∎

Corollary 1.9. *Let the conditions of Theorem 1.8 be satisfied. Then root functions of the problem (1.1)–(1.2) are n-fold complete in the space $\mathcal{H}_d$ (see Theorem 1.7).*

In the case where $n_p = 0$, the proof is greatly simplified if Theorem 2.3.4 is applied instead of Theorem 2.3.5. Consider the problem

$$L(x, D, \lambda)u = \lambda^n u(x) + \sum_{k=1}^{n} \lambda^{n-k} \left(\sum_{|\alpha|=dk} a_{k\alpha}(x) D^\alpha u(x) + B_k u|_x \right) = 0,$$

$$x \in G, \tag{1.31}$$

$$L_p(x', D)u = \sum_{|\alpha|=m_p} b_{p\alpha}(x') D^\alpha u(x') + T_p u|_{x'} = 0,$$

$$x' \in \Gamma, \ p = 1, \ldots, m. \tag{1.32}$$

Theorem 1.10. *Let the following conditions be satisfied:*

(1) $n \geq 1$, $m \geq 1$, $n \leq 2m$, $m_p \geq 0$, $\max\{m_p\} - \min\{m_p\} \leq 2m - 1$;

(2) *the weight $d = 2m/n$ of the problem (1.31)–(1.32) is an integer;*

(3) $a_{k\alpha} \in C^{q-2m}(\overline{G})$, *operators B_k from $W_2^{dk}(G)$ into $L_2(G)$ and from $W_2^{q-2m+dk}(G)$ into $W_2^{q-2m}(G)$ are compact, where $q \geq \max\{2m, m_p + 1\}$; $\Gamma \in C^q$;*

(4) $b_{p\alpha} \in C^{q-m_p}(\overline{G})$, *operators T_p from $W_2^{m_p+1/2}(G)$ into $L_2(\Gamma)$ and from $W_2^q(G)$ into $W_2^{q-m_p-1/2}(\Gamma)$ are compact; $T_p = 0$, if $m_p = 0$;*

(5) *the system*

$$L_p = \sum_{|\alpha|=m_p} b_{p\alpha}(x') D^\alpha + T_p, \quad p = 1, \ldots, m,$$

is normal and has property III;

(6) *there exist rays ℓ_k with the angles between the neighboring rays less than $2m\pi/r$, such that conditions I and II are fulfilled on the rays ℓ_k when $n_p = 0$.*

Then the spectrum of the problem (1.31)–(1.32) is discrete and a system of root functions of the problem (1.31)–(1.32) is n-fold complete in the space

$$\mathcal{H} = \bigoplus_{k=0}^{n-1} \{u \mid u \in W_2^{\ell+d(n-k-1)}(G), \sum_{|\alpha|=m_p} b_{p\alpha}(x')D^\alpha u(x') + T_p u|_{x'} = 0,$$

$$x' \in \Gamma, \ m_p \leq \ell + d(n - k - 1) - 1, \ p = 1, \ldots, m\}$$

for an integer $\ell \in [max\{0, m_p - (2m-1)\}, \min\{m_p\}]$.

Proof. As in Theorem 1.8, the problem (1.31)–(1.32) can be rewritten in the form of a system of operator pencil equations (1.30). Let us apply Theorem 2.3.5 to (1.30). All its conditions except 5 have been checked. The density of the set

$$\{u \mid u \in W_2^{\ell+dk}(G), \sum_{|\alpha|=m_p} b_{p\alpha}(x')D^\alpha u(x') + T_p u|_{x'} = 0,$$

$$x' \in \Gamma, \ m_p \leq \ell + dk - 1, \ p = 1, \ldots, m\}$$

in the space

$$\{u \mid u \in W_2^{\ell+d(k-1)}(G), \sum_{|\alpha|=m_p} b_{p\alpha}(x')D^\alpha u(x') + T_p u|_{x'} = 0,$$

$$x' \in \Gamma, \ m_p \leq \ell + d(k-1) - 1, \ p = 1, \ldots, m\}$$

follows from Remark 1.6.

Density of the set

$$\{u \mid u \in W_2^{\ell+d}(G), \sum_{|\alpha|=m_p} b_{p\alpha}(x')D^\alpha u(x') + T_p u|_{x'} = 0,$$

$$x' \in \Gamma, \ m_p \leq \ell + d - 1, \ p = 1, \ldots, m\}$$

in the space $W_2^\ell(G)$, for $\ell \leq \min\{m_p\}$ also follows from Remark 1.6. ∎

Consider the spectral problem

$$A(x, D, \lambda)u = (\lambda^2 - \frac{\partial^2}{\partial x_1^2} - \cdots - \frac{\partial^2}{\partial x_r^2})^m u(x)$$

$$+ \sum_{k+|\alpha|\leq 2m-1} \lambda^k a_{k\alpha}(x)D^\alpha u(x) = 0, \quad x \in G,$$

$$B_p(x', D, \lambda)u = \frac{\partial^{p-1}u(x')}{\partial \mu(x')^{p-1}} + \sum_{k+|\alpha|\leq p-2} \lambda^k b_{pk\alpha}(x')D^\alpha u(x') = 0,$$

$$x' \in \Gamma, \quad p = 1, \ldots, m,$$

(1.33)

where $\mu(x') \in C^\infty(\Gamma, \mathbb{R}^r)$, and is a non-tangent unit vector to Γ at the point $x' \in \Gamma$.

For $\lambda \in \ell(0, \varphi)$, $\varphi \neq \pi/2$, $\varphi \neq -\pi/2$, $\sigma \in \mathbb{R}^r$, $|\sigma| + |\lambda| \neq 0$, we have

$$A_0(x, \sigma, \lambda) = (\lambda^2 + \sigma_1^2 + \cdots + \sigma_r^2)^m \neq 0,$$

i.e., condition I is fulfilled on any ray, which is inconsistent with the imaginary semi-axes.

Let the vector σ' be tangent, and σ be normal, to Γ at the point $x' \in \Gamma$ and $\lambda \in \ell(0, \varphi)$, $\varphi \neq \pm\pi/2$. From

$$A_0(x', \sigma' + \tau\sigma, \lambda) = (\lambda^2 + \|\sigma'\|^2 + \|\sigma\|^2 \tau)^m = 0,$$

for $0 \leq \arg \lambda < \pi/2$ and $-\pi < \arg \lambda < -\pi/2$ we have

$$\tau_k^+ = i\|\sigma\|^{-1}(\lambda^2 + \|\sigma'\|^2)^{1/2}. \quad k = 1, \ldots, m,$$

where $z^{1/2} = |z|^{1/2} e^{i \arg z/2}$, $-\pi < \arg z \leq \pi$. Hence,

$$M^+(x', \sigma', \sigma, \lambda, \tau) = [\tau - i\|\sigma\|^{-1}(\lambda^2 + \|\sigma'\|^2)^{1/2}]^m.$$

Let σ'^s, $s = 1, \ldots, r-1$ be the orthonormal basis consisting of tangent vectors. Since

$$\mu = \sum_{s=1}^{r-1}(\mu, \sigma'^s)\sigma'^s + (\mu, \sigma)\sigma,$$

then

$$\cos(\mu, x_j) = (\mu, x_j) = \sum_{s=1}^{r-1}(\mu, \sigma'^s)\sigma_j'^s + (\mu, \sigma)\sigma_j.$$

Hence

$$B_{p0}(x', \sigma' + \tau\sigma) = \left[\sum_{j=1}^{r}(\sigma_j' + \tau\sigma_j)\cos(\mu, x_j)\right]^{p-1}$$

$$= \left[\sum_{j=1}^{r}(\sigma_j' + \tau\sigma_j)\left(\sum_{s=1}^{r-1}(\mu, \sigma'^s)\sigma_j'^s + (\mu, \sigma)\sigma_j\right)\right]^{p-1}$$

$$= ((\mu, \sigma)\|\sigma\|^2)^{p-1}\tau^{p-1} + \sum_{k=0}^{p-2}b_{pk}(x', \sigma', \sigma)\tau^k,$$

$$p = 1, \ldots, m.$$

From here, in turn, follows

$$\det |b_{pk}(x', \sigma', \sigma)|_{\substack{p=1,\ldots,m \\ k=0,\ldots,m-1}} = ((\mu, \sigma)\|\sigma\|^2)^{\sum_{p=1}^{m}(p-1)} \neq 0,$$

i.e., condition II$'$ is fulfilled.

So, for problem (1.33) conditions I and II are fulfilled. Hence, from Theorem 1.8 follows:

Theorem 1.11. *Let $a_{k\alpha} \in C^{2m}(\overline{G})$, $b_{pk\alpha} \in C^{4m-m_p}(\overline{G})$, $\Gamma \in C^{4m}$. Then the spectrum of problem (1.33) is discrete. There exists a finite number of eigenvalues outside the angles $|\arg \lambda \pm \pi/2| < \varepsilon$ and a system of root functions of problem (1.33) that is 2m-fold complete in the space*

$$\mathcal{H} = \{u \mid u = (u_1, \ldots, u_{2m}) \in \overset{2m-1}{\underset{k=0}{\oplus}} W_2^{2m-k-1}(G),$$

$$\frac{\partial^{p-1} u_s(x')}{\partial \mu(x')^{p-1}} + \sum_{k+|\alpha| \le p-2} b_{pk\alpha}(x') D^\alpha u_{k+s}(x') = 0,$$

$$x' \in \Gamma, \quad s = 1, \ldots, 2m - p, \quad p = 1, \ldots, m\}.$$

Corollary 1.12. *The root functions of problem (1.33) are 2m-fold complete in the space $L_2(G) \oplus \cdots \oplus L_2(G)$.*

Proof. It follows from facts that $\overline{C_0^\infty(G) \oplus \cdots \oplus C_0^\infty(G)} = L_2(G) \oplus \cdots \oplus L_2(G)$ and $C_0^\infty(G) \oplus \cdots \oplus C_0^\infty(G) \subset \mathcal{H}$, and from Theorem 1.11. ∎

Consider the problem (1.31)-(1.32) in the case $n = 1$

$$L(x, D, \lambda)u = \lambda u(x) + \sum_{|\alpha|=2m} a_\alpha(x) D^\alpha u(x) + Bu|_x = 0, \quad x \in G, \tag{1.34}$$

$$L_p(x', D)u = \sum_{|\alpha|=m_p} b_{p\alpha}(x') D^\alpha u(x') + T_p u|_{x'} = 0, \quad x' \in \Gamma, \ p = 1, \ldots, m. \tag{1.35}$$

Theorem 1.13. *Let the following conditions be satisfied:*

(1) $m_p \ge 0$, $\max\{m_p\} - \min\{n_p\} \le 2m - 1$;

(2) $a_\alpha \in C^{q-2m}(\overline{G})$, *operator B from $W_2^{2m}(G)$ into $L_2(G)$ and from $W_2^q(G)$ into $W_2^{q-2m}(G)$ is compact, where $q \ge \max\{2m, m_p + 1\}$; $\Gamma \in C^q$;*

(3) $b_{p\alpha} \in C^{q-m_p}(\overline{G})$, *operators T_p from $W_2^{m_p+1/2}(G)$ into $L_2(\Gamma)$ and from $W_2^q(G)$ into $W_2^{q-m_p-1/2}(\Gamma)$ are compact; $T_p = 0$, if $m_p = 0$;*

(4) *system (1.35) is normal;*

(5) *there exist rays ℓ_k with the angles between the neighboring rays less than $2m\pi/r$, such that for $x \in \overline{G}$, $\sigma \in \mathbb{R}^r$, $\lambda \in \ell_k$, $|\sigma| + |\lambda| \ne 0$,*

$$\lambda + \sum_{|\alpha|=2m} a_\alpha(x) \sigma^\alpha \ne 0;$$

(6) x' *is any point on Γ, the vector σ' is tangent and σ is a normal vector to Γ at the point $x' \in \Gamma$. Consider the following ordinary differential problem*

$$[\lambda + \sum_{|\alpha|=2m} a_\alpha(x')(\sigma' - i\sigma\frac{d}{dy})^\alpha]u(y) = 0, \quad y \ge 0, \ \lambda \in \ell_k, \tag{1.36}$$

$$\sum_{|\alpha|=m_k} b_{k\alpha}(x')(\sigma' - i\sigma\frac{d}{dy})^\alpha u(y)|_{y=0} = h_k, \quad k = 1, \ldots, m; \tag{1.37}$$

it is required that the problem (1.36)–(1.37) has one and only one solution, including all its derivatives, tending to zero as $y \to \infty$ for any numbers $h_k \in \mathbb{C}$.

Then the spectrum of the problem (1.34)–(1.35) is discrete, and a system of root functions of the problem (1.34)–(1.35) is complete in the spaces $W_2^\ell(G)$ and $W_2^{\ell+2m}(G, L_p u = 0, p = 1, \ldots, m)$, where the integer $\ell \in [\max\{0, m_p - (2m - 1)\}, \min\{m_p\}]$.[13]

Proof. As in Theorem 1.8 the problem (1.34)–(1.35) can be rewritten in the form

$$L(\lambda)u = \lambda u + Au = 0, \tag{1.38}$$

$$L_p u = A_{p0} u = 0, \quad p = 1, \ldots, m. \tag{1.39}$$

Let us apply Theorem 2.3.5 to (1.38)–(1.39). All conditions except condition 5 of Theorem 2.3.5 have been checked in Theorem 1.8. Then density of the set

$$\{u \mid u \in W_2^{\ell+2m}(G), \ \sum_{|\alpha|=m_p} b_{p\alpha}(x')D^\alpha u(x') + T_p u|_{x'} = 0,$$

$$x' \in \Gamma, \ p = 1, \ldots, m\}$$

in the space $W_2^\ell(G)$, for $\ell \leq \min\{m_p\}$, follows from Theorem 1.5, i.e., condition 5 of Theorem 2.3.5 is satisfied too. ∎

4.2. Irregular elliptic boundary value problems for the Laplace equation

Irregular elliptic problems are such that they do not lend themselves to description in the framework of one book. We shall try to show here that, even for the Laplace equation in a rectangle, there exist non-coercive and non-classical differential boundary value problems.

4.2.1. A boundary value problem for the Laplace equation with non-classical spectral asymptotics. Let $\{\varphi_n\}_1^\infty$ be a complete orthonormal system in H and let $f \in L_2((0,1), H)$. Then almost everywhere on $(0,1)$ the equality

$$f(x) = \sum_{k=1}^\infty (f(x), \varphi_k)\varphi_k$$

holds, where the series converges in H.

---

[13]This theorem strengthens the classical result of S. Agmon [1], since in [1] it is supposed that $m_p \leq 2m - 1$ and operators B and T_p are differential.

Lemma 2.1. *Let* $f \in L_2((0,1), H)$. *Then*

$$\int_0^1 \|f(x)\|^2 dx = \sum_{k=1}^{\infty} \int_0^1 |(f(x), \varphi_k)|^2 dx.$$

Proof. Denote $f_k(x) = (f(x), \varphi_k)$. Then by virtue of the Parceval equality we have

$$\|f(x)\|^2 = \sum_{k=1}^{\infty} |f_k(x)|^2$$

almost everywhere on $(0,1)$. Applying the dominated convergence theorem one obtains

$$\int_0^1 \|f(x)\|^2 dx = \int_0^1 \sum_{k=1}^{\infty} |f_k(x)|^2 dx = \lim_{n \to \infty} \sum_{k=1}^{n} \int_0^1 |f_k(x)|^2 dx$$

$$= \sum_{k=1}^{\infty} \int_0^1 |f_k(x)|^2 dx. \blacksquare$$

Consider the spectral boundary value problem in the square $\Omega = [0, 2\pi] \times [0, 2\pi]$

$$-\frac{\partial^2 u(x,y)}{\partial x^2} - \frac{\partial^2 u(x,y)}{\partial y^2} = \lambda u(x,y), \tag{2.1}$$

$$u(0,y) = 0, \quad \frac{\partial u(2\pi, y)}{\partial x} + i \frac{\partial u(2\pi, y)}{\partial y} = 0,$$

$$u(x,0) = u(x, 2\pi), \quad \frac{\partial u(x,0)}{\partial y} = \frac{\partial u(x, 2\pi)}{\partial y}. \tag{2.2}$$

Theorem 2.2. *For the problem* (2.1)–(2.2) *there exists a sequence of eigenvalues* λ_k *tending to zero.*

Proof. Consider in the space $L_2(0, 2\pi)$ an operator A defined by the equalities

$$D(A) = W_2^2((0, 2\pi), \quad u(0) = u(2\pi), \quad u'(0) = u'(2\pi)),$$
$$Au = -u''(y).$$

Eigenvalues of the operator A are numbers $\mu_k(A) = k^2$, $k = 0, \ldots, \infty$, to which, for $k > 0$, a pair of eigenfunctions corresponds

$$u_{k1}(y) = 1/\sqrt{2\pi} e^{iky}, \quad u_{k2}(y) = 1/\sqrt{2\pi} e^{-iky}, \quad k = 1, \ldots, \infty.$$

In the space $L_2(0, 2\pi)$ consider also an operator B defined by the equalities

$$D(B) = W_2^1((0, 2\pi), u(0) = u(2\pi)),$$
$$Bu = iu'(y).$$

To eigenvalues $\mu_k(B) = -k$, $k = 0,\ldots,\infty$, eigenfunctions $u_{k1}(y)$ correspond and to eigenvalues $\mu_{-k}(B) = k$, $k = 1,\ldots,\infty$, eigenfunctions $u_{k2}(y)$ correspond.

The problem (2.1)–(2.2) is equivalent to the problem

$$- u''(x) + Au(x) = \lambda u(x),$$
$$u(0) = 0, \quad u'(2\pi) + Bu(2\pi) = 0,$$

where $u(x)$ is a vector valued function with values from $L_2(0, 2\pi)$. Taking into account the spectral expansion of the operators A and B we obtain

$$-u''(x) + \sum_{k=1}^{\infty} k^2[(u(x), u_{k1})u_{k1} + (u(x), u_{k2})u_{k2}] = \lambda u(x),$$

$$u(0) = 0, \quad u'(2\pi) + \sum_{k=1}^{\infty}[-k(u(2\pi), u_{k1})u_{k1} + k(u(2\pi), u_{k2})u_{k2}] = 0.$$

For the Fourier coefficients $\widetilde{u}_{k1}(x) = (u(x), u_{k1})$ we have the problem

$$- \widetilde{u}_{k1}''(x) + k^2\widetilde{u}_{k1}(x) = \lambda\widetilde{u}_{k1}(x), \tag{2.3}$$

$$\widetilde{u}_{k1}(0) = 0, \qquad \widetilde{u}_{k1}'(2\pi) - k\widetilde{u}_{k1}(2\pi) = 0. \tag{2.4}$$

Eigenvalues of the problem (2.3)–(2.4) are roots of the equation

$$e^{4\pi\sqrt{k^2-\lambda}}(\sqrt{k^2 - \lambda} - k) + \sqrt{k^2 - \lambda} + k = 0. \tag{2.5}$$

Let $\varepsilon > 0$. Show that there exists k for which the equation (2.5) has a solution $\lambda_k \in (0, \varepsilon]$. Consider the function

$$f_k(\lambda) = e^{4\pi\sqrt{k^2-\lambda}}(\sqrt{k^2 - \lambda} - k) + \sqrt{k^2 - \lambda} + k$$

on the segment $[0, \varepsilon]$. We have $f_k(0) = 2k > 0$, but

$$\lim_{k\to\infty} f_k(\varepsilon) = -\infty.$$

Then we apply the Cauchy theorem to the function $f_k(\lambda)$. ∎

4.2.2. Non-coerciveness of the boundary value problem. The boundary value conditions (2.2) are non-classical in many senses. For example, let us show that the following boundary value problem in $\Omega = [0, 2\pi] \times [0, 2\pi]$

$$-\frac{\partial^2 u(x, y)}{\partial x^2} - \frac{\partial^2 u(x, y)}{\partial y^2} + \lambda u(x, y) = f(x, y),$$

$$u(0, y) = 0, \quad \frac{\partial u(2\pi, y)}{\partial x} + i\frac{\partial u(2\pi, y)}{\partial y} = 0, \tag{2.6}$$

$$u(x, 0) = u(x, 2\pi), \quad \frac{\partial u(x, 0)}{\partial y} = \frac{\partial u(x, 2\pi)}{\partial y}.$$

is non-coercive for $\lambda > 0$.

Theorem 2.3. *For $\lambda > 0$ and $f \in L_2(\Omega)$ problem (2.6) has a unique solution and for the solution the following estimate*

$$\|u\|_{L_2(\Omega)} \le C(\lambda)\|f\|_{L_2(\Omega)}$$

holds, and cannot be improved.

Proof. Problem (2.6) is equivalent to the problem

$$
\begin{aligned}
& -u''(x) + Au(x) + \lambda u(x) = f(x), \\
& u(0) = 0, \quad u'(2\pi) + Bu(2\pi) = 0,
\end{aligned}
\tag{2.7}
$$

in the space $L_2(0, 2\pi)$. For the Fourier coefficients $\tilde{u}_{k1}(x) = (u(x), u_{k1})$ we obtain the following problem

$$
\begin{aligned}
& -\tilde{u}''_{k1}(x) + (k^2 + \lambda)\tilde{u}_{k1}(x) = \tilde{f}_{k1}(x), \\
& \tilde{u}_{k1}(0) = 0, \quad \tilde{u}'_{k1}(2\pi) - k\tilde{u}_{k1}(2\pi) = 0,
\end{aligned}
$$

where $\tilde{f}_{k1}(x) = (f(x), u_{k1})$. It is easy to see, that

$$
\begin{aligned}
\tilde{u}_{k1}(x) &= 2^{-1}(k^2 + \lambda)^{-1/2}[C_1 e^{-\sqrt{k^2+\lambda}x} + C_2 e^{-\sqrt{k^2+\lambda}(2\pi-x)} \\
&+ \int_0^x e^{-\sqrt{k^2+\lambda}(x-y)}\tilde{f}_{k1}(y)\,dy + \int_x^{2\pi} e^{-\sqrt{k^2+\lambda}(y-x)}\tilde{f}_{k1}(y)\,dy],
\end{aligned}
\tag{2.8}
$$

where C_1, C_2 are solutions of the system

$$
\begin{aligned}
& C_1 + e^{-2\pi\sqrt{k^2+\lambda}}C_2 + \int_0^{2\pi} e^{-\sqrt{k^2+\lambda}y}\tilde{f}_{k1}(y)\,dy = 0, \\
& -(\sqrt{k^2 + \lambda} + k)e^{-2\pi\sqrt{k^2+\lambda}}C_1 + (\sqrt{k^2 + \lambda} - k)C_2 \\
& -(\sqrt{k^2 + \lambda} + k)\int_0^{2\pi} e^{-\sqrt{k^2+\lambda}(2\pi-y)}\tilde{f}_{k1}(y)\,dy = 0.
\end{aligned}
$$

Evidently

$$
\begin{aligned}
C_1 &= D^{-1}(k)\{-(\sqrt{k^2 + \lambda} - k)\int_0^{2\pi} e^{-\sqrt{k^2+\lambda}y}\tilde{f}_{k1}(y)\,dy \\
&- (\sqrt{k^2 + \lambda} + k)e^{-2\pi\sqrt{k^2+\lambda}}\int_0^{2\pi} e^{-\sqrt{k^2+\lambda}(2\pi-y)}\tilde{f}_{k1}(y)\,dy\}, \tag{2.9} \\
C_2 &= D^{-1}(k)\{(\sqrt{k^2 + \lambda} + k)\int_0^{2\pi} e^{-\sqrt{k^2+\lambda}(2\pi-y)}\tilde{f}_{k1}(y)\,dy \\
&- (\sqrt{k^2 + \lambda} + k)e^{-2\pi\sqrt{k^2+\lambda}}\int_0^{2\pi} e^{-\sqrt{k^2+\lambda}y}\tilde{f}_{k1}(y)\,dy\}. \tag{2.10}
\end{aligned}
$$

On the other hand, for $\lambda > 0$ and $k \to \infty$ we have

$$D(k) = \begin{vmatrix} 1 & e^{-2\pi\sqrt{k^2+\lambda}} \\ -(\sqrt{k^2+\lambda}+k)e^{-2\pi\sqrt{k^2+\lambda}} & \sqrt{k^2+\lambda}-k \end{vmatrix}$$

$$= \sqrt{k^2+\lambda} - k + (\sqrt{k^2+\lambda}+k)e^{-4\pi\sqrt{k^2+\lambda}} \sim \lambda/2k. \tag{2.11}$$

Taking into account (2.9)–(2.11) in (2.8) we obtain

$$\|\widetilde{u}_{k1}\|_{L_2(0,2\pi)} \leq C\|\widetilde{f}_{k1}\|_{L_2(0,2\pi)}. \tag{2.12}$$

For the Fourier coefficients $\widetilde{u}_{k2}(x) = (u(x), u_{k2})$ of a solution $u(x)$ to the problem (2.7) we have the problem

$$- \widetilde{u}''_{k2}(x) + (k^2 + \lambda)\widetilde{u}_{k2}(x) = \widetilde{f}_{k2}(x),$$
$$\widetilde{u}_{k2}(0) = 0, \qquad \widetilde{u}'_{k2}(2\pi) + k\widetilde{u}_{k2}(2\pi) = 0,$$

where $\widetilde{f}_{k2}(x) = (f(x), u_{k2})$. The same arguments bring us to the estimate

$$\|\widetilde{u}_{k2}\|_{L_2(0,2\pi)} \leq Ck^{-2}\|\widetilde{f}_{k2}\|_{L_2(0,2\pi)}. \tag{2.13}$$

From (2.12) and (2.13), by virtue of Lemma 2.1, follows the estimate

$$\|u\|^2_{L_2(\Omega)} = \int_0^{2\pi} \|u(x)\|^2_{L_2(0,2\pi)} \, dx = \sum_{k=0}^{\infty} \|\widetilde{u}_{k1}\|^2_{L_2(0,2\pi)}$$

$$+ \sum_{k=1}^{\infty} \|\widetilde{u}_{k2}\|^2_{L_2(0,2\pi)} \leq C\|f\|^2_{L_2(\Omega)}.$$

There is no stronger estimate than that in Theorem 2.3. More precisely, the estimate

$$\|u\|_{W_2^s(\Omega)} \leq C\|f\|_{L_2(\Omega)}$$

with some $s > 0$ is not true since in this case problem (2.6) would have a discrete spectrum. ∎

Remark 2.4. *From the paper by S. Ya. Yakubov and B. A. Aliev [83] it follows that if in (2.6) instead of the condition*

$$\frac{\partial u(2\pi, y)}{\partial x} + i\frac{\partial u(2\pi, y)}{\partial y} = 0$$

we consider the condition

$$\frac{\partial u(2\pi, y)}{\partial x} + b\frac{\partial u(2\pi, y)}{\partial y} = 0,$$

where b is a real number, then the problem becomes coercive.

4.2.3. A boundary value problem for the Laplace equation with an infinite dimensional kernel. An elliptic system, for which the Dirichlet problem in a circle has an infinite dimensional kernel, was constructed by A. V. Bitsadze [8]. It is shown here that, for the Laplace equation, there exist boundary value problems with an infinite dimensional kernel for which boundary conditions contain only differential operators.

Consider the following boundary value problem in the square $\Omega = [0, 2\pi] \times [0, 2\pi]$

$$-\frac{\partial^2 u(x,y)}{\partial x^2} - \frac{\partial^2 u(x,y)}{\partial y^2} = 0, \tag{2.14}$$

$$\frac{\partial u(0,y)}{\partial x} + i\frac{\partial u(0,y)}{\partial y} = 0, \qquad \frac{\partial u(2\pi,y)}{\partial x} + i\frac{\partial u(2\pi,y)}{\partial y} = 0,$$

$$u(x,0) = u(x,2\pi), \qquad \frac{\partial u(x,0)}{\partial y} = \frac{\partial u(x,2\pi)}{\partial y}. \tag{2.15}$$

Theorem 2.5. *The problem* (2.14)–(2.15) *has an infinite dimensional kernel.*

Proof. Consider in the space $L_2(0, 2\pi)$ an operator A defined by the equalities

$$D(A) = W_2^2((0, 2\pi), u(0) = u(2\pi), u'(0) = u'(2\pi)),$$
$$Au = -u''(y).$$

Eigenvalues of the operatror A are numbers $\mu_k(A) = k^2$, $k = 0, \ldots, \infty$, to which, for $k > 0$, a pair of eigenfunctions correspond

$$u_{k1}(y) = 1/\sqrt{2\pi}e^{iky}, \qquad u_{k2}(y) = 1/\sqrt{2\pi}e^{-iky}, \qquad k = 1, \ldots, \infty.$$

In the space $L_2(0, 2\pi)$ consider also an operator B defined by the equalities

$$D(B) = W_2^1((0, 2\pi), u(0) = u(2\pi)),$$
$$Bu = iu'(y).$$

To eigenvalues $\mu_k(B) = -k$, $k = 0, \ldots, \infty$, eigendunctions $u_{k1}(y)$ correspond and to eigenvalues $\mu_{-k}(B) = k$, $k = 1, \ldots, \infty$, eigenfunctions $u_{k2}(y)$ correspond.

The problem (2.14)–(2.15) is equivalent to the problem

$$-u''(x) + Au(x) = 0,$$
$$u'(0) + Bu(0) = 0, \qquad u'(2\pi) + Bu(2\pi) = 0,$$

where $u(x)$ is a vector-valued function with values from $L_2(0, 2\pi)$. Taking into account the spectral expansion of the operators A and B we obtain

$$-u''(x) + \sum_{k=1}^{\infty} k^2[(u(x), u_{k1})u_{k1} + (u(x), u_{k2})u_{k2}] = 0,$$

$$u'(0) + \sum_{k=1}^{\infty}[-k(u(0), u_{k1})u_{k1} + k(u(0), u_{k2})u_{k2}] = 0,$$

$$u'(2\pi) + \sum_{k=1}^{\infty}[-k(u(2\pi), u_{k1})u_{k1} + k(u(2\pi), u_{k2})u_{k2}] = 0.$$

For the Fourier coefficients $\tilde{u}_{k1}(x) = (u(x), u_{k1})$ we have the problem

$$-\tilde{u}_{k1}''(x) + k^2\tilde{u}_{k1}(x) = 0, \tag{2.16}$$

$$L_1\tilde{u}_{k1} = \tilde{u}_{k1}'(0) - k\tilde{u}_{k1}(0) = 0,$$
$$L_2\tilde{u}_{k1} = \tilde{u}_{k1}'(2\pi) - k\tilde{u}_{k1}(2\pi) = 0. \tag{2.17}$$

A general solution to equation (2.16) has the form

$$\tilde{u}_{k1}(x) = C_1 e^{kx} + C_2 e^{-kx}.$$

The characteristic determinant of the problem (2.16)–(2.17) has the form

$$D = \begin{vmatrix} L_1(e^{kx}) & L_1(e^{-kx}) \\ L_2(e^{kx}) & L_2(e^{-kx}) \end{vmatrix} = \begin{vmatrix} 0 & -2k \\ 0 & -2k \end{vmatrix} = 0.$$

So, functions of the form

$$u(x, y) = Ce^{kx}e^{iky}, \quad k = 1, \ldots, \infty$$

are solutions to the problem (2.14)–(2.15). ∎

And here the question arises: maybe the problem is pathological? Let us show that this is not the case.

Consider the boundary value problem

$$-\frac{\partial^2 u(x, y)}{\partial x^2} - \frac{\partial^2 u(x, y)}{\partial y^2} + \lambda u(x, y) = f(x, y),$$

$$\frac{\partial u(0, y)}{\partial x} + i\frac{\partial u(0, y)}{\partial y} = 0, \qquad \frac{\partial u(2\pi, y)}{\partial x} + i\frac{\partial u(2\pi, y)}{\partial y} = 0, \tag{2.18}$$

$$u(x, 0) = u(x, 2\pi), \qquad \frac{\partial u(x, 0)}{\partial y} = \frac{\partial u(x, 2\pi)}{\partial y}.$$

Theorem 2.6. For $\lambda > 0$ and $f \in L_2(\Omega)$ problem (2.18) has a unique solution and for the solution the estimate

$$\|u\|_{L_2(\Omega)} \leq C(\lambda)\|f\|_{L_2(\Omega)}$$

holds.

Proof. Problem (2.18) is equivalent to the problem

$$
\begin{aligned}
-u''(x) + Au(x) + \lambda u(x) &= f(x), \\
u'(0) + Bu(0) = 0, \qquad u'(2\pi) &+ Bu(2\pi) = 0,
\end{aligned}
\tag{2.19}
$$

in the space $L_2(0, 2\pi)$. For the Fourier coefficients $\tilde{u}_{k1}(x) = (u(x), u_{k1})$ we obtain the following problem

$$
\begin{aligned}
-\tilde{u}''_{k1}(x) + (k^2 + \lambda)\tilde{u}_{k1}(x) &= \tilde{f}_{k1}(x), \\
\tilde{u}'_{k1}(0) - k\tilde{u}_{k1}(0) = 0, \qquad \tilde{u}'_{k1}(2\pi) &- k\tilde{u}_{k1}(2\pi) = 0,
\end{aligned}
\tag{2.20}
$$

where $\tilde{f}_{k1}(x) = (f(x), u_{k1})$. It is easy to see, that

$$
\begin{aligned}
\tilde{u}_{k1}(x) = 2^{-1}(k^2 + \lambda)^{-1/2}[&C_1 e^{-\sqrt{k^2+\lambda}\,x} + C_2 e^{-\sqrt{k^2+\lambda}(2\pi-x)} \\
&+ \int_0^x e^{-\sqrt{k^2+\lambda}(x-y)}\tilde{f}_{k1}(y)\,dy + \int_x^{2\pi} e^{-\sqrt{k^2+\lambda}(y-x)}\tilde{f}_{k1}(y)\,dy],
\end{aligned}
\tag{2.21}
$$

where C_1, C_2 are solutions of the system

$$
\begin{aligned}
&-(\sqrt{k^2+\lambda} + k)C_1 + (\sqrt{k^2+\lambda} - k)e^{-2\pi\sqrt{k^2+\lambda}}C_2 \\
&+ (\sqrt{k^2+\lambda} - k)\int_0^{2\pi} e^{-\sqrt{k^2+\lambda}\,y}\tilde{f}_{k1}(y)\,dy = 0, \\
&-(\sqrt{k^2+\lambda} + k)e^{-2\pi\sqrt{k^2+\lambda}}C_1 + (\sqrt{k^2+\lambda} - k)C_2 \\
&- (\sqrt{k^2+\lambda} + k)\int_0^{2\pi} e^{-\sqrt{k^2+\lambda}(2\pi-y)}\tilde{f}_{k1}(y)\,dy = 0.
\end{aligned}
$$

Evidently

$$
\begin{aligned}
C_1 = D^{-1}(k)\{&-(\sqrt{k^2+\lambda} - k)^2 \int_0^{2\pi} e^{-\sqrt{k^2+\lambda}\,y}\tilde{f}_{k1}(y)\,dy \\
&- \lambda e^{-2\pi\sqrt{k^2+\lambda}} \int_0^{2\pi} e^{-\sqrt{k^2+\lambda}(2\pi-y)}\tilde{f}_{k1}(y)\,dy\}, \\
C_2 = D^{-1}(k)\{&-(\sqrt{k^2+\lambda} + k)^2 \int_0^{2\pi} e^{-\sqrt{k^2+\lambda}(2\pi-y)}\tilde{f}_{k1}(y)\,dy \\
&- \lambda e^{-2\pi\sqrt{k^2+\lambda}} \int_0^{2\pi} e^{-\sqrt{k^2+\lambda}\,y}\tilde{f}_{k1}(y)\,dy\},
\end{aligned}
\tag{2.22}
$$

where $D(k) = -\lambda + \lambda e^{-4\pi\sqrt{k^2+\lambda}}$. Taking into account (2.22) in (2.21) we have

$$\|\widetilde{u}_{k1}\|_{L_2(0,2\pi)} \leq C\|\widetilde{f}_{k1}\|_{L_2(0,2\pi)}.$$

For the Fourier coefficients $\widetilde{u}_{k2}(x) = (u(x), u_{k2})$ of a solution $u(x)$ of the problem (2.19) we obtain the problem

$$-\widetilde{u}_{k2}''(x) + (k^2 + \lambda)\widetilde{u}_{k2}(x) = \widetilde{f}_{k2}(x),$$
$$\widetilde{u}_{k2}'(0) + k\widetilde{u}_{k2}(0) = 0, \qquad \widetilde{u}_{k2}'(2\pi) + k\widetilde{u}_{k2}(2\pi) = 0$$

where $\widetilde{f}_{k2}(x) = (f(x), u_{k2})$. The above arguments bring us to the estimate

$$\|\widetilde{u}_{k2}\|_{L_2(0,2\pi)} \leq Ck^{-2}\|\widetilde{f}_{k2}\|_{L_2(0,2\pi)}.$$

By virtue of Lemma 2.1 we obtain the estimate

$$\|u\|_{L_2(\Omega)}^2 = \int_0^{2\pi} \|u(x)\|_{L_2(0,2\pi)}^2 \, dx$$
$$= \sum_{k=0}^{\infty} \|\widetilde{u}_{k1}\|_{L_2(0,2\pi)}^2 + \sum_{k=1}^{\infty} \|\widetilde{u}_{k2}\|_{L_2(0,2\pi)}^2 \leq C\|f\|_{L_2(\Omega)}^2.$$

There is no stronger estimate than that in Theorem 2.6. ∎

Chapter 5

Differential-operator equations

The role and importance of the Fourier method in mathematical physics are well known. Unfortunately, the method can be applied as a rule only in cases when the corresponding spectral problem is selfadjoint. This chapter does not answer the question about the applicability of the Fourier method when the corresponding spectral problem is essentially non-selfadjoint. Instead, we show how solutions to problems for differential-operator equations can be approximated by linear combinations of elementary solutions.

5.1. Differential-operator equations on the whole axis

In the theory of differential-operator equations both spaces of continuously differentiable vector-valued functions with value from a Banach space, and spaces of vector-valued functions with derivatives in the Sobolev sense, are important. Here we will add a number of new results to the information given in Chapter 1.

5.1.1. Compactness of embeddings. Let E and F be Banach spaces and the embedding $E \subset F$ be continuous. Consider the Banach space

$$W_p^1((0,1), E, F) = L_p((0,1), E) \cap W_p^1((0,1), F).$$

Theorem 1.1. Let the embedding $E \subset F$ be compact. Then for $1 < p \leq \infty$, $1 \leq q < \infty$ the embedding

$$W_p^1((0,1), E, F) \subset L_q((0,1), F)$$

is compact.

Proof. Let M be a bounded set in $W_p^1((0,1), E, F)$. Let us show that M is a precompact set in $L_q((0,1), F)$. Take $f \in M$ and extend $f(x)$ by zero outside $[0,1]$. Consider the averaging function of $f(x)$

$$f_h(x) = 1/2h \int_{x-h}^{x+h} f(y) \, dy.$$

Let us show that for a fixed h the set

$$M_h = \{f_h(\cdot) \mid f \in M\}$$

is precompact in $L_q((0,1), F)$. From

$$\|f_h(x)\|_E \leq 1/2h \int_{x-h}^{x+h} \|f(y)\|_E \, dy \leq (2h)^{-1/p} \|f\|_{L_p((0,1),E)},$$

it follows that for a fixed h the set

$$E_h = \{u \mid u = f_h(x), \ f \in M, \ x \in [0,1]\}$$

is bounded in E, hence it is precompact in F. For $0 < x'' - x' < h$ we have

$$\|f_h(x') - f_h(x'')\|_E \leq (2h)^{-1}[\int_{x'-h}^{x''-h} \|f(y)\|_E \, dy + \int_{x'+h}^{x''+h} \|f(y)\|_E \, dy]$$

$$\leq (2h)^{-1} |x'' - x'|^{1/p'} \|f\|_{L_p((0,1),E)},$$

where $1/p + 1/p' = 1$. It follows that the set M_h consists of equicontinuous functions. By virtue of [31, Ch.2, §8.7] the set is precompact in $C([0,1], F)$ and hence in $L_q((0,1), F)$.   It is obvious that

$$\|f_h - f\|_{L_q((0,1),F)}^q \leq (2h)^{-q} \int_0^1 \| \int_{x-h}^{x+h} (f(y) - f(x)) \, dy \|_F^q \, dx$$

$$= (2h)^{-q}[\int_0^h \| \int_{x-h}^{x+h} (f(y) - f(x)) \, dy \|_F^q \, dx$$

$$+ \int_h^{1-h} \| \int_{x-h}^{x+h} (f(y) - f(x)) \, dy \|_F^q \, dx$$

$$+ \int_{1-h}^1 \| \int_{x-h}^{x+h} (f(y) - f(x)) \, dy \|_F^q \, dx] = J_1 + J_2 + J_3.$$

$$(1.1)$$

By virtue of [72, p.42] the embeddings

$$W_p^1((0,1), E, F) \subset W_p^1((0,1), F) \subset C([0,1], F)$$

are continuous. Then

$$J_1 \leq 2h\|f\|_{C([0,1],F)} \leq 2h\|f\|_{W_p^1((0,1),E,F)}.$$

Similarly,

$$J_3 \leq 2h\|f\|_{W_p^1((0,1),E,F)}.$$

On the other hand,

$$J_2 = (2h)^{-q} \int_h^{1-h} \|\int_{x-h}^{x+h} \int_x^y f'(t) \, dt dy\|_F^q \, dx \leq \int_h^{1-h} (\int_x^{x+h} \|f'(t)\|_F \, dt)^q \, dx$$

$$\leq h^{q/p'} \int_h^{1-h} (\int_x^{x+h} \|f'(t)\|_F^p \, dt)^{q/p} \, dx \leq h^{q/p'} \|f\|_{W_p^1((0,1),E,F)}^q.$$

Hence, by virtue of (1.1)

$$\lim_{h \to 0} \|f_h - f\|_{L_q((0,1),F)} = 0$$

uniformly relative to M. Hence, for any $\varepsilon > 0$ there exists $M_h(\varepsilon)$ such that $\|f_h(\varepsilon) - f\|_{L_q((0,1),F)} < \varepsilon/2$ for all $f \in M$. Then a finite $\varepsilon/2$ net for $M_{h(\varepsilon)}$ will be an ε net for M. ∎

Lemma 1.2. *Let the following conditions be satisfied:*

(1) *the embedding $E \subset F$ is compact;*

(2) *for almost all $x \in [0,1]$ an operator $A(x)$ from E into F is compact; for any $\varepsilon > 0$ and for almost all $x \in [0,1]$*

$$\|A(x)u\|_F \leq \varepsilon\|u\|_E + C(\varepsilon)\|u\|_F, \quad u \in E; \tag{1.2}$$

(3) *for $u \in E$ the function $A(x)u$ is measurable on $[0,1]$ in F.*

Then the operator $G: u(x) \to A(x)u(x)$ from $W_p^1((0,1), E, F)$ into $L_p((0,1), F)$ is compact.

Proof. After raising (1.2) to power p and integrating we obtain

$$\|A(\cdot)u(\cdot)\|_{L_p((0,1),F)} \leq \varepsilon\|u\|_{L_p((0,1),E)} + C(\varepsilon)\|u\|_{L_p((0,1),F)},$$

where $u \in L_p((0,1), E)$. Hence, for $u \in W_p^1((0,1), E, F)$ and any $\varepsilon > 0$

$$\|Gu\|_{L_p((0,1),F)} \leq \varepsilon\|u\|_{W_p^1((0,1),E,F)} + C(\varepsilon)\|u\|_{L_p((0,1),F)}.$$

Then by virtue of Theorem 1.1 and Lemma 2.2.6, the operator G from $W_p^1((0,1), E, F)$ into $L_p((0,1), F)$ is compact ∎

5.1.2. Coercive equations of the higher order. Consider the equation

$$L(D)u = D^n u(x) + A_1 D^{n-1}u(x) + \cdots + A_n u(x) = f(x), \quad x \in \mathbb{R}, \tag{1.3}$$

with closed operators A_k acting in a Hilbert space H, where $D^j u(x) = \dfrac{d^j u(x)}{dx^j}$ is a generalized derivative of the function $u(x)$ with values from H.

Let there exist Hilbert spaces H_k, $k = 0, \ldots, n$, for which continuous embeddings $H_n \subset H_{n-1} \subset \cdots \subset H_0 = H$ take place.

Introduce the space $W^n_{p,\gamma}(\mathbb{R}, H_n, \ldots, H_0)$, $1 < p < \infty$, $\gamma \in \mathbb{R}$, of functions with the norm

$$\|u\|_{W^n_{p,\gamma}(\mathbb{R}, H_n, \ldots, H_0)} = \left(\sum_{k=0}^{n} \int_{-\infty}^{\infty} \|D^{n-k}(e^{-\gamma x} u(x))\|_{H_k}^p \, dx \right)^{1/p}$$

Let us denote

$$W^0_{p,\gamma}(\mathbb{R}, H) = L_{p,\gamma}(\mathbb{R}, H), \qquad L_{p,0}(\mathbb{R}, H) = L_p(\mathbb{R}, H),$$
$$W^n_{p,0}(\mathbb{R}, H_n, \ldots, H_0) = W^n_p(\mathbb{R}, H_n, \ldots, H_0).$$

Assume that the operators A_k from H_k into H are bounded. The operator pencil

$$L(\lambda) = \lambda^n I + \lambda^{n-1} A_1 + \cdots + A_n, \tag{1.4}$$

defined on $D(L) = H_n$ is called a *characteristic operator pencil* of equation (1.3).

Denote by Ff the Fourier transform of the function $f(x)$ with values from H.

Denote by $B(H)$ the Banach space of linear bounded operators acting in H with the norm equal to the operator norm.

The mapping $\sigma \to T(\sigma)$: $\mathbb{R} \to B(H)$ is called a *Fourier multiplier of type* (p, p), if

$$\|F^{-1}TFf\|_{L_p(\mathbb{R}, H)} \le C\|f\|_{L_p(\mathbb{R}, H)}, \quad f \in L_p(\mathbb{R}, H).$$

It is known [13, p.1181] that if the mapping $\sigma \to T(\sigma)$: $\mathbb{R} \to B(H)$ is continuously differentiable and

$$\|T(\sigma)\| \le C, \qquad \|T'(\sigma)\| \le C|\sigma|^{-1}, \quad \sigma \in \mathbb{R},$$

the $T(\sigma)$ is a Fourier multiplier of type (p, p).

Theorem 1.3. *Let the following conditions be satisfied:*

(1) *there exist Hilbert spaces H_k, $k = 0, \ldots, n$, for which the continuous embeddings $H_n \subset H_{n-1} \subset \cdots \subset H_0 = H$ are fulfilled and $\overline{H_n}|_H = H$;*

(2) *operators A_k from H_k into H are bounded;*

(3) *for the characteristic operator pencil (1.4) the coercive estimate*

$$\sum_{k=0}^{n} |\lambda|^{n-k} \|L^{-1}(\lambda)\|_{B(H,H_k)} \leq C, \quad \mathrm{Re}\lambda = \gamma,$$

holds.

Then the operator $\mathbb{L} : u \to L(D)u = \sum_{k=0}^{n} A_k D^{n-k} u(x)$, where $A_0 = I$, from $W_{p,\gamma}^n(\mathbb{R}, H_n, \ldots, H_0)$ onto $L_{p,\gamma}(\mathbb{R}, H)$ is an isomorphism.

The solution to (1.3) is given by the formula

$$u(x) = -\frac{i}{\sqrt{2\pi}} \int_{\gamma-i\infty}^{\gamma+i\infty} e^{\lambda x} L^{-1}(\lambda) \widetilde{f}(-i\lambda) \, d\lambda, \qquad (1.5)$$

where $\widetilde{f}(z) = (Ff)(z)$.

Proof. Let $u(x)$ be a solution to (1.3) that belongs to $W_{p,\gamma}^n(\mathbb{R}, H_n, \ldots, H_0)$. Let us multiply equation (1.3) by $e^{-\gamma x}$ and then apply the Fourier transform

$$\frac{1}{\sqrt{2\pi}} \sum_{k=0}^{n} \int_{-\infty}^{\infty} e^{-i(\sigma-i\gamma)x} A_k D^{n-k} u(x) \, dx = \frac{1}{\sqrt{2\pi}} \int_{-\infty}^{\infty} e^{-i(\sigma-i\gamma)x} f(x) \, dx. \qquad (1.6)$$

Since

$$\frac{1}{\sqrt{2\pi}} \int_{-\infty}^{\infty} e^{-i(\sigma-i\gamma)x} A_k D^{n-k} u(x) \, dx = (i\sigma + \gamma)^{n-k} A_k \widetilde{u}(\sigma - i\gamma),$$

then from (1.6) we obtain

$$\sum_{k=0}^{n} (i\sigma + \gamma)^{n-k} A_k \widetilde{u}(\sigma - i\gamma) = \widetilde{f}(\sigma - i\gamma). \qquad (1.7)$$

Since the operator $L(\lambda)$ in H is invertible on the line $\mathrm{Re}\lambda = \gamma$, then from (1.7) follows

$$\widetilde{u}(\sigma - i\gamma) = L^{-1}(i\sigma + \gamma) \widetilde{f}(\sigma - i\gamma).$$

On the other hand, from

$$e^{-\gamma x} \widetilde{u}(\sigma) = \frac{1}{\sqrt{2\pi}} \int_{-\infty}^{\infty} e^{-i\sigma x} e^{-\gamma x} u(x) \, dx$$

$$= \frac{1}{\sqrt{2\pi}} \int_{-\infty}^{\infty} e^{-i(\sigma-i\gamma)x} u(x) \, dx = \widetilde{u}(\sigma - i\gamma)$$

it follows that

$$e^{-\gamma x} u(x) = \frac{1}{\sqrt{2\pi}} \int_{-\infty}^{\infty} e^{i\sigma x} \widetilde{u}(\sigma - i\gamma) \, d\sigma.$$

Hence

$$\begin{aligned}
u(x) &= \frac{1}{\sqrt{2\pi}} \int_{-\infty}^{\infty} e^{(i\sigma+\gamma)x} \widetilde{u}(\sigma - i\gamma) \, d\sigma \\
&= \frac{1}{\sqrt{2\pi}} \int_{-\infty}^{\infty} e^{(i\sigma+\gamma)x} L^{-1}(i\sigma + \gamma) \widetilde{f}(\sigma - i\gamma) \, d\sigma \\
&= -\frac{i}{\sqrt{2\pi}} \int_{\gamma-i\infty}^{\gamma+i\infty} e^{\lambda x} L^{-1}(\lambda) \widetilde{f}(-i\lambda) \, d\lambda.
\end{aligned}$$

Thus formula (1.5) is established. The boundedness of the operator $\mathbb{L}$ is obvious. It remains to be shown that if $f \in L_{p,\gamma}(\mathbb{R}, H)$, then the function $u(x)$ given by (1.5) belongs to $W_{p,\gamma}^n(\mathbb{R}, H_n, \ldots, H_0)$ and is the solution to (1.3). We have

$$\begin{aligned}
\|u\|_{W_{p,\gamma}^n(\mathbb{R},H_n,\ldots,H_0)}^p &= \sum_{k=0}^{n} \int_{-\infty}^{\infty} \|F^{-1} F D^{n-k}(e^{-\gamma x} u(x))\|_{H_k}^p \, dx \\
&= \sum_{k=0}^{n} \int_{-\infty}^{\infty} \|F^{-1}(i\sigma)^{n-k} \widetilde{u}(\sigma - i\gamma)\|_{H_k}^p \, d\sigma.
\end{aligned}$$

Consider selfadjoint, positively defined operators S_k in H such that $H(S_k) = H_k$, $k = 0, \ldots, n$ [44, Ch.1, §2.1]. Then

$$\begin{aligned}
\|u\|_{W_{p,\gamma}^n(\mathbb{R},H_n,\ldots,H_0)}^p &= \sum_{k=0}^{n} \int_{-\infty}^{\infty} \|F^{-1}(i\sigma)^{n-k} S_k L^{-1}(i\sigma + \gamma) \widetilde{f}(\sigma - i\gamma)\|^p \, d\sigma \\
&= \sum_{k=0}^{n} \int_{-\infty}^{\infty} \|F^{-1}(i\sigma)^{n-k} S_k L^{-1}(i\sigma + \gamma) F(e^{-\gamma x} f(x))\|^p \, d\sigma.
\end{aligned} \tag{1.8}$$

Show that functions $T_k(\sigma) = (i\sigma)^{n-k} S_k L^{-1}(i\sigma + \gamma)$, $k = 0, \ldots, n$ are the Fourier multipliers of type (p, p) in H. From the theorem conditions follows

$$\|T_k(\sigma)\| \leq C. \tag{1.9}$$

Since

$$\begin{aligned}
T_k'(\sigma) &= (n - k) i^{n-k} \sigma^{n-k-1} S_k L^{-1}(i\sigma + \gamma) \\
&\quad - i^{n-k+1} \sigma^{n-k} S_k L^{-1}(i\sigma + \gamma) L'(i\sigma + \gamma) L^{-1}(i\sigma + \gamma),
\end{aligned}$$

then

$$\begin{aligned}
\|T_k'(\sigma)\| &\leq C(|\sigma|^{-1} + |\sigma|^{n-k} \|S_k\|_{B(H_k,H)} \|L^{-1}(i\sigma + \gamma)\|_{B(H,H_k)} \\
&\quad \times \sum_{j=1}^{n-1} |\sigma|^{n-j-1} \|A_j\|_{B(H_j,H)} \|L^{-1}(i\sigma + \gamma)\|_{B(H,H_j)} \leq C|\sigma|^{-1}.
\end{aligned} \tag{1.10}$$

By virtue of the Mikhlin-Schwartz theorem [13, p.1181] from (1.9) and (1.10) it follows that the functions $T_k(\sigma)$ are Fourier multipliers of type (p, p) in H. Then (1.8) implies

$$\|u\|_{W_{p,\gamma}^n(\mathbb{R}, H_n, \dots, H_0)} \leq C \|e^{-\gamma \cdot} f(\cdot)\|_{L_p(\mathbb{R}, H)} \leq C \|f\|_{L_{p,\gamma}(\mathbb{R}, H)}.$$

On the other hand, from (1.5) we have

$$L(D)u = -\frac{i}{\sqrt{2\pi}} \int\limits_{\gamma - i\infty}^{\gamma + i\infty} e^{\lambda x} \widetilde{f}(-i\lambda) \, d\lambda = \frac{1}{\sqrt{2\pi}} \int_{-\infty}^{\infty} e^{(i\sigma + \gamma)x} \widetilde{f}(\sigma - i\gamma) \, d\sigma = f(x).$$

The theorem has been proved. ∎

Remark 1.4. *In the theory of partial differential equations for regular elliptic boundary value problems the conditions of Theorem 1.3 are fulfilled.*

Consider the equation

$$L(D)u = D^n u(x) + (a_1 A + A_1) D^{n-1} u(x) + \cdots + (a_n A^n + A_n) u(x) = f(x),$$

$$x \in \mathbb{R}, \tag{1.11}$$

with closed operator coefficients A and A_k in a Hilbert space H, where a_k are complex numbers, $a_n \neq 0$. The roots of the *characteristic equation*

$$\omega^n + a_1 \omega^{n-1} + \cdots + a_n = 0$$

will be denoted by ω_j, $j = 1, \dots, n$.

Lemma 1.5. *Let A be an operator in H and*

$$\|R(\lambda, A)\| \leq C |\lambda|^{-1}, \quad \lambda \in \ell(\gamma \omega_j^{-1}, \pm \pi/2 - \arg \omega_j), \ |\lambda| \to \infty.$$

Then for the operator pencil

$$L_0(\lambda) = \lambda^n I + \lambda^{n-1} a_1 A + \cdots + a_n A^n$$

the coercive estimate

$$\sum_{k=0}^{n} |\lambda|^{n-k} \|A^k L^{-1}(\lambda)\| \leq C, \quad \operatorname{Re} \lambda = \gamma, |\lambda| \to \infty$$

is valid.

Proof. It follows from Lemma 2.3.14 under $\eta = 0$, $a = \gamma$, $\varphi = \pm \pi/2$. ∎

We turn $D(A^k)$ into a Hilbert space $H(A^k)$ with respect to the norm

$$\|u\|_{H(A^k)} = (\|A^k u\|^2 + \|u\|^2)^{1/2}.$$

Theorem 1.6. *Let the following conditions be satisfied:*

(1) *A is an operator in H and*

$$\|R(\lambda, A)\| \le C|\lambda|^{-1}, \quad \lambda \in \ell(\gamma \omega_j^{-1}, \pm\pi/2 - \arg\omega_j), \ |\lambda| \to \infty;$$

(2) *A_k are operators in H, $D(A_k) \supset D(A^k)$ and for any $\varepsilon > 0$*

$$\|A_k u\| \le \varepsilon \|A^k u\| + C(\varepsilon)\|u\|, \quad u \in D(A^k), \ k = 1,\ldots,n;$$

(3) *for every λ for which $\mathrm{Re}\,\lambda = \gamma$ the operator*

$$L(\lambda) = \lambda^n I + \lambda^{n-1}(a_1 A + A_1) + \cdots + (a_n A^n + A_n)$$

in H is invertible.[14]

Then the operator

$$\mathbb{L} : u \to L(D)u = D^n u(x) + \sum_{k=1}^{n} (a_k A^k + A_k) D^{n-k} u(x)$$

from $W_{p,\gamma}^n(\mathbb{R}, H(A^n)), \ldots, H)$ onto $L_{p,\gamma}(\mathbb{R}, H)$ is an isomorphism.

Proof. From condition 1, by virtue of Lemma 1.5, follows the estimate

$$\sum_{k=0}^{n} |\lambda|^{n-k} \|L_0^{-1}(\lambda)\|_{B(H,H(A^k))} \le C, \quad \mathrm{Re}\,\lambda = \gamma, \ |\lambda| \to \infty,$$

where $L_0(\lambda) = \lambda^n I + \lambda^{n-1} a_1 A + \cdots + a_n A^n$. Then, by Lemma 2.3.7

$$\sum_{k=0}^{n} |\lambda|^{n-k} \|L^{-1}(\lambda)\|_{B(H,H(A^k))} \le C, \quad \mathrm{Re}\,\lambda = \gamma, \ |\lambda| \to \infty. \qquad (1.12)$$

By virtue of condition 3 estimate (1.12) holds under all $\mathrm{Re}\,\lambda = \gamma$. So, Theorem 1.3 is applicable to our operator from which the statement of Theorem 1.6 follows. ∎

---

[14]In this case $D(L(\lambda)) = \bigcap_{k=1}^{n} D(A^k) = D(A^n)$.

Let H and H_n be Hilbert spaces, for which the continuous embedding $H_n \subset H$ takes place. Let us introduce the space $W_p^n(\mathbb{R}, H_n, H)$, $1 < p < \infty$, of functions with the norm

$$\|u\|_{W_p^n(\mathbb{R}, H_n, H)} = \|u\|_{L_p(\mathbb{R}, H_n)} + \|u^{(n)}\|_{L_p(\mathbb{R}, H)}.$$

Theorem 1.7. *Let B be an operator in H, $\overline{D(B)} = H$ and*

$$\|(B + sI)^{-1}\| \leq C(1 + s)^{-1}, \quad s \geq 0.$$

Then the operator $D^j : u(x) \to u^{(j)}(x)$ from $W_p^{2n}(\mathbb{R}, H(B), H)$ into

$$W_p^{2n-j}(\mathbb{R}, H(B^{1-j/2n}), \dots, H)$$

is bounded.

Proof. Consider the operator pencil $L(\lambda) = \lambda^{2n} I + aB$, where $a > 0$ if $n = 2k$, and $a < 0$ if $n = 2k + 1$. By virtue of the momentum inequality [39, p.115]

$$\|B^\alpha u\| \leq C\|Bu\|^\alpha \|u\|^{1-\alpha}, \quad u \in D(B),$$

we have

$$\sum_{k=0}^{2n} |\lambda|^{2n-k} \|B^{k/2n} L^{-1}(\lambda)\| \leq C, \quad \operatorname{Re}\lambda = 0.$$

Let $u \in W_p^{2n}(\mathbb{R}, H(B), H)$. Then $f = u^{(2n)} + aBu \in L_p(\mathbb{R}, H)$. Hence by virtue of Theorem 1.3 a solution to the equation

$$u^{(2n)}(x) + aBu(x) = f(x), \quad x \in \mathbb{R}$$

belongs to the space $W_p^{2n}(\mathbb{R}, H(B), \dots, H)$ and

$$\|u\|_{W_p^{2n}(\mathbb{R}, H(B), \dots, H)} \leq C\|f\|_{L_p(\mathbb{R}, H)}. \quad \blacksquare$$

Theorem 1.8. *Let the following conditions be satisfied:*

(1) *there exist Hilbert spaces H_k, $k = 0, \dots, n$, such that continuous embeddings $H_n \subset H_{n-1} \subset \cdots \subset H_0 = H$ are fulfilled and $\overline{H_n}|_H = H$;*

(2) *there exist operators A_k bounded from H_k into H;*

(3) *for the operator pencil*

$$L(\lambda) = \lambda^n I + \lambda^{n-1} A_1 + \cdots + A_n,$$

the coercive estimate

$$\sum_{k=0}^{n} |\lambda|^{n-k} \|L^{-1}(\lambda)\|_{B(H,H_k)} \leq C, \quad \operatorname{Re}\lambda = 0$$

holds.

Then the operator $D^j : u(x) \to u^{(j)}(x)$ from $W_p^n(\mathbb{R}, H_n, \ldots, H)$ into

$$L_p(\mathbb{R}, (H_k, H_m)_{(n-k-j)/(m-k),q}), \quad m > k, n - m \leq j \leq n - k, 1 \leq q \leq \infty$$

is bounded.

Proof. By virtue of inequality [72, p.25] we have

$$\|u\|_{(H_k,H_m)\alpha,q} \leq C\|u\|_{H_m}^{\alpha}\|u\|_{H_k}^{1-\alpha}, \quad u \in H_k \cap H_m.$$

Then for $m > k$, for a solution to the equation

$$L(\lambda)u = (\lambda^n I + \lambda^{n-1} A_1 + \cdots + A_n)u = f,$$

we have

$$|\lambda|^j \|u\|_{(H_k,H_m)_{(n-k-j)/(m-k),q}} \leq C|\lambda|^{(n-k-j)(n-m)/(m-k)}\|u\|_{H_m}^{(n-k-j)/(m-k)}$$
$$\times |\lambda|^{(m-n+j)(n-k)/(m-k)}\|u\|_{H_k}^{(m-n+j)/(m-k)}.$$

By virtue of the Young inequality (1.2.13)

$$|\lambda|^j \|u\|_{(H_k,H_m)_{(n-k-j)/(m-k),q}} \leq C|\lambda|^{n-m}\|u\|_{H_m} + |\lambda|^{n-k}\|u\|_{H_k}.$$

By virtue of condition 3

$$|\lambda|^j \|L^{-1}(\lambda)\|_{B(H,(H_k,H_m)_{(n-k-j)/(m-k),q})} \leq C, \quad \operatorname{Re}\lambda = 0.$$

Let $u \in W_p^n(\mathbb{R}, H_n, \ldots, H)$. Then $f = u^{(n)} + A_1 u^{(n-1)} + \cdots + A_n u \in L_p(\mathbb{R}, H)$. By virtue of Theorem 1.3, for a solution to the equation

$$u^{(n)}(x) + A_1 u^{(n-1)}(x) + \cdots + A_n u(x) = f(x),$$

we have the following estimate

$$\|D^j u\|_{L_p(\mathbb{R},(H_k,H_m)_{(n-k-j)/(m-k),q})} \leq C\|f\|_{L_p(\mathbb{R},H)}.$$

Hence,

$$\|D^j u\|_{L_p(\mathbb{R},(H_k,H_m)_{(n-k-j)/(m-k),q})} \leq \|u\|_{W_p^n(\mathbb{R},H_n,\ldots,H)}. \quad \blacksquare$$

Corollary 1.9. *Theorems 1.7 and 1.8 are also true if instead of $\mathbb{R}$ we consider* $(0, 1)$. *In this case we have to use a continuation operator from* $(0, 1)$ *onto* $\mathbb{R}$.

5.2. The Cauchy problem for parabolic differential-operator equations

In spite of the fact that a great number of monographs, including monographs by J. L. Lions [43], S. G. Krein [39], E. Hille and R. S. Phillips [23], S. Ya. Yakubov [85], M. Sova [67], V. I. Gorbachuk and M. L. Gorbachuk [20], A. Pazy [54], H. O. Fattorini [15], and H. Tanabe [70], are devoted to the theory of differential operator equations in a Banach space, this field of science is far from being completely investigated. Out of this extensive material we shall present only what is necessary in 5.3 and 5.4.

5.2.1. Homogeneous equations with a constant operator in a Banach space. A solution to the Cauchy problem

$$u'(t) = Au(t),$$
$$u(0) = u_0, \tag{2.1}$$

with a closed operator A in a Banach space E, is connected with the construction of the function e^{tA}. Many papers are devoted to this question (see, for example, [23], [39]). We shall present only one class of such function. In all cases below the contours are oriented counterclockwise.

Lemma 2.1. Let the following conditions be satisfied:

(1) Γ is a piecewise smooth contour that coincides with the rays $\arg \lambda = \pm(\pi/2 + \alpha)$, $\alpha > 0$ for large $|\lambda|$;

(2) a function $g(\lambda)$ is analytic from the left side of Γ and for some $k \in \mathbb{R}$

$$|g(\lambda)| \le C|\lambda|^k, \quad \pi/2 + \alpha \le \arg \lambda \le 3\pi/2 - \alpha, \quad |\lambda| \to \infty.$$

Then[15]

$$\int_\Gamma e^{\lambda t} g(\lambda) \, d\lambda = 0, \quad |\arg t| < \alpha.$$

Proof. Denote by Γ_R the arc $\lambda = Re^{i\varphi}$, $\varphi \in [\pi/2 + \alpha, 3\pi/2 - \alpha]$. Then by the Cauchy theorem

$$\int_\Gamma e^{\lambda t} g(\lambda) \, d\lambda = \lim_{R \to \infty} \int_{\substack{\lambda \in \Gamma \\ |\lambda| \le R}} e^{\lambda t} g(\lambda) \, d\lambda = \lim_{R \to \infty} \int_{\Gamma_R} e^{\lambda t} g(\lambda) \, d\lambda.$$

Hence, for $|\arg t| \le \gamma < \alpha$

$$|\int_\Gamma e^{\lambda t} g(\lambda) \, d\lambda| \le C_R \lim_{R \to \infty} \int_{|\lambda| \le R} e^{-|\lambda| \, |t| \sin(\alpha - \gamma)} |\lambda|^k |d\lambda| = 0. \quad \blacksquare$$

---

[15] The angle $|\arg t| < \alpha$ does not contain zero.

Theorem 2.2. *Let the estimate*

$$\|R(\lambda, A)\| \le C|\lambda|^{-\beta}, \quad |\arg \lambda| \le \pi/2 + \alpha, \ |\lambda| \to \infty,$$

hold for an operator A in E for some $\alpha > 0$ and $\beta \in (0, 1]$. Then the function

$$U(t) = -\frac{1}{2\pi i} \int_\Gamma e^{\lambda t} R(\lambda, A) \, d\lambda, \quad |\arg t| < \alpha, \tag{2.2}$$

where Γ coincides with the rays $\arg \lambda = \pm(\pi/2 + \alpha)$ for large $|\lambda|$ and is completely contained in the operator A resolvent set, has the following properties:

  (1) *the mapping $t \to U(t)$: $|\arg t| < \alpha \to B(E)$ is analytic;*
  (2) $\|U^{(k)}(t)\| \le C_k(\gamma)|t|^{\beta - k - 1}$, $|\arg t| \le \gamma < \alpha$, $t \to 0$;
  (3) $\lim\limits_{\substack{t \to 0 \\ |\arg t| \le \gamma}} U(t)u = u, u \in D(A)$;
     and if $\beta = 1$ then $\lim\limits_{\substack{t \to 0 \\ |\arg t| \le \gamma}} U(t)u = u, \ u \in \overline{D(A)}$;
  (4) $U(t + \tau) = U(t)U(\tau)$, $|\arg t| < \alpha$, $|\arg \tau| < \alpha$;
  (5) $U^{(k)}(t) = A^k U(t)$, $|\arg t| < \alpha$;

Proof. Let us take as Γ the arc

$$\lambda = Re^{i\varphi}, \quad \varphi \in [-\pi/2 - \alpha, \pi/2 + \alpha],$$

with the rays

$$\lambda = (R + r)e^{\pm i(\pi/2 + \alpha)}, \quad r \ge 0.$$

For large enough R such contour Γ is completely contained in the operator A resolvent set. Denote the arc of the circle $|\lambda| = R$ by Γ_1, and the remaining part of Γ by Γ_2. Then

$$U(t) = -\frac{1}{2\pi i} \int_{\Gamma_1} e^{\lambda t} R(\lambda, A) \, d\lambda - \frac{1}{2\pi i} \int_{\Gamma_2} e^{\lambda t} R(\lambda, A) \, d\lambda.$$

The existence of the first integral for all $t \in \mathbb{C}$ is obvious. The second integral for $|\arg t| \le \gamma < \alpha$ exists by virtue of the estimate

$$\|e^{\lambda t} R(\lambda, A)\| = e^{\text{Re}\lambda t} \|R(\lambda, A)\| \le e^{|\lambda| \, |t| \cos(\pi/2 + \alpha - \gamma)} |\lambda|^{-\beta}$$
$$\le C|\lambda|^{-\beta} e^{-|\lambda| \, |t| \sin(\alpha - \gamma)}.$$

Analyticity of the mapping $t \to U(t)$: $|\arg t| < \alpha \to B(E)$ follows from the existence of integrals

$$U^{(k)}(t) = -\frac{1}{2\pi i} \int_{\Gamma} \lambda^k e^{\lambda t} R(\lambda, A) \, d\lambda,$$

which, in turn, follows from the estimate

$$\|\lambda^k e^{\lambda t} R(\lambda, A)\| \le C|\lambda|^{k-\beta} e^{-|\lambda| \, |t| \sin(\alpha-\gamma)}$$

on Γ_2, and on Γ_1 its existence is obvious.

For $|\arg t| \le \gamma < \alpha$, $t \to 0$ we have

$$\|U^{(k)}(t)\| \le C(1 + \int_{\substack{\arg \lambda = \pm(\pi/2+\alpha) \\ |\lambda| \ge R}} |\lambda|^{k-\beta} e^{-|\lambda| \, |t| \sin(\alpha-\gamma)} |d\lambda|).$$

Substituting $|\lambda| \, |t| \sin(\alpha - \gamma) = \tau$ we obtain

$$\|U^{(k)}(t)\| \le C_k(1 + |t|^{\beta-k-1} \int_0^\infty \tau^{k-\beta} e^{-\tau} \, d\tau),$$

from which under $k - \beta > -1$ property 2 of the theorem follows.

If $\beta = 1$ and $k = 0$ then the last integral diverges. In this case we obtain the estimate in property 2 as follows

$$\|U(t)\| \le C(e^{R|t|} + \int_{R|t|\sin(\alpha-\gamma)} \tau^{-1} e^{-\tau} \, d\tau).$$

Under small enough $|t|$ one can choose $R = 1/|t|$. Then

$$\|U(t)\| \le C(e + \int_{\sin(\alpha-\gamma)}^\infty \tau^{-1} e^{-\tau} \, d\tau) \le C.$$

Let $u \in D(A)$, $s \in \rho(A)$ and $s > R$. Denote $(A - sI)u = v$. Then $u = R(s, A)v$ and

$$U(t)u = -\frac{1}{2\pi i} \int_{\Gamma} e^{\lambda t} R(\lambda, A) R(s, A) v \, d\lambda$$

$$= -\frac{1}{2\pi i} \int_{\Gamma} e^{\lambda t} \frac{R(\lambda, A) - R(s, A)}{\lambda - s} v \, d\lambda.$$

Since by virtue of Lemma 2.1

$$\int_{\Gamma} e^{\lambda t} \frac{R(s, A)}{\lambda - s} v \, d\lambda = 0,$$

then

$$U(t)u = -\frac{1}{2\pi i} \int_\Gamma e^{\lambda t} \frac{R(\lambda, A)}{\lambda - s} v \, d\lambda.$$

Applying the Cauchy formula we obtain

$$\lim_{\substack{t \to 0 \\ |\arg t| \le \gamma}} U(t)u = -\frac{1}{2\pi i} \int_\Gamma \frac{R(\lambda, A)}{\lambda - s} v \, d\lambda = R(s, A)v = u, \quad u \in D(A). \qquad (2.3)$$

Since for $\beta = 1$ we have $\|U(t)\| \le C$, then from (2.3) by virtue of [39, p.9] it follows that

$$\lim_{\substack{t \to 0 \\ |\arg t| \le \gamma}} U(t)u = u, \quad u \in \overline{D(A)}.$$

So, property 3 of the theorem is completely proved.

Let Γ_1 be a contour obtained from the contour Γ by a shift to the right so that Γ_1 and Γ are disjoint. Integral (2.2) can be transformed into the integral on Γ_1 by virtue of the function under the integral (2.2) sign being analytic. Then for $|\arg t| < \alpha$ and $|\arg \tau| < \alpha$ we have

$$U(t)U(\tau) = \frac{1}{(2\pi i)^2} \int_{\Gamma_1} \int_\Gamma e^{\lambda t} e^{\mu t} R(\lambda, A) R(\mu, A) \, d\lambda d\mu$$

$$= \frac{1}{(2\pi i)^2} \int_{\Gamma_1} \int_\Gamma e^{\lambda t} e^{\mu t} \frac{R(\lambda, A) - R(\mu, A)}{\lambda - \mu} \, d\lambda d\mu$$

$$= \frac{1}{(2\pi i)^2} \int_\Gamma e^{\lambda t} R(\lambda, A) \Big(\int_{\Gamma_1} \frac{e^{\mu t}}{\lambda - \mu} \, d\mu \Big) d\lambda$$

$$- \frac{1}{(2\pi i)^2} \int_{\Gamma_1} e^{\mu t} R(\mu, A) \Big(\int_\Gamma \frac{e^{\lambda t}}{\lambda - \mu} \, d\lambda \Big) d\mu$$

By the Cauchy formula

$$\int_{\Gamma_1} \frac{e^{\mu t}}{\lambda - \mu} \, d\mu = -2\pi i e^{\lambda \tau}, \quad |\arg \tau| < \alpha, \ \lambda \in \Gamma,$$

and by Lemma 2.1

$$\int_\Gamma \frac{e^{\lambda t}}{\lambda - \mu} \, d\lambda = 0, \quad |\arg t| < \alpha, \ \mu \in \Gamma_1.$$

So, for $|\arg t| < \alpha$ and $|\arg \tau| < \alpha$

$$U(t)U(\tau) = \frac{1}{2\pi i} \int_\Gamma e^{\lambda(t+\tau)} R(\lambda, A) \, d\lambda = U(t + \tau),$$

i.e., property 4 is proved.

Since $(A - \lambda I)R(\lambda, A) = I$, then $\lambda R(\lambda, A) = AR(\lambda, A) - I$. Hence

$$U'(t) = -\frac{1}{2\pi i} \int_\Gamma \lambda e^{\lambda t} R(\lambda, A) \, d\lambda = -\frac{1}{2\pi i} \int_\Gamma A e^{\lambda t} R(\lambda, A) \, d\lambda$$
$$+ \frac{1}{2\pi i} \int_\Gamma e^{\lambda t} \, d\lambda, \quad |\arg t| < \alpha.$$

By virtue of Lemma 2.1

$$\int_\Gamma e^{\lambda t} \, d\lambda = 0, \quad |\arg t| < \alpha.$$

Hence,

$$U'(t) = AU(t), \quad |\arg t| < \alpha.$$

We can now prove property 5 by induction on k. Suppose, for $k = m$ property 5 holds. Then by virtue of the closedness of A under $|\arg t| < \alpha$

$$U^{(m+1)}(t) = [U^{(m)}(t)]' = [A^m U(t)]' = A^m U'(t) = A^{m+1} U(t). \quad \blacksquare$$

A family of bounded operators $U(t)$ depending on the parameter $t \in (0, \infty)$ is called a *semigroup* if

$$U(t + \tau) = U(t)U(\tau), \quad t > 0, \ \tau > 0.$$

The semigroup defined by (2.2) will be denoted by e^{tA}. So, if operator A satisfies the condition of Theorem 2.2 then it generates the semigroup according to the formula

$$e^{tA} = -\frac{1}{2\pi i} \int_\Gamma e^{\lambda t} R(\lambda, A) \, d\lambda, \quad |\arg t| < \alpha. \tag{2.4}$$

Moreover, the semigroup $U(t) = e^{tA}$ has properties 1–5 of Theorem 2.2.

Remark 2.3. Let A be an operator in E and

$$\|R(\lambda, A)\| \leq M|\lambda|^{-1}, \quad |\arg \lambda| \leq \pi/2, \ |\lambda| \to \infty.$$

Then for some $\alpha > 0$

$$\|R(\lambda, A)\| \leq C|\lambda|^{-1}, \quad |\arg \lambda| \leq \pi/2 + \alpha, \ |\lambda| \to \infty.$$

Proof. Expand the resolvent $R(\lambda, A)$ at a point s, where $\arg s = \varphi$, in the Taylor series

$$R(\lambda, A) = R(s, A) \sum_{k=0}^{\infty} [R(s, A)]^k (\lambda - s)^k.$$

In the circle $|\lambda - s| < M^{-1}|s|q$ with $q < 1$ we have

$$\|R(\lambda, A)\| \le M|s|^{-1} \sum_{k=0}^{\infty} (M|s|^{-1})^k (M^{-1}|s|q)^k = M|s|^{-1}(1-q)^{-1}.$$

Since $M^{-1}|s|q > |\lambda - s| \ge |\lambda| - |s|$, then $|s|^{-1} < (M^{-1}q+1)|\lambda|^{-1}$. So, in the circle $|\lambda - s| < M^{-1}|s|q$, $|s| \to \infty$, we have

$$\|R(\lambda, A)\| \le M(1-q)^{-1}(M^{-1}q+1)|\lambda|^{-1} \le C|\lambda|^{-1}.$$

Since the circles $|\lambda - s| < M^{-1}|s|q$ cover the angles $|\arg \lambda - \varphi| < \alpha$, where $\alpha = \arcsin M^{-1}q$, then for $\alpha = \arcsin M^{-1}q$ the necessary estimate is fulfilled. ∎

Let us show that problem (2.1) is well-posed, i.e., existence, uniqueness and continuous dependence of a solution on the data can be proved. First we shall prove the uniqueness of a solution to problem (2.1).

Let A in E be closed. Transform $D(A)$ into the Banach space $E(A)$ with the norm $\|u\|_{E(A)} = (\|Au\|^2 + \|u\|^2)^{1/2}$. Introduce the space

$$W_p^1((0,T), E(A), E) = L_p((0,T), E(A)) \cap W_p^1((0,T), E)$$

of functions with the norm

$$\|u\|_{W_p^1((0,T),E(A),E)} = \|Au\|_{L_p((0,T),E)} + \|u'\|_{L_p((0,T),E)}$$

and the set of functions

$$W_{p,lok}^1((0,T), E(A), E) = L_{p,lok}((0,T), E(A)) \cap W_{p,lok}^1((0,T), E).$$

Theorem 2.4. *Let an operator A in E have the resolvent for sufficiently large positive λ and*

$$\varlimsup_{\lambda \to \infty} \frac{\ln \|R(\lambda, A)\|}{\lambda} = 0.$$

Then a solution to problem (2.1) that belongs to the space $C([0,T], E) \cap W_{p,lok}^1((0,T), E(A), E)$ is unique.

Proof. Let $u \in C([0,T], E) \cap W_{p,lok}^1((0,T), E(A), E)$ be a solution to the equation $u'(t) = Au(t)$. Then by integrating by parts, for any $\varepsilon > 0$ we obtain

$$\int_{\varepsilon}^{T-\varepsilon} e^{\lambda(T-\tau)} u(\tau) \, d\tau = -\frac{e^{\lambda(T-\tau)}}{\lambda} u(\tau)\big|_{\varepsilon}^{T-\varepsilon} + \frac{1}{\lambda} \int_{\varepsilon}^{T-\varepsilon} e^{\lambda(T-\tau)} Au(\tau) \, d\tau.$$

Since A is closed, then the last integral sign can be removed. Hence

$$\int_\varepsilon^{T-\varepsilon} e^{\lambda(T-\tau)} u(\tau)\, d\tau = -\frac{e^{\lambda\varepsilon}}{\lambda} u(T-\varepsilon)$$

$$+ \frac{e^{\lambda(T-\varepsilon)}}{\lambda} u(\varepsilon) + \frac{1}{\lambda} A \int_\varepsilon^{T-\varepsilon} e^{\lambda(T-\tau)} u(\tau)\, d\tau.$$

Then passing to the limit for $\varepsilon \to 0$ we obtain

$$(A - \lambda I) \int_0^T e^{\lambda(T-\tau)} u(\tau)\, d\tau = u(T) - e^{\lambda T} u(0).$$

If $u(0) = 0$, then

$$R(\lambda, A)u(T) = \int_0^T e^{\lambda(T-\tau)} u(\tau)\, d\tau = \int_0^T e^{\lambda s} u(T-s)\, ds.$$

Hence,

$$\varlimsup_{\lambda\to\infty} \frac{\ln\| \int_0^T e^{\lambda s} u(T-s)\, ds\|}{\lambda} = \varlimsup_{\lambda\to\infty} \frac{\ln\|R(\lambda, A)u(T)\|}{\lambda} = 0.$$

Then, by virtue of [71, ch.XI, §11.10] we have $u(T - s) = 0$ on $[0, T]$. ∎

5.2.2. Nonhomogeneous equations with a constant operator in a Banach space. Consider Banach spaces

1) $C_\mu((0,T], E) = \{f(\cdot) \mid f \in C((0,T], E),\ \|f\|_{C_\mu((0,T],E)}$
   $= \sup\limits_{t\in(0,T]} \|t^\mu f(t)\| < \infty\},\ \mu \geq 0$;

2) $C_\mu^\gamma((0,T], E) = \{f(\cdot) \mid f \in C((0,T], E),\ \|f\|_{C_\mu^\gamma((0,T],E)}$
   $= \sup\limits_{t\in(0,T]} \|t^\mu f(t)\| + \sup\limits_{0<t<t+h\leq T} \|f(t+h) - f(t)\| h^{-\gamma} t^\mu < \infty\}$
   $\gamma \in (0,1],\ \mu \geq 0$;

   and the linear space

3) $C^1((0,T], E(A), E) = \{f(\cdot) \mid f \in C((0,T], E(A)) \cap C^1((0,T], E)\}$.

   Consider in the Banach space E the Cauchy problem

$$u'(t) = Au(t) + f(t),$$
$$u(0) = u_0. \tag{2.5}$$

Theorem 2.5. *Let the following conditions be satisfied:*

(1) *for the operator A in E under some $\alpha > 0$ and $\beta \in (0,1]$*

$$\|R(\lambda, A)\| \leq M|\lambda|^{-\beta}, \quad |\arg\lambda| \leq \pi/2 + \alpha,\ |\lambda| \to \infty;$$

(2) $f \in C_\mu^\gamma((0,T], E)$ *for some* $\gamma \in (1-\beta, 1],\ \mu \in [0, \beta)$;

(3) $u_0 \in D(A)$.

Then problem (2.5) has a unique solution in $C([0,T], E) \cap C^1((0,T], E(A), E)$, and the solution can be represented in the form

$$u(t) = e^{tA} u_0 + \int_0^t e^{(t-\tau)A} f(\tau) \, d\tau. \tag{2.6}$$

Moreover, for $t \in (0,T]$ the estimates

$$\|u(t)\| \leq C(\|Au_0\| + \|u_0\| + \|f\|_{C_\mu((0,t],E)}),$$
$$\|u'(t)\| + \|Au(t)\| \leq Ct^{-1}(\|Au_0\| + \|u_0\| + \|f\|_{C_\mu^\gamma((0,t],E)}), \tag{2.7}$$

are valid.

Proof. Let us show that the function (2.6) is a solution to problem (2.5). From the continuity of $e^{(t-\tau)A} f(\tau)$ in E by τ on $(0,t)$ and the estimate

$$\|e^{(t-\tau)A} f(\tau)\| \leq \tau^{-\mu} \|e^{(t-\tau)A}\| \sup_{\tau \in (0,t]} \|\tau^\mu f(\tau)\|$$
$$\leq C\tau^{-\mu}(t-\tau)^{-1+\beta} \|f\|_{C_\mu((0,t],E)} \tag{2.8}$$

it follows that the function $e^{(t-\tau)A} f(\tau)$ can be integrated on $(0,t)$.

Show that the function $u(t)$ is continuous in 0 and $\lim_{t \to +0} u(t) = u_0$. By virtue of (2.8) from (2.6) we have

$$\|u(t) - u_0\| \leq \|(e^{tA} - I)u_0\|$$
$$+ C \int_0^t \tau^{-\mu}(t-\tau)^{-1+\beta} \, d\tau \, \|f\|_{C_\mu((0,t],E)}. \tag{2.9}$$

By virtue of condition 3 and property 3 of Theorem 2.2 it follows that the first summand in (2.9) tends to 0 as $t \to +0$. Substituting $\tau = ts$ we obtain

$$\int_0^t \tau^{-\mu}(t-\tau)^{-1+\beta} \, d\tau = t^{-\mu+\beta} \int_0^1 s^{-\mu}(1-s)^{-1+\beta} \, ds,$$

from which, by virtue of $\mu < \beta$, it follows that the second summand in (2.9) also tends to 0 if $t \to +0$.

Consider functions

$$g_h(t) = \int_h^{t-h} e^{(t-\tau)A} f(\tau) \, d\tau$$

on the segment $[2h_0, \tau]$ for $0 < h < h_0 < T/2$. By virtue of (2.8)

$$\lim_{h \to 0} g_h(t) = g(t) = \int_0^t e^{(t-\tau)A} f(\tau) \, d\tau$$

uniformly in $[2h_0, T]$. On the other hand, for $t \in [2h_0, T]$

$$g'_h(t) = \int\limits_h^{t-h} Ae^{(t-\tau)A} f(\tau) \, d\tau + e^{hA} f(t-h)$$

$$= \int\limits_h^{t-h} Ae^{(t-\tau)A}[f(\tau) - f(t)] \, d\tau + \int\limits_h^{t-h} Ae^{(t-\tau)A} f(t) \, d\tau$$

$$+ e^{hA} f(t-h) = \int\limits_h^{t-h} Ae^{(t-\tau)A}[f(\tau) - f(t)] \, d\tau$$

$$- e^{(t-\tau)A} f(t)|_h^{t-h} + e^{hA} f(t-h)$$

$$= \int\limits_h^{t-h} Ae^{(t-\tau)A}[f(\tau) - f(t)] \, d\tau - e^{hA}[f(t) - f(t-h)]$$

$$+ e^{(t-h)A} f(t). \tag{2.10}$$

Obviously,

$$\|Ae^{(t-\tau)A}[f(\tau) - f(t)]\| \le Ce^{\alpha t} \tau^{-\mu} (t-\tau)^{-2+\beta+\gamma} \|f\|_{C_\mu^\gamma((0,t],E)},$$
$$\|e^{hA}[f(t) - f(t-h)]\| \le Ce^{\alpha t} (t-h)^{-\mu} h^{-1+\beta+\gamma} \|f\|_{C_\mu^\gamma((0,t],E)}.$$

Since $-2 + \beta + \gamma > -1$ and $-1 + \beta + \gamma > 0$ then in (2.10), passing to the limit when $h \to 0$, we obtain that when $t > 0$

$$\lim_{h \to 0} g'_h(t) = \int_0^t Ae^{(t-\tau)A}[f(\tau) - f(t)] \, d\tau + e^{tA} f(t).$$

So, for $t > 0$ the function $g(t)$ is continuously differentiable and

$$g'(t) = \int_0^t Ae^{(t-\tau)A}[f(\tau) - f(t)] \, d\tau + e^{tA} f(t).$$

Hence, under $t > 0$ the function $u(t)$ is continuously differentiable and the following formula

$$u'(t) = Ae^{tA} u_0 + \int_0^t Ae^{(t-\tau)A}[f(\tau) - f(t)] \, d\tau + e^{tA} f(t) \tag{2.11}$$

holds.

Suppose that the operator A in E is invertible. Then

$$u(t) = e^{tA}u_0 + \int_0^t e^{(t-\tau)A} f(\tau) \, d\tau$$

$$= e^{tA}u_0 + \int_0^t e^{(t-\tau)A}[f(\tau) - f(t)] \, d\tau$$

$$+ \int_0^t e^{(t-\tau)A} f(t) \, d\tau = e^{tA}u_0$$

$$+ \int_0^t e^{(t-\tau)A}[f(\tau) - f(t)] \, d\tau$$

$$+ (-A^{-1}e^{(t-\tau)A} f(t))|_0^t = e^{tA}u_0$$

$$+ \int_0^t e^{(t-\tau)A}[f(\tau) - f(t)] \, d\tau - A^{-1}f(t) + A^{-1}e^{tA} f(t).$$

Hence, under $t > 0$ we have

$$Au(t) = Ae^{tA}u_0 + \int_0^t Ae^{(t-\tau)A}[f(\tau) - f(t)] \, d\tau - f(t) + e^{tA} f(t). \qquad (2.12)$$

From (2.11) and (2.12) it follows that $u(t)$ is the required solution to problem (2.5). Let us show that estimates (2.7) hold. Obviously,

$$\|u(t)\| \le \|e^{tA}A^{-1}\| \, \|Au_0\| + \int_0^t \|e^{(t-\tau)A} f(\tau)\| \, d\tau$$

$$\le C(\|Au_0\| + \int_0^t \tau^{-\mu}(t - \tau)^{-1+\beta} \, d\tau \|f\|_{C_\mu((0,t],E)})$$

$$\le C(\|Au_0\| + t^{-\mu+\beta} \int_0^1 s^{-\mu}(1 - s)^{-1+\beta} \, ds \, \|f\|_{C_\mu((0,t],E)})$$

$$\le C(\|Au_0\| + \|f\|_{C_\mu((0,t],E)}).$$

On the other hand, from (2.11) and (2.12) we have

$$\|u'(t)\| + \|Au(t)\| \le C(\|e^{tA}\| \, \|Au_0\| + \int_0^t \|Ae^{(t-\tau)A}[f(\tau) - f(t)]\| \, d\tau$$

$$+ \|f(t)\| + \|e^{tA}\| \, \|f(t)\|) \le C(t^{-1+\beta}\|Au_0\|$$

$$+ \int_0^t \tau^{-\mu}(t - \tau)^{-2+\beta+\gamma} \, d\tau \, \|f\|_{C_\mu^\gamma((0,t],E)}$$

$$+ t^{-1-\mu+\beta}\|f\|_{C_\mu((0,t],E)}) \le C(t^{-1+\beta}\|Au_0\|$$

$$+ t^{-1-\mu+\beta+\gamma}\|f\|_{C_\mu^\gamma((0,t],E)} + t^{-1-\mu+\beta}\|f\|_{C_\mu((0,t],E)})$$

$$\le Ct^{-1}(\|Au_0\| + \|f\|_{C_\mu^\gamma((0,t],E)}).$$

If the operator A is not invertible, then by substituting $u(t) = e^{st}v(t)$, where $s \in \rho(A)$, we have the problem

$$v'(t) = (A - sI)v(t) + e^{-st}f(t), \qquad v(0) = u_0',$$

with the invertible operator $A - sI$.

From condition 1, by virtue of Theorem 2.4, follows the uniqueness of the solution to problem (2.5) in the class $C([0, T], E) \cap C^{-1}((0, T], E(A), E)$. ■

Remark 2.6. Under $\beta = 1$ conditions 1 and 3 of Theorem 2.5 can be weakened, i.e., one can suppose that

1′. $\|R(\lambda, A)\| \leq M|\lambda|^{-1}$, $|\arg \lambda| \leq \pi/2$, $|\lambda| \to \infty$;

3′. $u_0 \in \overline{D(A)}$.

In this case estimates (2.7) for $t \in (0, T]$ have the following form

$$\|u(t)\| \leq C(\|u_0\| + \|f\|_{C_\mu((0,t],E)}),$$
$$\|u'(t)\| + \|Au(t)\| \leq Ct^{-1}(\|Au_0\| + \|u_0\| + \|f\|_{C_\mu^\gamma((0,t],E)}),$$

where $\gamma \in (0, 1]$, $\mu \in [0, 1)$.

Proof. To prove this, it is enough to apply Remark 2.3 and property 3 of Theorem 2.2 when $\beta = 1$. ■

Consider in a Banach space E the Cauchy problem for a nonhomogeneous equation with a perturbed constant operator

$$u'(t) = Au(t) + Bu(t) + f(t), \qquad (2.13)$$
$$u(0) = u_0.$$

Theorem 2.7. Let the following conditions be satisfied:

(1) for the operator A in E for some $\alpha > 0$ and $\beta \in (0, 1]$

$$\|R(\lambda, A)\| \leq C|\lambda|^{-\beta}, \quad |\arg \lambda| \leq \pi/2 + \alpha, \ |\lambda| \to \infty;$$

(2) B is an operator in E, $D(B) \supset D(A)$ and for any $\varepsilon > 0$

$$\|Bu\| \leq \varepsilon\|Au\|^\beta\|u\|^{1-\beta} + C(\varepsilon)\|u\|, \quad u \in D(A);$$

(3) $f \in C_\mu^\gamma((0, T], E)$ for some $\gamma \in (1 - \beta, 1]$, $\mu \in [0, \beta)$;

(4) $u_0 \in D(A)$.

Then problem (2.13) has a unique solution in $C([0,T],E) \cap C^1((0,T],E(A),E)$ and for $t \in (0,T]$ estimates (2.7) are valid.

Proof. By virtue of Lemma 2.2.8 the operator $A + B$ satisfies condition 1 of Theorem 2.5. Further, it is necessary to apply Theorem 2.5 to problem (2.13). ∎

Remark 2.8. Under $\beta = 1$ conditions 1 and 4 of Theorem 2.7 can be substituted by conditions $1'$ and $3'$ of Remark 2.6.

5.2.3. Nonhomogeneous equations with a variable operator in a Hilbert space.

Items 2.1 and 2.2 investigate the Cauchy problem for parabolic differential operator equations of the first order with a constant operator in the class of classical solutions. Let us show that in a Hilbert space one can prove not only the well-posed solvability of such problems, but also the coercive solvability.

Consider in a Hilbert space H the Cauchy problem

$$u'(t) = Au(t) + Bu(t) + f(t),$$

$$u(0) = u_0.$$

Theorem 2.9. Let the following conditions be satisfied:

(1) the operator A in H has a dense domain of definition and

$$\|R(\lambda, A)\| \le M|\lambda|^{-1}, \quad |\arg \lambda| \le \pi/2, \ |\lambda| \to \infty;$$

(2) B is an operator in H, $D(B) \supset D(A)$ and for any $\varepsilon > 0$

$$\|Bu\| \le \varepsilon \|Au\| + C(\varepsilon)\|u\|, \quad u \in D(A).$$

Then the operator $\mathbb{L} : u \to (u'(t) - Au(t) - Bu(t), u(0))$ from $W_p^1((0,T),H(A),H)$ onto $L_p((0,T),H) \dotplus (H,H(A))_{1-1/p,p}$, when $p \in (1,\infty)$, is an isomorphism.

Proof. Using the Banach theorem it is enough to prove that $\mathbb{L}$ is an algebraic isomorphism. By virtue of the definition $W_p^1((0,T),H(A),H)$, from $u \in W_p^1((0,T),H(A),H)$ follows $u'(\cdot) - Au(\cdot) - Bu(\cdot) \in L_p((0,T),H)$, and by virtue of the traces theorem [72, p.44], from $u \in W_p^1((0,T),H(A),H)$ follows $u(0) \in (H,H(A))_{1-1/p,p}$. By virtue of Lemma 2.2.8 the estimate

$$\|R(\lambda, A + B)\| \le C|\lambda|^{-1}, \quad |\arg \lambda| \le \pi/2, \ |\lambda| \to \infty$$

holds.

Then by virtue of Theorem 2.4 the mapping $\mathbb{L}$ is injective. Let us prove that it is surjective, i.e., for any pair of $(f, u_0) \in L_p((0, T), H) \dot{+} (H, H(A))_{1-1/p,p}$ there exists a solution to the Cauchy problem

$$u'(t) - Au(t) - Bu(t) = f(t), \qquad u(0) = u_0, \tag{2.14}$$

that belongs to $W_p^1((0, T), H(A), H)$. When $s \in \mathbb{C}$ a solution to the problem

$$u'(t) - (A + B - sI)u(t) = f(t), \qquad u(0) = u_0$$

can be represented in the form of a sum $u(t) = u_1(t) + u_2(t)$, where $u_1(t)$ is the restriction on $[0, T]$ of the solution to the equation

$$u_1'(t) - (A + B - sI)u_1(t) = \widehat{f}(t), \quad t \in \mathbb{R}, \tag{2.15}$$

where $\widehat{f}(t) = f(t)$ for $t \in [0, T]$ and $\widehat{f}(t) = 0$ for $t \notin [0, T]$, and $u_2(t)$ is a solution to the problem

$$u_2'(t) - (A + B - sI)u_2(t) = 0, \qquad u_2(0) = u_0 - u_1(0). \tag{2.16}$$

Form conditions 1–2, by virtue of Lemma 2.2.8, it follows that for large enough s

$$\|R(\lambda, A + B - sI)\| \le C|\lambda|^{-1}, \quad |\arg \lambda| \le \pi/2. \tag{2.17}$$

Then, by virtue of Theorem 1.6, problem (2.15) has a unique solution $u_1 \in W_p^1(\mathbb{R}, H(A), H)$. From (2.17), by virtue of Remark 2.3 and Theorem 2.2, it follows that there exists a semigroup $e^{t(A+B-sI)}$ with properties 1–5 of Theorem 2.2 under $\beta = 1$. On the other hand, by virtue of the traces theorem [72, p.44] we have $u_1(0) \in (H, H(A))_{1-1/p,p}$. Now it is easy to see that the function

$$u_2(t) = e^{t(A+B-sI)}(u_0 - u_1(0))$$

belongs to the space $W_p^1((0, T), H(A), H)$ and is a solution to the Cauchy problem (2.16). Indeed, by virtue of [72, p.96]

$$\|u_2\|_{W_p^1((0,T),H(A+B-sI),H)}^p \le C \int_0^1 \|(A + B - sI)e^{t(A+B-sI)}(u_0 - u_1(0))\|^p \, dt$$

$$\le C\|u_0 - u_1(0)\|_{(H,H(-A-B+sI))_{1-1/p,p}}^p.$$

Now it is enough to consider $H(-A - B + sI) = H(A)$. Using the substitution $u(t) = e^{-st}v(t)$ we also establish the existence of the solution to problem (2.14), that belongs to $W_p^1((0, T), H(A), H)$. ∎

Now consider in a Hilbert space H the Cauchy problem for an equation with variable operators

$$u'(t) = A(t)u(t) + B(t)u(t) + f(t),$$
$$u(0) = u_0. \tag{2.18}$$

Theorem 2.10. *Let the following conditions be satisfied:*

(1) *for* $t \in [0, T]$ *the operator* $A(t)$ *in* H *has a dense domain of definition,* $D(A(t)) = D(A(0))$, *and*

$$\|R(\lambda, A(t))\| \le M|\lambda|^{-1}, \quad |\arg \lambda| \le \pi/2, \ |\lambda| \to \infty;$$

(2) *for* $t \in [0, T]$ $B(t)$ *is an operator in* H, $D(B(t)) \supset D(A(t))$ *and for any* $\varepsilon > 0$

$$\|B(t)u\| \le \varepsilon\|A(t)u\| + C(\varepsilon)\|u\|, \quad u \in D(A(t)).$$

(3) *functions* $t \to A(t) : [0, T] \to B(H(A), H)^{16}$ *and* $t \to B(T) : [0, T] \to B(H(A), H)$ *are continuous.*

Then the operator

$$\mathbb{L} : u \to (u'(t) - A(t)u(t) - B(t)u(t), u(0))$$

from $W_p^1((0, T), H(A), H)$ *onto* $L_p((0, T), H) \dot{+} (H, H(A))_{1-1/p,p}$ *is an isomorphism,* $p \in (1, \infty)$.

Proof. Without loss of generality one can assume that for $t \in [0, T]$ the operator $A(t)$ in H is invertible. Otherwise, instead of the operators $A(t)$ and $B(t)$ it is necessary to consider the operators $A_1(t) = A(t) - \lambda_0 I$ and $B_1(t) = B(t) + \lambda_0 I$ with some large enough λ. By virtue of the traces theorem [72, p.44] the operator $\mathbb{L}$ is bounded. Let us show that for $f \in L_p((0, T), H)$, $u_0 \in (H, H(A))_{1-1/p,p}$ the problem (2.18) has a unique solution from $W_p^1((0, T), H(A), H)$.

Let $\tau \in [0, T]$. Consider the problem

$$u'(t) = A(t)u(t) + B(t)u(t) + f(t), \qquad u(\tau) = \psi, \tag{2.19}$$

which can be rewritten in the form

$$u'(t) = [A(\tau) + B(\tau)]u(t) + [A(t) - A(\tau)]u(t)$$
$$+ [B(t) - B(\tau)]u(t) + f(t), \tag{2.20}$$
$$u(\tau) = \psi.$$

---

[16] $H(A) = H(A(0))$. Obviously, $H(A(0)) = H(A(t))$.

The following notations will be used:

$$E = W_p^1((\tau, \tau + h), H(A), H), \qquad F = L_p((\tau, \tau + h, H) \dotplus (H, H(A))_{1-1/p,p},$$
$$\mathbb{L}_1 u = (u'(t) - [A(\tau) + B(\tau)]u(t), u(\tau)),$$
$$\mathbb{L}_2 u = (-[A(t) - A(\tau)]u(t) - [B(t) - B(\tau)]u(t), 0).$$

From Theorem 2.9 it follows that the operator $u \to \mathbb{L}_1 u : E \to F$ is invertible. Let us show that, for small enough $h > 0$ and independent of τ, the estimate

$$\|\mathbb{L}_2 u\|_F \leq \beta \|\mathbb{L}_1 u\|_F, \qquad \beta < 1, \ u \in E$$

is valid.

Indeed,

$$\|L_2 u\|_F^p = \| - [A(\cdot) - A(\tau)]u(\cdot) - [B(\cdot) - B(\tau)]u(\cdot)\|_{L_p((\tau, \tau+h, H)}^p$$
$$\leq C \int\limits_\tau^{\tau+h} (\|[A(t) - A(\tau)]A^{-1}(\tau)\|^p + \|[B(t) - B(\tau)]A^{-1}(\tau)\|^p)$$
$$\times \|A(\tau)u(t)\|^p \ dt \leq C \max_{\tau \leq t \leq \tau+h} (\|[A(t) - A(\tau)]A^{-1}(\tau)\|^p$$
$$+ \|[B(t) - B(\tau)]A^{-1}(\tau)\|^p) \|u\|_{L_p((\tau, \tau+h), H(A))}^p$$
$$\leq C\varepsilon \|u\|_{W_p^1((\tau, \tau+h), H(A), H)}^p \leq C\varepsilon \|\mathbb{L}_1 u\|_F^p.$$

Let us now choose $\varepsilon > 0$ such that $C\varepsilon < 1$. Hence it is shown that one can choose $h > 0$ so small that problem (2.20), and therefore problem (2.19) is coercive solvable. In addition for the solution to problem (2.19) the uniform estimate in $\tau \in [0, T]$

$$\|u\|_{W_p^1((\tau, \tau+h), H(A), H)} \leq C(\|\psi\|_{(H, H(A))_{1-1/p,p}} + \|f\|_{L_p((\tau, \tau+h), H)}),$$

is valid, where C does not depend on h and τ. By virtue of the traces theorem [72, p.44] we have $u(t) \in (H, H(A))_{1-1/p,p}$. Then the method of step continuation is applied and it can be established that for any pair $(f, u_0) \in L_p((0, T), H) \dotplus (H, H(A))_{1-1/p,p}$ there exists a unique solution to problem (2.18) that belongs to $W_p^1((0, T), H(A), H)$. ∎

5.3. Boundary value problems for elliptic differential-operator equations

The completeness of elementary solutions for some classes of boundary value problems was proved by S. Agmon and L. Nirenberg [2], however they did not address the completeness of elementary solutions of boundary value problems for the biharmonic equation. This question was addressed by Y. A. Ustinov and V. I. Yudovich [74] (for the multidimensional case) and by D. D. Joseph [25,26], and G. Geymonat and P. Grisvard [18] (for the 2-dimensional case). The best possible results were obtained in [18] and [25]. In these papers the Fourier series expansion for the solution to the problem was found. in this paragraph we will prove the completeness of elementary solutions for a more general class of boundary value problems for elliptic equations.

5.3.1. Fredholm property of a boundary value problem for elliptic differential-operator equations on [0,1]. It is known [39, p.114] that if A is an operator in a Hilbert space H with a dense domain of definition $D(A)$ and

$$\|(A+sI)^{-1}\| \le C(1+s)^{-1}, \qquad s \ge 0,$$

then there exist fractional powers $A^\alpha, \alpha \in \mathbb{R}$. We turn $D(A^\alpha)$ into a Hilbert space $H(A^\alpha)$ with respect to the norm

$$\|u\|^2_{H(A^\alpha)} = \|A^\alpha u\|^2 + \|u\|^2, \qquad u \in D(A^\alpha).$$

Let us introduce the space $W_p^2((0,1), H(A), H)$ of functions with the norm

$$\|u\|_{W_p^2((0,1),H(A),H)} = \|Au\|_{L_p((0,1),H)} + \|u''\|_{L_p((0,1),H)}.$$

Let H, H_4 be Hilbert spaces with continuous embedding $H_4 \subset H_0 = H$. Consider the space $W_p^4((0,1), H_4, H)$ of functions with the norm

$$\|u\|_{W_p^4((0,1),H_4,H)} = \|u\|_{L_p((0,1),H_4)} + \|u''''\|_{L_p((0,1),H)}.$$

Consider in the Hilbert space H a boundary value problem in $[0,1]$ for the 4-th order elliptic equation

$$L(D)u = u''''(t) + A_2 u''(t) + A_4 u(t) + \sum_{k \le 3} B_k(t) u^{(k)}(t)$$

$$= f(t), \qquad t \in [0,1], \tag{3.1}$$

$$L_k u = \alpha_k u^{(m_k)}(0) + \beta_k u^{(m_k)}(1) + \sum_{j \le m_k - 1} \sum_{s \le N_{kj}} T_{kj} u^{(j)}(t_{kjs})$$

$$= \varphi_k, \qquad k = 1, \dots, 4, \tag{3.2}$$

where $0 \le m_1, m_2 \le 1$, $m_3 = m_1 + 2$, $m_4 = m_2 + 2$, and α_k and β_k are complex numbers, $t_{kjs} \in [0,1]$.

Theorem 3.1. *Let the following conditions be satisfied:*

(1) *there exist Hilbert spaces* H, H_4 *for which the compact embedding* $H_4 \subset H_0 = H$ *takes place and* $\overline{H_4}|_H = H$;

(2) *operators* A_k, $k = 2, 4$ *from* H_k *into* H *are bounded, where* $H_2 = (H, H_4)_{1/2,2}$;

(3) *the characteristic operator pencil* $L_0(\lambda) = \lambda^4 I + \lambda^2 A_2 + A_4$ *is coercive for* $\mathrm{Re}\lambda = 0$, $|\lambda| \to \infty$, *i.e.,*

$$|\lambda|^2 \|L_0^{-1}(\lambda)f\|_{H_2} + \|L_0^{-1}(\lambda)f\|_{H_4} \le C\|f\|_H, \quad f \in H,$$
$$\mathrm{Re}\lambda = 0, \ |\lambda| \to \infty;$$

$$|\lambda|^4 \|L_0^{-1}(\lambda)f\|_{H_2} + |\lambda|^2 \|L_0^{-1}(\lambda)f\|_{H_4} \le C\|f\|_{H_2}, \quad f \in H_2,$$
$$\mathrm{Re}\lambda = 0, \ |\lambda| \to \infty;$$

(4)
$$\begin{vmatrix} \alpha_1(-1)^{m_1} & 0 & \beta_1 & 0 \\ 0 & \alpha_3(-1)^{m_1} & 0 & \beta_3 \\ \alpha_2(-1)^{m_2} & 0 & \beta_2 & 0 \\ 0 & \alpha_4(-1)^{m_2} & 0 & \beta_4 \end{vmatrix} \ne 0;$$

(5) *operators* $B_k(t)$ *from* $(H, H_4)_{1-k/4,2}$ *into* H *act compactly for almost all* $t \in [0,1]$; *for any* $\varepsilon > 0$ *and for almost all* $t \in [0,1]$

$$\|B_k(t)u\|_H \le \varepsilon\|u\|_{(H,H_4)_{1-k/4,2}} + C(\varepsilon)\|u\|_H, \qquad u \in H;$$

functions $B_k(t)u$, *for* $u \in (H, H_4)_{1-k/4,2}$, *are measurable on* $[0,1]$ *in* H;

(6) *operators* T_{kj} *from*

$$(H, H_4)_{(4p-jp-1)/4p,p} \quad \text{into} \quad (H, H_4)_{(4p-m_kp-1)/4p,p}$$

are compact.

Then the operator

$$L : u \to (L(D)u, L_1u, L_2u, L_3u, L_4u)$$

from $W_p^4((0,1), H_4, H)^{17}$ *into* $L_p((0,1), H) \overset{4}{\underset{k=1}{\dotplus}} (H, H_4)_{(4p-m_kp-1)/4p,p}$ *is fredholm.*

Proof. Consider the problem

$$L_0(D)u = u''''(t) + A_2u''(t) + A_4u(t) = f(t), \qquad t \in [0,1], \tag{3.3}$$
$$L_{k0}u = \alpha_k u^{(m_k)}(0) + \beta_k u^{(m_k)}(1) = \varphi_k, \qquad k = 1, \dots, 4. \tag{3.4}$$

---

[17]By virtue of Theorem 1.8 the embedding $W_p^4((0,1), H_4, H) \subset W_p^2((0,1), (H, H_4)_{1/2,2})$ is continuous. Since $(H, H_4)_{1/2,2} = H_2$ then $Au'' \in L_p((0,1), H)$.

By the substitution

$$v(t) = \begin{pmatrix} v_1(t) \\ v_2(t) \end{pmatrix} = \begin{pmatrix} u(t) \\ u''(t) \end{pmatrix}$$

the problem (3.3)–(3.4) is reduced to the equivalent problem

$$v''(t) = Gv(t) + \Phi(t), \qquad t \in [0, 1],$$
$$a_k v^{(m_k)}(0) + b_k v^{(m_k)}(1) = \Phi_k, \qquad k = 1, 2,$$

(3.5)

where,

$$G = \begin{pmatrix} 0 & I \\ -A_4 & -A_2 \end{pmatrix}, \quad a_k = \begin{pmatrix} \alpha_k I & 0 \\ 0 & \alpha_{k+2}I \end{pmatrix}, \quad b_k = \begin{pmatrix} \beta_k I & 0 \\ 0 & \beta_{k+2}I \end{pmatrix},$$

$$\Phi(t) = \begin{pmatrix} 0 \\ f(t) \end{pmatrix}, \quad \Phi_k = \begin{pmatrix} \varphi_k \\ \varphi_{k+2} \end{pmatrix}.$$

We consider the operator G in the space $\mathcal{H} = H_2 \oplus H$. Let $D(G) = H_4 \oplus H_2$ and $F = (f_1, f_2) \in \mathcal{H} = H_2 \oplus H$. From the first equation of the system

$$(\lambda^2 I - G)v = F$$

(3.6)

we find

$$v_2 = \lambda^2 v_1 - f_1.$$

Substituting this expression into the second equation of system (3.6) we have

$$\lambda^2(\lambda^2 v_1 - f_1) = -A_4 v_1 - A_2(\lambda^2 v_1 - f_1) + f_2.$$

Hence,

$$L_0(\lambda)v_1 = \lambda^2 f_1 + A_2 f_1 + f_2,$$

i.e.,

$$v_1 = \lambda^2 L_0^{-1}(\lambda)f_1 + L_0^{-1}(\lambda)A_2 f_1 + L_0^{-1}(\lambda)f_2.$$

(3.7)

Consequently

$$v_2 = \lambda^4 L_0^{-1}(\lambda)f_1 + \lambda^2 L_0^{-1}(\lambda)A_2 f_1 - f_1 + \lambda^2 L_0^{-1}(\lambda)f_2.$$

(3.8)

Then from (3.7), (3.8) and condition 3 we have

$$\|G(\lambda^2 I - G)^{-1}F\|_{\mathcal{H}} \le C(\|v_2\|_{H_2} + \|A_4 v_1 + A_2 v_2\|_H)$$
$$\le C(|\lambda|^4 \|L_0^{-1}(\lambda)f_1\|_{H_2} + |\lambda|^2 \|L_0^{-1}(\lambda)A_2 f_1\|_{H_2}$$
$$+ |\lambda|^2 \|L_0^{-1}(\lambda)f_2\|_{H_2} + |\lambda|^2 \|L_0^{-1}(\lambda)f_1\|_{H_4}$$
$$+ \|L_0^{-1}(\lambda)A_2 f_1\|_{H_4} + \|L_0^{-1}(\lambda)f_2\|_{H_4} + \|f_1\|_{H_2})$$
$$\le C(\|f_1\|_{H_2} + \|A_2 f_1\|_H + \|f_2\|_H)$$
$$\le C\|F\|_{\mathcal{H}}, \qquad \text{Re}\lambda = 0, \ |\lambda| \to \infty.$$

From this and from the identity

$$\lambda^2(\lambda^2 I - G)^{-1} = G(\lambda^2 I - G)^{-1} + I$$

we have

$$\|(\lambda^2 I - G)^{-1}\| \leq C(1 + |\lambda|^2)^{-1}, \qquad \operatorname{Re}\lambda = 0, \ |\lambda| \to \infty.$$

So for sufficiently large $s_0 > 0$ one can show that

$$\|[(G + s_0 I) + \mu I]^{-1}\| \leq C(1 + |\mu|)^{-1}, \ \mu \geq 0.$$

Hence, by virtue of [39, p.119] there exist operators $G_0^{1/2}$, $e^{-tG_0^{1/2}}$ and

$$\|(G_0^{1/2} + \lambda I)^{-1}\| \leq C(1 + |\lambda|)^{-1}, \qquad \operatorname{Re}\lambda \geq 0, \tag{3.9}$$

$$\|e^{-tG_0^{1/2}}\| \leq Ce^{-\omega t}, \qquad t \geq 0, \tag{3.10}$$

where $G_0 = G + s_0 I$. Let us prove that for any $F \in L_p((0,1), \mathcal{H})$ and any $\Phi_k \in (\mathcal{H}, \mathcal{H}(G))_{(2p - m_k p - 1)/2p, p}$ the problem

$$\begin{aligned}
v''(t) &= (G_0^{1/2} + sI)^2 v(t) + \Phi(t), &\qquad t \in [0,1], \\
a_k v^{(m_k)}(0) + b_k v^{(m_k)}(1) &= \Phi_k, &\qquad k = 1, 2,
\end{aligned} \tag{3.11}$$

when $s \to \infty$, has a unique solution that belongs to the space

$$W_p^2((0,1), \mathcal{H}(G), \mathcal{H}).$$

Let us now show that a solution to the problem (3.11) is represented in the form of the sum $v(t) = v_1(t) + v_2(t)$, where $v_1(t)$ is the restriction on [0,1] of the solution $\widehat{v}_1(t)$ to the equation

$$\widehat{v}_1''(t) - (G_0^{1/2} + sI)^2 \widehat{v}_1(t) = \widehat{\Phi}(t), \qquad t \in \mathbb{R}, \tag{3.12}$$

where $\widehat{\Phi}(t) = \Phi(t)$ if $t \in [0,1]$, and $\widehat{\Phi}(t) = 0$ if $t \notin [0,1]$, and $v_2(t)$ is the solution to the problem

$$\begin{aligned}
v_2''(t) &= (G_0^{1/2} + sI)^2 v_2(t), &\qquad t \in [0,1], &\tag{3.13} \\
a_k v_2^{(m_k)}(0) + b_k v_2^{(m_k)}(1) &= \Phi_k, &\qquad k = 1, 2. &\tag{3.14}
\end{aligned}$$

Consider the characteristic operator pencil of equation (3.12), i.e.,

$$\mathbb{L}(\lambda) = \lambda^2 I - (G_0^{1/2} + sI)^2.$$

From (3.9) for $s \geq 0$ and $\mu \in \mathbb{R}$ we have

$$\|L^{-1}(i\mu)\| = \|[(G_0^{1/2} + sI)^2 + \mu^2 I]^{-1}\|$$

$$\leq \|(G_0^{1/2} + sI - i\mu I)^{-1}\| \|(G_0^{1/2} + sI + i\mu I)^{-1}\| \leq C|\mu|^{-2},$$

and

$$\|G_0 L^{-1}(i\mu)\| \leq \|G_0^{1/2}(G_0^{1/2} + sI - i\mu I)^{-1}\|$$

$$\times \|G_0^{1/2}(G_0^{1/2} + sI + i\mu I)^{-1}\| \leq C.$$

Let us now apply Theorem 1.3 to equation (3.12). Set $\mathcal{H}_2 = \mathcal{H}(G)$, $\mathcal{H}_1 = (\mathcal{H}, \mathcal{H}_2)_{1/2,2}$, $A_1 = 0$, and $A_2 = (G_0^{1/2} + sI)^2$. By virtue of (1.2.7) and the Young inequality (1.2.13), for $F \in \mathcal{H}$ we have

$$|\mu| \|L^{-1}(i\mu)F\|_{\mathcal{H}_1} \leq C(|\mu|^2 \|L^{-1}(i\mu)F\|_{\mathcal{H}})^{1/2} \|L^{-1}(i\mu)F\|_{\mathcal{H}_2}^{1/2}$$

$$\leq C(|\mu|^2 \|L^{-1}(i\mu)F\|_{\mathcal{H}} + \|G_0 L^{-1}(i\mu)F\|_{\mathcal{H}}) \leq C\|F\|_{\mathcal{H}}.$$

So, for $\gamma = 0$ conditions 1–3 of Theorem 1.3 are satisfied. Hence, equation (3.12), by virtue of Theorem 1.3, has a solution $\hat{v}_1 \in W_p^2(\mathbb{R}, \mathcal{H}_2, \mathcal{H}_1, \mathcal{H})$. Then $v_1 \in W_p^2((0,1), \mathcal{H}_2, \mathcal{H})$.

Now, let us prove that when $s \to \infty$ and when

$$\Phi_k \in (\mathcal{H}, \mathcal{H}(G))_{(2p-m_k-1)/2p,p}$$

then the problem (3.13)–(3.14) has a unique solution $v_2(t)$ that belongs to $W_p^2((0,1), \mathcal{H}(G), \mathcal{H})$. We now show that an arbitrary solution to (3.13) belonging to $W_p^2((0,1), \mathcal{H}(G), \mathcal{H})$ has the form

$$v_2(t) = e^{-t(G_0^{1/2} + sI)} g_1 + e^{-(1-t)(G_0^{1/2} + sI)} g_2, \tag{3.15}$$

where $g_k \in (\mathcal{H}, \mathcal{H}(G))_{1-1/2p,p}$. From (3.13) we have

$$(D - G_0^{1/2} - sI)(D + G_0^{1/2} + sI)v_2(t) = 0,$$

where $D = D_t = \frac{\partial}{\partial t}$. Denote

$$w(t) = (D + G_0^{1/2} + sI)v_2(t). \tag{3.16}$$

Then $w \in W_p^1((0,1), \mathcal{H}(G_0^{1/2}), \mathcal{H})$ and $(D - G_0^{1/2} - sI)w(t) = 0$. Hence,

$$w(t) = e^{-(1-t)(G_0^{1/2}+sI)}w(1), \tag{3.17}$$

where, by virtue of [72, p.44],

$$w(1) \in (\mathcal{H}(G_0^{1/2}), H)_{1/p,p} = (\mathcal{H}, \mathcal{H}(G_0^{1/2}))_{1-1/p,p}.$$

From (3.16) and (3.17) we have

$$\begin{aligned}
v_2(t) &= e^{-t(G_0^{1/2}+sI)}v_2(0) \\
&\quad + \int_0^t e^{-(t-\tau)(G_0^{1/2}+sI)}e^{-(1-\tau)(G_0^{1/2}+sI)}w(1)\,d\tau \\
&= e^{-t(G_0^{1/2}+sI)}v_2(0) \\
&\quad + 1/2(G_0^{1/2}+sI)^{-1}\{e^{-(1-t)(G_0^{1/2}+sI)} \\
&\quad - e^{-t(G_0^{1/2}+sI)}e^{-(G_0^{1/2}+sI)}\}w(1), \tag{3.18}
\end{aligned}$$

where, by virtue of [72, p.44], $v_2(0) \in (\mathcal{H}(G), \mathcal{H})_{1/2p,p}$. By virtue of [72, p.101] the operator $G_0^{1/2}$ from $(\mathcal{H}, \mathcal{H}(G))_{1-1/2p,p}$ onto

$$(\mathcal{H}, \mathcal{H}(G_0^{1/2}))_{1-1/p,p} = (\mathcal{H}, \mathcal{H}(G))_{(p-1)/2p,p}$$

is an isomorphism. Hence (3.18) has the form (3.15).

The function (3.15) satisfies condition (3.14) if

$$\begin{aligned}
(G_0^{1/2} + sI)^{m_k}\{a_k(-1)^{m_k} g_1 + a_k e^{-(G_0^{1/2}+sI)}g_2 \\
+ b_k(-1)^{m_k}e^{-(G_0^{1/2}+sI)}g_1 + b_k g_2\} = \Phi_k, \qquad k = 1,2. \tag{3.19}
\end{aligned}$$

By virtue of condition 4 the system (3.19) has a unique solution when $s \to \infty$ and

$$g_k = C_1^k(G_0^{1/2} + sI)^{-m_1}\Phi_1 + C_2^k(G_0^{1/2} + sI)^{-m_2}\Phi_2, \qquad k = 1,2,$$

where C_j^k are bounded operators commutative with G. Since

$$\Phi_k \in (\mathcal{H}, \mathcal{H}(G))_{(2p-m_kp-1)/2p,p}$$

then $g_k \in (\mathcal{H}, \mathcal{H}(G))_{1-1/2p,p}$. Let us show that the operator that corresponds to problem (3.5), i.e., the operator

$$\mathbb{L}_0 : v \to ((D^2 - G)v(t), a_1 v^{(m_1)}(0) + b_1 v^{(m_1)}(1), a_2 v^{(m_2)}(0) + b_2 v^{(m_2)}(1))$$

from $W_p^2((0,1), \mathcal{H}(G), \mathcal{H})$ into

$$L_p((0,1), \mathcal{H}) \dot{+} (\mathcal{H}, \mathcal{H}(G))_{(2p-m_1p-1)/2p,p} \dot{+} (\mathcal{H}, \mathcal{H}(G))_{(2p-m_2p-1)/2p,p}$$

is fredholm. Obviously

$$\mathbb{L}_0 = \mathbb{L}_s + \mathbb{K}_s, . \tag{3.20}$$

where

$$\mathbb{L}_s : v \to ([D^2 - (G_0^{1/2} + sI)^2]v(t), \ a_k v^{(m_k)}(0) + b_k v^{(m_k)}(1), \ k = 1, 2),$$

$$\mathbb{K}_s v = ((2sG_0^{1/2} + s_0 I + s^2 I)v(t), 0, 0).$$

We have, that the operator $\mathbb{L}_s$ from $W_p^2((0,1), \mathcal{H}(G), \mathcal{H})$ into

$$L_p((0,1), \mathcal{H}) \dot{+} (\mathcal{H}, \mathcal{H}(G))_{(2p-m_1p-1)/2p,p} \dot{+} (\mathcal{H}, \mathcal{H}(G))_{(2p-m_2p-1)/2p,p}$$

is an isomorphism as $s \to \infty$.

From (3.7)–(3.8) it follows that

$$(\lambda^2 I - G)^{-1} = \begin{pmatrix} \lambda^2 L^{-1}(\lambda) + L^{-1}(\lambda)A_2 & L^{-1}(\lambda) \\ \lambda^4 L^{-1}(\lambda) + \lambda^2 L^{-1}(\lambda)A_2 - I & \lambda^2 L^{-1}(\lambda) \end{pmatrix}.$$

By virtue of conditions 1 and 3 the operator $(\lambda^2 I - G)^{-1}$ in $\mathcal{H} = H_2 \oplus H$ is compact. Hence, the embedding $\mathcal{H}(G) \subset \mathcal{H}$ is compact. From the momentum inequality [39, p.115]

$$\|G_0^{1/2} v\| \le C \|G_0 v\|^{1/2} \|v\|^{1/2}, \qquad v \in D(G),$$

and from the Young inequality (1.2.13) it follows that

$$\|G_0^{1/2} v\| \le \varepsilon \|G_0 v\| + C(\varepsilon) \|v\|, \qquad \varepsilon > 0.$$

Then, by virtue of Lemma 1.2, the operator $\mathbb{K}_s$ from $W_p^2((0,1), \mathcal{H}(G), \mathcal{H})$ into

$$L_p((0,1), \mathcal{H}) \dot{+} (\mathcal{H}, \mathcal{H}(G))_{(2p-m_1p-1)/2p,p} \dot{+} (\mathcal{H}, \mathcal{H}(G))_{(2p-m_2p-1)/2p,p}$$

is compact. By virtue of the perturbation theorem of fredholm operators [28, p.238] from (3.20) it follows that the operator $\mathbb{L}_0$ is fredholm.

We have $(\mathcal{H}, \mathcal{H}(G))_{q,p} = (H_2 \oplus H, \ H_4 \oplus H_2)_{q,p} = (H_2, \ H_4)_{q,p} \oplus (H, H_2)_{q,p}$. Since $H_2 = (H, H_4)_{1/2,2}$, then by virtue of [72, p.66, 105]

$$(H, H_4)_{(4p-m_kp-1)/4p,p} = (H_2, H_4)_{(2p-m_kp-1)/2p,p}, \qquad k = 1, 2.$$

Since $m_{k+2} = m_k + 2$, $k = 1, 2$, then by virtue of [72, p.66, 105]

$$(H, H_4)_{(4p-m_{k+2}p-1)/4p,p} = (H, H_2)_{(2p-m_kp-1)/2p,p}, \qquad k = 1, 2.$$

Hence, the operator that corresponds to the problem (3.3)–(3.4)

$$L_0 : u \to (L_0(D)u, L_{10}u, L_{20}u, L_{30}u, L_{40}u)$$

from $W_p^4((0,1), H_4, H)$ into $L_p((0,1), H) \overset{4}{\underset{k=1}{+}} (H, H_4)_{(4p-m_kp-1)/4p,p}$ is fredholm. It is enough now to note that the operator L has the form

$$L = L_0 + T, \tag{3.21}$$

where $T : u \to (\sum_{k \le 3} B_k(t)u^{(k)}(t), \sum_{j \le m_k - 1} \sum_{s \le N_{kj}} T_{kj}u^{(j)}(t_{kjs}), \quad k = 1, \dots, 4)$. Using Theorem 1.8 (more exactly Corollary 1.9) about intermediate derivatives (when $p = 2$ see [44, p.28]), Lemma 1.2 and the trace theorem [72, p.44] we prove that the operator T from $W_p^4((0,1), H_4, H)$ into

$$L_p((0,1), H) \overset{4}{\underset{k=1}{+}} (H, H_4)_{(4p-m_kp-1)/4p,p}$$

is compact. It is enough now to apply the perturbation theorem of fredholm operators [28, p.238] to operator (3.21). ■

Theorem 3.2. *Let the following conditions be satisfied:*

(1) A_4 *is a selfadjoint operator in H; $(A_4u, u) \ge c^2(u, u), c \ne 0, u \in D(A_4)$; the embedding $H(A_4) \subset H$ is compact;*

(2) A_2 *and A_2^* are operators in H; $\mathrm{Re}(A_2u, u) \le 0, u \in D(A_4^{1/2})$;*

$\mathrm{Re}(A_2^*u, u) \le 0, \quad u \in D(A_4^{1/2}); A_2A_4^{-1/2} \in B(H); A_2^*A_4^{-1/2} \in B(H)$;

(3)

$$\begin{vmatrix} \alpha_1(-1)^{m_1} & 0 & \beta_1 & 0 \\ 0 & \alpha_3(-1)^{m_1} & 0 & \beta_3 \\ \alpha_2(-1)^{m_2} & 0 & \beta_2 & 0 \\ 0 & \alpha_4(-1)^{m_2} & 0 & \beta_4 \end{vmatrix} \ne 0;$$

(4) *operators $B_k(t)$ from $H(A_4^{1-k/4})$ into H act compactly for almost all $t \in [0,1]$; for any $\varepsilon > 0$ and for almost all $t \in [0,1]$*

$$\|B_k(t)u\|_H \le \varepsilon\|u\|_{H(A_4^{1-k/4})} + C(\varepsilon)\|u\|, \quad u \in H;$$

functions $B_k(t)u$, when $u \in H(A_4^{1-k/4})$, are measurable on $[0,1]$ in H;

(5) operators T_{kj} from $(H, H(A_4))_{(4p-jp-1)/4p,p}$ into

$$(H, H(A_4))_{(4p-m_kp-1)/4p,p}$$

are compact.

Then the operator

$$L : u \to (L(D)u, L_1 u, L_2 u, L_3 u, L_4 u)$$

from $W_p^4((0,1), H(A_4), H)$ into

$$L_p((0,1), H) \overset{4}{\underset{k=1}{+}} (H, H(A_4))_{(4p-m_kp-1)/4p,p}$$

is fredholm.

Proof. Repeat the proof of Theorem 3.1, but for the operator

$$G = \begin{pmatrix} 0 & I \\ -A_4 & -A_2 \end{pmatrix}$$

in the space $\mathcal{H} = H(A_4^{1/2}) \oplus H$ the inequality

$$\|(G - \lambda I)^{-1}\| \le C(1 + |\lambda|)^{-1}, \qquad \lambda \le 0$$

must be established in the following form. For $v \in D(G)$, $\varphi \in D(G^*)$ we have

$$(Gv, \varphi)_{\mathcal{H}} = (A_4^{1/2} v_2, A_4^{1/2} \varphi_1) + (-A_4 v_1 - A_2 v_2, \varphi_2) = (v, G^* \varphi)_{\mathcal{H}}.$$

Let $\varphi_2 \in D(A_4^{1/2}) \subset D(A_2^*)$, $\varphi_1 \in D(A_4)$. Then

$$-(A_4^{1/2} v_1, A_4^{1/2} \varphi_2) + (v_2, A_4 \varphi_1) - (v_2, A_2^* \varphi_2) = (v, G^* \varphi)_{\mathcal{H}}.$$

Hence, from this it follows that $D(G^*) \supset D(A_4) \oplus D(A_4^{1/2})$ and

$$G^* = \begin{pmatrix} 0 & -I \\ A_4 & -A_2^* \end{pmatrix}.$$

Since the restriction of the operator G^* on $D(A_4) \oplus D(A_4^{1/2})$ is invertible then $D(G^*) = D(A_4) \oplus D(A_4^{1/2})$. Obviously for $v \in D(G) = D(A_4) \oplus D(A_4^{1/2})$

$$(Gv, v)_{\mathcal{H}} = (A_4^{1/2} v_2, A_4^{1/2} v_1) - (A_4 v_1 + A_2 v_2, v_2)$$

$$= 2i \mathrm{Im}(A_4 v_2, v_1) - (A_2 v_2, v_2)$$

and for $\varphi \in D(G^*) = D(A_4) \oplus D(A_4^{1/2})$

$$(G^*\varphi, \varphi)_{\mathcal{H}} = -(A_4^{1/2}\varphi_2, A_4^{1/2}\varphi_1) + (A_4\varphi_1 - A_2^*\varphi_2, \varphi_2)$$
$$= -2i Im(A_4\varphi_2, \varphi_1) - (A_2^*\varphi_2, \varphi_2).$$

Then

$$Re(Gv, v)_{\mathcal{H}} \geq 0, \ v \in D(G),$$

$$Re(G^*\varphi, \varphi)_{\mathcal{H}} \geq 0, \ \varphi \in D(G^*).$$

Consequently, by virtue of [39, p.85] and G^{-1} being bounded, the operator $G - \lambda I$ for $\lambda \leq 0$ has an inverse and we have the required estimate. ∎

5.3.2. An isomorphism generated by a boundary value problem for elliptic differential-operator equations on a semi-axis. Consider in a Hilbert space H a boundary value problem in $[0, \infty)$ for the 4-th order elliptic equation

$$L(D)u = u''''(t) + A_2 u''(t) + A_4 u(t) = f(t), \qquad t > 0, \qquad (3.22)$$

$$L_1 u = \alpha u(0) + \beta u'(0) = \varphi_1,$$
$$L_2 u = \alpha u''(0) + \beta u'''(0) = \varphi_2, \qquad (3.23)$$

where α and β are complex numbers.

Theorem 3.3. *Let the following conditions be satisfied:*

(1) *there exist Hilbert spaces H, H_4 for which the continuous embedding $H_4 \subset H_0 = H$ takes place and $\overline{H_4}|_H = H$;*

(2) *operators A_k, $k = 2, 4$ from H_k into H act boundedly, where $H_2 = (H, H_4)_{1/2,2}$;*

(3) *the characteristic operator pencil $L(\lambda) = \lambda^4 I + \lambda^2 A_2 + A_4$ is coercive when $Re\lambda = 0$, i.e.,*

$$|\lambda|^2 \|L^{-1}(\lambda)f\|_{H_2} + \|L^{-1}(\lambda)f\|_{H_4} \leq C\|f\|_H,$$
$$f \in H, \ Re\lambda = 0;$$

$$|\lambda|^4 \|L^{-1}(\lambda)f\|_{H_2} + |\lambda|^2 \|L^{-1}(\lambda)f\|_{H_4} \leq C\|f\|_{H_2},$$
$$f \in H_2, \ Re\lambda = 0;$$

(4) *at least one of two numbers α and β is not equal to zero; $Re\alpha\beta^{-1} \leq 0$ when $\beta \neq 0$;*

Then the operator

$$L : u \to (L(D)u, L_1 u, L_2 u)$$

from $W_p^4((0, \infty), H_4, H)$ into

$$L_p((0, \infty), H) \dotplus (H, H_4)_{(4p-mp-1)/4p,p} \dotplus (H, H_4)_{(2p-mp-1)/4p,p},$$

where $m = 0$ if $\beta = 0$ and $m = 1$ if $\beta \neq 0$, is an isomorphism.

Proof. By virtue of [72, p.44] and condition 2 the operator L acts linearly and continuously from $W_p^4((0, \infty), H_4, H)$ into

$$L_p((0, \infty), H) \dotplus (H, H_4)_{(4p-mp-1)/4p,p} \dotplus (H, H_4)_{(2p-mp-1)/4p,p}.$$

Let us prove that for any $f \in L_p((0, \infty), H)$ and any

$$\varphi_1 \in (H, H_4)_{(4p-mp-1)/4p,p}, \quad \varphi_2 \in (H, H_4)_{(2p-mp-1)/4p,p}$$

the problem (3.22)–(3.23) has a unique solution that belongs to

$$W_p^4((0, \infty), H_4, H).$$

Let us show that a solution to the problem (3.22)–(3.23) is represented in the form of the sum $u(t) = u_1(t) + u_2(t)$, where $u_1(t)$ is the restriction on $[0, \infty)$ of the solution $\widehat{u}_1(t)$ to the equation

$$\widehat{u}_1''''(t) + A_2 \widehat{u}_1''(t) + A_4 \widehat{u}_1(t) = \widehat{f}(t), \qquad t \in \mathbb{R}, \tag{3.24}$$

where $\widehat{f}(t) = f(t)$ if $t \in [0, \infty)$ and $\widehat{f}(t) = 0$ if $t \in (-\infty, 0)$, and $u_2(t)$ is the solution of the problem

$$u_2''''(t) + A_2 u_2''(t) + A_4 u_2(t) = 0, \qquad t > 0, \tag{3.25}$$

$$\alpha u_2(0) + \beta u_2'(0) = -L_1 u_1 + \varphi_1,$$
$$\alpha u_2''(0) + \beta u_2'''(0) = -L_2 u_1 + \varphi_2. \tag{3.26}$$

Apply Theorem 1.3 to equation (3.24). Let $H_1 = (H, H_4)_{1/4,2}$, $H_3 = (H, H_4)_{3/4,2}$, $A_1 = 0$, $A_3 = 0$. From condition 3 and the equation

$$L(\lambda)u = \lambda^4 u + \lambda^2 A_2 u + A_4 u = f$$

we have

$$|\lambda|^4 \|u\| \le \|f\| + |\lambda|^2 \|A_2 u\| + \|A_4 u\| \le C \|f\|, \quad f \in H, \ \operatorname{Re}\lambda = 0.$$

Since $H_2 = (H, H_4)_{1/2,2}$ then, by virtue of (1.2.7),

$$\|u\|_{H_k} = \|u\|_{(H,H_4)_{k/4,2}} \leq C\|u\|_H^{1-k/4}\|u\|_{H_4}^{k/4}, \qquad k = 1, 2, 3.$$

Using the Young inequality (1.2.13) we have

$$|\lambda|^{4-k}\|u\|_{H_k} \leq C(|\lambda|^4\|u\|_H)^{1-k/4}\|u\|_{H_4}^{k/4} \leq C(|\lambda|^4\|u\|_H + \|u\|_{H_4}).$$

Then conditions 1–3 of Theorem 1.3 are satisfied. Hence, equation (3.24), by virtue of Theorem 1.3 has a solution $\hat{u}_1 \in W_p^4(\mathbb{R}, H_4, \ldots, H)$. Then $u_1 \in W_p^4((0, \infty), H_4, H)$.

Let us now prove, that for any

$$\varphi_1 \in (H, H_4)_{(4p-mp-1)/4p,p}, \qquad \varphi_2 \in (H, H_4)_{(2p-mp-1)/4p,p}$$

the problem (3.25)–(3.26) has a unique solution $u_2(t)$ that belongs to $W_p^4((0, \infty), H_4, H)$. By the substitution

$$v(t) = \begin{pmatrix} v_1(t) \\ v_2(t) \end{pmatrix} = \begin{pmatrix} u_2(t) \\ u_2''(t) \end{pmatrix}$$

the problem (3.25)–(3.26) is reduced to the equivalent problem

$$v''(t) = Gv(t), \quad t > 0, \tag{3.27}$$

$$\alpha v(0) + \beta v'(0) = \Phi_0, \tag{3.28}$$

where

$$G = \begin{pmatrix} 0 & I \\ -A_4 & -A_2 \end{pmatrix}, \qquad \Phi_0 = \begin{pmatrix} -L_1 u_1 + \varphi_1 \\ -L_2 u_1 + \varphi_2 \end{pmatrix}.$$

We consider the operator G in the space $\mathcal{H} = H_2 \oplus H$. Let $D(G) = H_4 \oplus H_2$. From the proof of Theorem 3.1 the following estimate

$$\|(\lambda^2 I - G)^{-1}\| \leq C(1 + |\lambda|^2)^{-1}, \qquad \mathrm{Re}\,\lambda = 0,$$

follows. Hence, by virtue of [39, p.119] there exists an operator $e^{-tG^{1/2}}$ and for some $\omega > 0$

$$\|e^{-tG^{1/2}}\| \leq Ce^{-\omega t}, \qquad t \geq 0.$$

Let us show that an arbitrary solution (3.27) that belongs to $W_p^2((0, \infty), \mathcal{H}(G), \mathcal{H})$ has the form

$$v(t) = e^{-tG^{1/2}}g, \tag{3.29}$$

where $g \in (\mathcal{H}, \mathcal{H}(G))_{1-1/2p,p}$. By virtue of [39, p.265]

$$v(t) = e^{-tG^{1/2}} v(0).$$

Since $v \in W_p^2((0, \infty), \mathcal{H}(G), \mathcal{H})$, then using [72, p.44] we have

$$v(0) \in (\mathcal{H}, \mathcal{H}(G))_{1-1/2p,p},$$

i.e., (3.29) has been proved.

Function (3.29) satisfies condition (3.28) if

$$\alpha g - \beta G^{1/2} g = \Phi_0.$$

By virtue of [72, p.44]

$$L_1 u_1 \in (H, H_4)_{(4p-mp-1)/4p,p}, \quad L_2 u_1 \in (H, H_4)_{(2p-mp-1)/4p,p}.$$

Obviously we have

$$(\mathcal{H}, \mathcal{H}(G))q, p = (H_2 \oplus H, H_4 \oplus H_2)_{q,p} = (H_2, H_4)_{q,p} \oplus (H, H_2)_{q,p}.$$

Since $H_2 = (H, H_4)_{1/2,2}$, then by virtue of [72, p.66, 105]

$$(H, H_4)_{(4p-mp-1)/4p,p} = (H_2, H_4)_{(2p-mp-1)/2p,p},$$

$$(H, H_4)_{(4p-(m+2)p-1)/4p,p} = (H, H_2)_{(2p-mp-1)/2p,p}.$$

Then, from condition 4, when $\beta = 0$, it follows that a solution to equation (3.27) has the form (3.29). By virtue of [39, p.118] and condition 4, when $\beta \neq 0$, the problem (3.27)–(3.28) has a unique solution in $W_p^2((0, \infty), \mathcal{H}(G), \mathcal{H})$ and this solution is given by the formula

$$v(t) = e^{-tG^{1/2}} (\alpha - \beta G^{1/2})^{-1} \Phi_0. \tag{3.30}$$

By virtue of [72, p.101] the operator $G^{1/2}$ from

$$(\mathcal{H}, \mathcal{H}(G^{1/2}))_{1-1/p,p} = (\mathcal{H}, \mathcal{H}(G))_{1-1/2p,p}$$

onto $(\mathcal{H}, \mathcal{H}(G))_{(p-1)/2p,p}$ is an isomorphism. Then

$$(\alpha - \beta G^{1/2})^{-1} \Phi_0 \in (\mathcal{H}, \mathcal{H}(G))_{1-1/2p,p},$$

i.e., (3.30) has the form (3.29). ∎

Let us now show that condition 3 can be replaced by a condition for the coefficients. It is known [39, p.114] that if A is an operator in a Hilbert space H with a dense domain of definition $D(A)$ and

$$\|(A + sI)^{-1}\| \leq C(1 + s)^{-1}, \qquad s \geq 0, \tag{3.31}$$

then positive fractional powers A^α, $\alpha \in (0,1)$ are defined by the formula

$$A^\alpha u = \frac{\sin \pi \alpha}{\pi} \int_0^\infty s^{\alpha-1}(A + sI)^{-1}Au \, ds, \qquad u \in D(A). \tag{3.32}$$

This formula could have been applied to yield the positive fractional powers of A on $D(A)$. It remains only to be observed that the entire operator A^α is obtained by closure from its restriction to $D(A)$.

Lemma 3.4. Let the following conditions be satisfied:

(1) A is an operator in H, $\overline{D(A)} = H$ and estimate (3.31) holds;
(2) $\text{Re}(Au, u) \geq 0$, $u \in D(A)$;

Then

$$\text{Re}(A^\alpha u, u) \geq 0, \qquad u \in D(A^\alpha).$$

Proof. Formula (3.32) follows that for $u \in D(A)$

$$\text{Re}(A^\alpha u, u) = \frac{\sin \pi \alpha}{\pi} \int_0^\infty s^{\alpha-1}\text{Re}(A(A + sI)^{-1}u, u) \, ds.$$

Denote $v = (A + sI)^{-1}u$. Then $u = (A + sI)v$ and $\text{Re}(A(A + sI)^{-1}u, u) = \text{Re}(Av, (A + sI)v) = \text{Re}(Av, Av) + s\text{Re}(Av, v) \geq 0$. ∎

Theorem 3.5. Let the following conditions be satisfied:

(1) A_4 is a selfadjoint operator in H; $(A_4u, u) \geq c^2(u, u)$, $c \neq 0$, $u \in D(A_4)$;
(2) A_2 and A_2^* are operators in H; $\text{Re}(A_2u, u) \leq 0$, $u \in D(A_4^{1/2})$; $\text{Re}(A_2^*u, u) \leq 0$, $u \in D(A_4^{1/2})$; $A_2A_4^{-1/2} \in B(H)$; $A_2^*A_4^{-1/2} \in B(H)$;
(3) at least one of two numbers α and β is not equal to zero; $\text{Re}\alpha\beta^{-1} \leq 0$ when $\beta \neq 0$;

Then the operator

$$L : u \to (L(D)u, L_1u, L_2u)$$

from $W_p^4((0,\infty), H(A_4), H)$ into

$$L_p((0,\infty), H) \dot{+} (H, H(A_4))_{(4p-mp-1)/4p,p} \dot{+} (H, H(A_4))_{(2p-mp-1)/4p,p},$$

where $m = 0$ if $\beta = 0$ and $m = 1$ if $\beta \neq 0$, is an isomorphism.

Proof. Repeat the proof of Theorem 3.3, but for the operator

$$G = \begin{pmatrix} 0 & I \\ -A_4 & -A_2 \end{pmatrix},$$

in the space $\mathcal{H} = H(A_4^{1/2}) \oplus H$, the proof of the inequality

$$\|(G - \lambda I)^{-1}\| \leq C(1 + |\lambda|)^{-1}, \qquad \lambda \leq 0, \tag{3.33}$$

must be taken from Theorem 3.2. We show that, in this case, (3.29) can be proved otherwise. From (3.27) we have

$$(D - G^{1/2})(D + G^{1/2})v(t) = 0,$$

where $D = d/dt$. Denote

$$w(t) = (D + G^{1/2})v(t).$$

Then

$$(D - G^{1/2})w(t) = 0. \tag{3.34}$$

Since $v \in W_p^2((0, \infty), \mathcal{H}(G), \mathcal{H})$, then $w \in W_p^1((0, \infty), \mathcal{H}(G^{1/2}), \mathcal{H})$. A function that belongs to $W_p^1((0, \infty), \mathcal{H}(G^{1/2}), \mathcal{H})$ and satisfies (3.34) is equal to zero. In Theorem 3.2 we proved

$$\mathrm{Re}(Gv, v)_{\mathcal{H}} \geq 0, \quad v \in D(G),$$

$$\mathrm{Re}(G^*v, v)_{\mathcal{H}} \geq 0, \quad v \in D(G^*).$$

Consequently, from (3.34) by virtue of Lemma 3.4, we have

$$\mathrm{Re}(w'(t), w(t)) = \mathrm{Re}(G^{1/2}w(t), w(t)),$$

$$\frac{d}{dt}\|w(t)\|^2 \geq 0,$$

$$\|w(t)\|^2 \geq \|w(0)\|^2,$$

which implies that $w(t) = 0$. So,

$$(D + G^{1/2})v(t) = 0, \qquad t \geq 0.$$

Hence, $v(t) = e^{-tG^{1/2}}v(0)$. Since $v \in W_p^2((0, \infty), \mathcal{H}(G), \mathcal{H})$, then by virtue of [72, p.44] $v(0) \in (\mathcal{H}, \mathcal{H}(G))_{1-1/2p, p}$. $\blacksquare$

5.4. Completeness of elementary solutions
of differential-operator equations

In cases when it is difficult to prove the applicability of the Fourier method, it is desirable at least to establish that a solution to an initial boundary value problem may be approximated by linear combinations of elementary solutions.

5.4.1. Completeness of elementary solutions of parabolic differential-operator equations. Consider in a Hilbert space H the Cauchy problem for the perturbed homogeneous equation of the first order

$$u'(t) = Au(t) + Bu(t), \tag{4.1}$$

$$u(0) = \varphi_0. \tag{4.2}$$

Let us find conditions that guarantee an approximation of the solution to the problem (4.1)–(4.2) by linear combinations of elementary solutions to equation (4.1).

By virtue of Lemma 2.0.1 the function of the form

$$u_i(t) = e^{\lambda_i t}\left(\frac{t^{k_i}}{k_i!}u_{i0} + \frac{t^{k_i-1}}{(k_i-1)!}u_{i1} + \cdots + u_{ik_i}\right) \tag{4.3}$$

becomes the *elementary solution* to equation (4.1) if and only if $u_{i0}, u_{i1}, \ldots, u_{ik_i}$ is a chain of root vectors of the operator $A + B$, corresponding to the eigenvalue λ_i.

Theorem 4.1. *Let the following conditions be satisfied:*

(1) A *is an operator in* H; $\overline{D(A)} = H$;

(2) *for some* $p > 0$ $J \in \sigma_p(H(A), H)$;

(3) *there exist rays* ℓ_k *with the angles between the neighboring rays less than* $\frac{\pi}{p}$ *and numbers* $\alpha > 0$, $\beta \in (0,1]$ *such that*

$$\|R(\lambda, A)\| \le C|\lambda|^{-\beta}, \quad |\arg\lambda| \le \frac{\pi}{2} + \alpha \text{ or } \lambda \in \ell_k, \ |\lambda| \to \infty;$$

(4) B *is an operator in* H, $D(B) \supset D(A)$ *and for any* $\varepsilon > 0$

$$\|Bu\| \le \varepsilon\|Au\|^\beta\|u\|^{1-\beta} + C(\varepsilon)\|u\|, \quad u \in D(A);$$

(5) $\varphi_0 \in D(A)$.

Then the problem (4.1)–(4.2) *has a unique solution*

$$u \in C([0,T], H) \cap C^1((0,T], H(A), H)$$

and there exist numbers C_{in} such that

$$\lim_{n \to \infty} \max_{t \in [0,T]} \|u(t) - \sum_{i=1}^{n} C_{in} u_i(t)\| = 0,$$

$$\lim_{n \to \infty} \sup_{t \in (0,T]} t(\|u'(t) - \sum_{i=1}^{n} C_{in} u_i'(t)\|$$

$$+ \|Au(t) - \sum_{i=1}^{n} C_{in} A u_i(t)\|) = 0,$$

where $u(t)$ is a solution to the problem (4.1)–(4.2) and $u_i(t)$ are the elementary solutions (4.3) to equation (4.1).

Proof. By virtue of Theorem 2.2.10 (more exactly, Theorem 2.3.9 when $n = 1$) a system of root vectors of the operator $A + B$ is complete in the space $H(A)$. Hence, there exist numbers C_{in} such that

$$\lim_{n \to \infty} (\|\varphi_0 - \sum_{i=1}^{n} C_{in} u_i(0)\| + \|A\varphi_0 - \sum_{i=1}^{n} C_{in} A u_i(0)\|) = 0. \qquad (4.4)$$

On the other hand, from Theorem 2.7 follow the estimates

$$\|u(t) - \sum_{i=1}^{n} C_{in} u_i(t)\| \le C(\|\varphi_0 - \sum_{i=1}^{n} C_{in} u_i(0)\|$$

$$+ \|A\varphi_0 - \sum_{i=1}^{n} C_{in} A u_i(0)\|), \qquad (4.5)$$

$$\|u'(t) - \sum_{i=1}^{n} C_{in} u_i'(t)\| + \|Au(t) - \sum_{i=1}^{n} C_{in} A u_i(t)\|$$

$$\le C t^{-1}(\|\varphi_0 - \sum_{i=1}^{n} C_{in} u_i(0)\| + \|A\varphi_0 - \sum_{i=1}^{n} C_{in} A u_i(0)\|). \qquad (4.6)$$

From (4.5) and (4.6) by virtue of (4.4) the theorem statement follows. ∎

Remark 4.2. *When $\beta = 1$ the problem (4.1)–(4.2) has a unique solution $u \in C^1([0,T], H(A), H)$ and there exist numbers C_{in} such that*

$$\lim_{n \to \infty} \max_{t \in [0,T]} (\|u'(t) - \sum_{i=1}^{n} C_{in} u_i'(t)\|$$

$$+ \|Au(t) - \sum_{i=1}^{n} C_{in} A u_i(t)\|) = 0.$$

Theorem 4.3. *Let the following conditions be satisfied:*

(1) A *is an operator in* H, $\overline{D(A)} = H$;

(2) *for some* $p > 0$ $\quad J \in \sigma_p(H(A), H)$;

(3) *there exist rays* ℓ_k *with the angles between the neighboring rays less than* π/p *such that*

$$\|R(\lambda, A)\| \le C|\lambda|^{-1}, \qquad |\arg \lambda| \le \frac{\pi}{2} \text{ or } \lambda \in \ell_k, \ |\lambda| \to \infty;$$

(4) B *is an operator on* H, $D(B) \supset D(A)$ *and the operator* BA^{-1} *is compact in* H;

(5) $\varphi_0 \in (H, H(A))_{1-1/q,q}$ *for some* $q \in (1, \infty)$.

Then the problem (4.1)–(4.2) *has a unique solution* $u \in W_q^1((0,T), H(A), H)$ *and there exist numbers* C_{in} *such that*

$$\lim_{n \to \infty} \int_0^T \left(\|u'(t) - \sum_{i=1}^n C_{in}u_i'(t)\|^q + \|Au(t) - \sum_{i=1}^n C_{in}Au_i(t)\|^q\right) dt = 0,$$

where $u(t)$ *is a solution to the problem* (4.1)–(4.2) *and* $u_i(t)$ *are the elementary solutions* (4.3) *to equation* (4.1).

Proof. From conditions 2 and 4, by virtue of Lemma 2.2.5, for any $\varepsilon > 0$ the following estimate

$$\|Bu\| \le \varepsilon\|Au\| + C(\varepsilon)\|u\|, \qquad u \in D(A)$$

holds. Then by virtue of Theorem 2.2.10 (more exactly, theorem 2.3.9 when $n = 1$) (4.4) holds. The relation (4.4) means that the linear span of the root vector system of the operator $A + B$ is complete in the space $H(A)$. On the other hand, by virtue of [72, p.39] the set $H(A)$ is dense in the space $(H, H(A))_{s,q}$. Hence, the linear span of the root vector system of the operator $A + B$ is dense in the space $(H, H(A))_{s,q}$, i.e., there exist numbers C_{in} such that

$$\lim_{n \to \infty} \|\varphi_0 - \sum_{i=1}^n C_{in}u_i(0)\|_{(H,H(A))_{s,q}} = 0.$$

Then by virtue of Theorem 2.9

$$\|u(\cdot) - \sum_{i=1}^n C_{in}u_i(\cdot)\|_{W_q^1((0,T),H(A),H)}$$

$$\le C\|\varphi_0 - \sum_{i=1}^n C_{in}u_i(0)\|_{(H,H(A))_{1-1/q,q}},$$

from which the theorem statement follows. ∎

5.4.2. Completeness of elementary solutions for elliptic differential-operator equations. Consider in a Hilbert space H a boundary value problem in $[0, \infty)$ for the 4-th order elliptic equation

$$L(D)u = u''''(t) + A_2 u''(t) + A_4 u(t) = 0, \qquad t > 0, \qquad (4.7)$$

$$u(0) = \varphi_1, \quad u''(0) = \varphi_2. \qquad (4.8)$$

Let us find conditions that guarantee an approximation of the solution to the problem (4.7)–(4.8) by linear combinations of elementary solutions to equation (4.7).

By virtue of Lemma 2.0.1 the function (4.3) is a solution to equation (4.7) if and only if $u_{i0}, u_{i1}, \ldots, u_{ik_i}$ is a chain of root vectors of the characteristic operator pencil

$$L(\lambda) = \lambda^4 I + \lambda^2 A_2 + A_4 \qquad (4.9)$$

and (4.3) is called an *elementary solution* to equation (4.7).

Theorem 4.4. *Let the following conditions be satisfied:*

(1) *there exist Hilbert spaces H, H_4 for which the compact embedding $H_4 \subset H_0 = H$ takes place and $\overline{H_4}|_H = H$;*

(2) *operators A_k, $k = 2, 4$ from H_k into H act boundedly, where $H_2 = (H, H_4)_{1/2,2}$;*

(3) *$J \in \sigma_q(H_4, H)$ for some $q > 0$;*

(4) *the characteristic operator pencil (4.9) is coercive when $\mathrm{Re}\lambda = 0$, i.e.,*

$$|\lambda|^2 \|L^{-1}(\lambda)f\|_{H_2} + \|L^{-1}(\lambda)f\|_{H_4} \le C\|f\|_H,$$
$$f \in H, \ \mathrm{Re}\lambda = 0;$$

$$|\lambda|^4 \|L^{-1}(\lambda)f\|_{H_2} + |\lambda|^2 \|L^{-1}(\lambda)f\|_{H_4} \le C\|f\|_{H_2},$$
$$f \in H_2, \ \mathrm{Re}\lambda = 0;$$

(5) *there exist rays ℓ_k with angles between the neighboring rays less than $\pi/4q$ and ω such that*

$$\|L^{-1}(\lambda)\|_{B(H,H_4)} \le C|\lambda|^\omega, \qquad \lambda \in \ell_k, \ |\lambda| \to \infty;$$

(6) *$\varphi_1 \in (H, H_4)_{1-1/4p,p}$, $\varphi_2 \in (H, H_4)_{(2p-1)/4p,p}$.*

Then the problem (4.7)–(4.8) *has a unique solution* $u \in W_p^4((0,\infty), H_4, H)$ *and there exist numbers* C_{in} *such that*

$$\lim_{n \to \infty} \sum_{k=0,2,4} \int_0^\infty \|u^{(k)}(t) - \sum_{i=1}^n C_{in} u_i^{(k)}(t)\|_{H_{4-k}}^p \, dt = 0, \qquad (4.10)$$

where $u(t)$ *is a solution to the problem* (4.7)–(4.8) *and* $u_i(t)$ *are elementary solutions* (4.3) *to equation* (4.7), *corresponding to the eigenvalue* λ_i *with* $\mathrm{Re}\lambda_i < 0$.

Proof. By virtue of Theorem 2.3.17 a system of root vectors of the pencil (4.9), corresponding to the eigenvalues λ_i with $\mathrm{Re}\lambda_i < 0$, is $(u(0), u''(0))$-complete in the spaces $H_4 \oplus H_2$ and $H_2 \oplus H$. Then the same system of root vectors is $(u(0), u''(0))$-complete in the spaces $(H_2 \oplus H, H_4 \oplus H_2)_{s,p}$ [72, p.39]. Obviously we have

$$(H_2 \oplus H, H_4 \oplus H_2)_{s,p} = (H_2, H_4)_{s,p} \oplus (H, H_2)_{s,p}.$$

Since $H_2 = (H, H_4)_{1/2,2}$, then by virtue of [72, p.66,105]

$$(H, H_4)_{1-1/4p,p} = (H_2, H_4)_{(2p-1)/2p,p},$$

$$(H, H_4)_{(2p-1)/4p,p} = (H, H_2)_{(2p-1)/2p,p}.$$

Hence, there exist numbers C_{in} such that

$$\lim_{n \to \infty} (\|\varphi_1 - \sum_{i=1}^n C_{in} u_i(0)\|_{(H,H_4)_{1-1/4p,p}}$$

$$+ \|\varphi_2 - \sum_{i=1}^n C_{in} u_i''(0)\|_{(H,H_4)_{(2p-1)/4p,p}}) = 0.$$

On the other hand, from Theorem 3.3 we have

$$\|u - \sum_{i=1}^n C_{in} u_i\|_{W_p^4((0,\infty),H_4,H)}$$

$$\leq C(\|\varphi_1 - \sum_{i=1}^n C_{in} u_i(0)\|_{(H,H_4)_{1-1/4p,p}}$$

$$+ \|\varphi_2 - \sum_{i=1}^n C_{in} u_i''(0)\|_{(H,H_4)_{(2p-1)/4p,p}}). \blacksquare$$

Let us now show that condition 4 can be replaced by some condition for coefficients.

Theorem 4.5. *Let the following conditions be satisfied:*

(1) A_4 *is a selfadjoint operator in* H; $(A_4 u, u) \geq c^2(u, u), c \neq 0,\ u \in D(A_4)$;

(2) A_2 *and* A_2^* *are operators in* H; $\mathrm{Re}(A_2 u, u) \leq 0,\ u \in D(A_4^{1/2})$;

$$\mathrm{Re}(A_2^* u, u) \leq 0, \quad u \in D(A_4^{1/2}); A_2 A_4^{-1/2} \in B(H);\ A_2^* A_4^{-1/2} \in B(H);$$

(3) *the embedding* $H(A_4) \subset H$ *is compact;* $J \in \sigma_q(H(A_4), H)$ *for some* $q > 0$;

(4) *there exist rays* ℓ_k *with angles between the neighboring rays less than* $\pi/4q$ *and* ω *such that*

$$\|L^{-1}(\lambda)\|_{B(H, H_4)} \leq C|\lambda|^\omega, \qquad \lambda \in \ell_k,\ |\lambda| \to \infty;$$

(5) $\varphi_1 \in (H, H_4)_{1-1/4p, p}, \quad \varphi_2 \in (H, H_4)_{(2p-1)/4p, p}.$

Then the problem (4.7)–(4.8) *has a unique solution* $u \in W_p^4((0, \infty), H(A_4), H)$ *and there exist numbers* C_{in} *such that* (4.10) *is satisfied.*

Proof. Repeat the proof of Theorem 4.4 but use Theorem 3.5 instead of Theorem 3.3. ∎

Chapter 6

Partial Differential Equations

The role and importance of the Fourier method for investigation of mathematical physics problems is well known. By this method we can investigate problems for which the corresponding spectral problem is selfadjoint. By the Fourier method a solution to the problem $u(t, x)$ is written in the form of the series

$$u(t, x) = \sum_{k=1}^{\infty} C_k u_k(t, x),$$

where $u_k(t, x)$ are elementary solutions to the considered problem. In the selfadjoint case the Hilbert theory of selfadjoint operators with a compact resolvent gives us information about the existence of elementary solutions and whether enough elementary solutions exist to write $u(t, x)$ in this series form.

Equations that are considered here belong to known types only in their principal parts. Algebraic conditions determining solvability or completeness of elementary solutions are found. Theorems of completeness of elementary solutions are new even for simple boundary conditions. Similar theorems were proved in the papers [25], [26], [18], [74].

6.1. Principally initial boundary value problems for parabolic equations

6.1.1. An isomorphism of principally initial boundary value problems for parabolic equations. In spite of the fact that there are many papers devoted to the question of completeness, the results suggested here are still new.

Consider in the domain $[0, T] \times [0, 1]$ an initial boundary value problem for the parabolic equation

$$L(t, x, D_x, D_t)u = D_t u(t, x) + a(t, x)D_x^{2m} u(t, x) + B(t)u(t, \cdot)|_x$$
$$= f(t, x), \quad (t, x) \in [0, T] \times [0, 1], \tag{1.1}$$

$$L_p(D_x)u = \alpha_p D_x^{m_p} u(t, 0) + \beta_p D_x^{m_p} u(t, 1) + \sum_{s=1}^{N_p} \delta_{ps} D_x^{m_p} u(t, x_{ps})$$
$$+ T_p u(t, \cdot) = 0, \quad t \in [0, T], \; p = 1, \ldots, 2m, \tag{1.2}$$

$$u(0, x) = u_0(x), \quad x \in [0, 1], \tag{1.3}$$

where $x_{ps} \in (0, 1)$, $m_p \leq 2m - 1$; $\alpha_p, \beta_p, \delta_{ps}$ are complex numbers, $D_t = \frac{\partial}{\partial t}, D_x = \frac{\partial}{\partial x}$.

Theorem 1.1. *Let the following conditions be satisfied:*

(1) $a \in C([0, T] \times [0, 1])$; $a(t, x) \neq 0$ for $(t, x) \in [0, T] \times [0, 1]$;

(2) $a(t, 0) = a(t, 1)$[18]; *for some fixed $\delta > 0$ we have $|\arg a(t, x)| \leq \pi/2 - \delta$ if m is an even number and $|\arg a(t, x)| \geq \pi/2 + \delta$ if m is odd;*

(3) *conditions (1.2) are m-regular with respect to numbers*

$$\omega_1 = 1, \; \omega_2 = e^{i\pi/m}, \ldots, \omega_{2m} = e^{i\pi(2m-1)/m},$$

i.e.,

$$\begin{vmatrix} \alpha_1 \omega_1^{m_1} & \cdots & \alpha_1 \omega_m^{m_1} & \beta_1 \omega_{m+1}^{m_1} & \cdots & \beta_1 \omega_{2m}^{m_1} \\ \vdots & \cdots & \vdots & \vdots & \cdots & \vdots \\ \alpha_{2m} \omega_1^{m_{2m}} & \cdots & \alpha_{2m} \omega_m^{m_{2m}} & \beta_{2m} \omega_{m+1}^{m_{2m}} & \cdots & \beta_{2m} \omega_{2m}^{m_{2m}} \end{vmatrix} \neq 0,$$

and for some $r \in [1, \infty)$ functionals T_p in $W_r^{m_p}(0, 1)$ are continuous;

(4) *for $t \in [0, T]$ an operator $B(t)$ from $W_2^{2m}(0, 1)$ into $L_2(0, 1)$ is compact; the function $t \to B(t)$ from $[0, T]$ into $B(W_2^{2m}(0, 1), L_2(0, 1))$ is continuous.*

Then the operator $\mathbb{L} : u \to (L(t, x, D_x, D_t)u, u(0, x))$ from

$$W_q^1((0, T), W_2^{2m}((0, 1), L_p u = 0, p = 1, \ldots, 2m), L_2(0, 1))$$

onto

$$L_{(q,2)}((0, T) \times (0, 1)) \dotplus (L_2(0, 1), W_2^{2m}((0, 1), L_p u = 0, p = 1, \ldots, 2m))_{1-1/q,q}$$

---

[18]See the footnote on p.100.

is an isomorphism when $q \in (1, \infty)$.

Proof. To prove Theorem 1.1 let us reduce the problem (1.1)–(1.3) to the Cauchy problem for an abstract parabolic equation of the first order.

Denote by $A(t)$ an operator in $L_2(0,1)$ with the domain of definition

$$D(A(t)) = W_2^{2m}((0,1), L_p u = 0, p = 1, \ldots, 2m),$$

independent of $t \in [0, T]$ and with the action law

$$A(t)u = -a(t,x)u^{(2m)}(x) + B(t)u|_x.$$

then the problem (1.1)–(1.3) may be rewritten in the form

$$\begin{aligned}
u'(t) &= A(t)u(t) + f(t), \\
u(0) &= u_0,
\end{aligned} \tag{1.4}$$

where $u(t) = u(t, \cdot)$, $f(t) = f(t, \cdot)$, $u_0 = u_0(\cdot)$ are functions with values in the Hilbert space $H = L_2(0,1)$. From Theorems 3.2.1 and 3.1.7 it follows that the operator $A(t)$ satisfies condition 1 of Theorem 5.2.10, i.e.,

$$\overline{D(A(t))} = L_2(0,1)$$

and

$$\|R(\lambda, A(t))\| \leq C|\lambda|^{-1}, \qquad |\arg \lambda| \leq \pi/2, \ |\lambda| \to \infty.$$

Condition 3 of Theorem 5.2.10 follows from conditions 1 and 4. So, for problem (1.4) all conditions of Theorem 5.2.10 are fulfilled and the statement of Theorem 1.1 follows. ∎

Let G be a bounded domain in the Euclidean space $\mathbb{R}^r$ with an $(r-1)$-dimensional boundary Γ. Consider in $[0, T] \times G$ a principally initial boundary value problem for the parabolic equation

$$L(t, x, D_x, D_t)u = D_t u(t, x) + \sum_{|\alpha|=2m} a_\alpha(t, x) D_x^\alpha u(t, x)$$

$$+ B(t)u(t, \cdot)|_x = f(t, x), \quad (t, x) \in [0, T] \times G, \tag{1.5}$$

$$L_p(x', D_x)u = \sum_{|\alpha|=m_p} b_{p\alpha}(x') D^\alpha u(t, x') + T_p u(t, \cdot)|_{x'}$$

$$= 0, \quad (t, x') \in [0, T] \times \Gamma, \ p = 1, \ldots, m, \tag{1.6}$$

$$u(0, x) = \varphi_0(x), \qquad x \in G, \tag{1.7}$$

where $m_p \leq 2m - 1$, $D_t = \frac{\partial}{\partial t}$, $D_x^\alpha = D_1^{\alpha_1} \cdots D_r^{\alpha_r}$, $D_k = -i\frac{\partial}{\partial x_k}$.

Theorem 1.2. *Let the following conditions be satisfied:*

(1) $a_\alpha \in C([0,T] \times \overline{G}), b_{p\alpha} \in C^{2m-m_p}(\overline{G}), \Gamma \in C^{2m}$;

(2) *when* $t \in [0,T], x \in \overline{G}, \sigma \in \mathbb{R}^r, |\arg\lambda| \leq \pi/2$, *and* $|\sigma| + |\lambda| \neq 0$ *then*

$$\lambda + \sum_{|\alpha|=2m} a_\alpha(t,x)\sigma^\alpha \neq 0,$$

where $\sigma^\alpha = \sigma_1^{\alpha_1} \cdots \sigma_r^{\alpha_r}$;

(3) *system (1.6) is normal, i.e.,* $m_i \neq m_k$ *if* $i \neq k$, *and for any vector* σ *normal to the boundary* Γ *at the point* $x' \in \Gamma$ *the following condition is fulfilled*

$$\sum_{|\alpha|=m_k} b_{k\alpha}(x')\sigma^\alpha \neq 0, \qquad k = 1, \ldots, m.$$

(4) *let* x' *be any point on* Γ; *let the vector* σ' *be tangent and the vector* σ *be normal to* Γ *at the point* $x' \in \Gamma$; *consider the following ordinary differential problem*

$$(\lambda + \sum_{|\alpha|=2m} a_\alpha(x')(\sigma' - i\sigma\frac{d}{dy})^\alpha)u(y) = 0, \qquad y \geq 0, \ |\arg\lambda| \leq \pi/2,$$

$$\sum_{|\alpha|=m_k} b_{k\alpha}(x')(\sigma' - i\sigma\frac{d}{dy})^\alpha u(y)|_{y=0} = h_k, \qquad k = 1, \ldots, m;$$

this problem should have one and only one solution, including all its derivatives, tending to zero as $y \to +\infty$ *for any numbers* $h_k \in \mathbb{C}$;

(5) *for* $t \in [0,T]$ *the operator* $B(t)$ *from* $W_2^{2m}(G)$ *into* $L_2(G)$ *is compact; the function* $t \to B(t)$ *from* $[0,T]$ *into* $B(W_2^{2m}(G), L_2(G))$ *is continuous;*

(6) *operators* T_p *from* $W_2^{m_p+1/2}(G)$ *into* $L_2(\Gamma)$ *and from* $W_2^{2m}(G)$ *into* $W_2^{2m-m_p-1/2}(\Gamma)$ *are compact;* $T_p = 0$, *if* $m_p = 0$;

Then the operator $\mathbb{L} : u \to (L(t,x,D_x,D_t)u, u(0,x))$ *from*

$$W_q^1((0,T), W_2^{2m}(G, L_p u = 0, p = 1, \ldots, m), L_2(G))$$

onto

$$L_{(q,2)}((0,T) \times G) \dotplus (L_2(G), W_2^{2m}(G, L_p u = 0, p = 1, \ldots, m))_{1-1/q,q}$$

is an isomorphism when $q \in (1,\infty)$.

Proof. Denote by $A(t)$ an operator in $L_2(G)$ with the domain of definition

$$D(A(t)) = W_2^{2m}(G, L_p u = 0, p = 1, \ldots, m)$$

independent of $t \in [0, T]$ and with the action law

$$A(t)u = -\sum_{|\alpha|=2m} a_\alpha(t, x)D_x^\alpha u(x) + B(t)u|_x.$$

Then the problem (1.5)–(1.7) may be rewritten in the form (1.4). The further proof is analogous to the proof of Theorem 1.1, but instead of Theorems 3.2.1 and 3.1.7 we have to use Theorems 4.1.5 and 4.1.1. ■

6.1.2. Completeness of elementary solutions of principally initial boundary value problems for parabolic equations. Consider in the domain $[0, T] \times [0, 1]$ an initial boundary value problem for the parabolic equation

$$D_t u(t, x) + a(x)D_x^{2m} u(t, x) + Bu(t, \cdot)|_x = 0, \quad (t, x) \in [0, T] \times [0, 1], \tag{1.8}$$

$$L_p u = \alpha_p D_x^{m_p} u(t, 0) + \beta_p D_x^{m_p} u(t, 1) + \sum_{s=1}^{N_p} \delta_{ps} D_x^{m_p} u(t, x_{ps})$$

$$+ T_p u(t, \cdot) = 0, \qquad t \in [0, T], \; p = 1, \ldots, 2m, \tag{1.9}$$

$$u(0, x) = u_0(x), \qquad x \in [0, 1], \tag{1.10}$$

where $x_{ps} \in (0, 1), m_p \leq 2m - 1; \alpha_p, \beta_p, \delta_{ps}$ are complex numbers, $D_t = \frac{\partial}{\partial t}, D_x = \frac{\partial}{\partial x}$, and the corresponding spectral problem is

$$\lambda u(x) + a(x)u^{(2m)}(x) + Bu|_x = 0, \qquad x \in [0, 1], \tag{1.11}$$

$$L_p u = 0, \qquad p = 1, \ldots, 2m. \tag{1.12}$$

By virtue of Lemma 2.0.1 the function of the form

$$u_i(t, x) = e^{\lambda_i t}\left(\frac{t^{k_i}}{k_i!}u_{i0}(x) + \frac{t^{k_i-1}}{(k_i - 1)!}u_{i1}(x) + \cdots + u_{ik_i}(x)\right) \tag{1.13}$$

becomes an *elementary solution* to the problem (1.8)–(1.9) if and only if a system of functions $u_{i0}(x), u_{i1}(x), \ldots, u_{ik_i}(x)$ is a chain of root functions of the problem (1.11)–(1.12), corresponding to the eigenvalue λ_i.

Theorem 1.3. *Let the following conditions be satisfied:*

(1) $a \in C[0, 1], a(x) \neq 0$ for $x \in [0, 1], a(0) = a(1)^{19}$;

(2) under a certain fixed $\delta > 0$, if m is an even number then $|\arg a(x)| \leq \pi/2 - \delta$ and if m is odd then $|\arg a(x)| \geq \pi/2 + \delta$;

(3) condition 3 of Theorem 1.1 is fulfilled;

(4) the operator B from $W_2^{2m}(0, 1)$ into $L_2(0, 1)$ is compact;

(5) $\varphi_0 \in W_2^{2m}((0, 1), L_p u = 0, p = 1, \ldots, 2m)$.

---

[19]See the footnote on p.100.

Then the problem (1.8)–(1.10) has a unique solution

$$u \in C^1([0,T], W_2^{2m}(0,1), L_2(0,1))$$

and there exist numbers C_{in} such that

$$\lim_{n \to \infty} \max_{t \in [0,T]} (\|D_t u(t,\cdot) - \sum_{i=1}^n C_{in} D_t u_i(t,\cdot)\|_{0,2}$$

$$+ \|u(t,\cdot) - \sum_{i=1}^n C_{in} u_i(t,\cdot)\|_{2m,2}) = 0,$$

where $u(t,x)$ is a solution to the problem (1.8)–(1.10) and $u_i(t,x)$ are elementary solutions (1.13) to the problem (1.8)–(1.9).

Proof. Apply Theorem 5.4.1 to the problem (1.8)–(1.10). Consider in $H = L_2(0,1)$ an operator A given by the equalities

$$D(A) = W_2^{2m}((0,1), L_p u = 0, p = 1, \ldots, 2m),$$
$$Au = -a(x)u^{(2m)}(x) + Bu|_x.$$

Then the problem (1.8)–(1.10) may be rewritten in the form

$$u'(t) = Au(t),$$
$$u(0) = u_0, \tag{1.14}$$

where $u(t) = u(t,\cdot), u_0 = u_0(\cdot)$ are functions with values in the Hilbert space $H = L_2(0,1)$. From theorems 3.2.1 and 3.1.7 it follows that for the operator A under some $\alpha > 0$ the following correlations hold:

$$\overline{D(A)} = L_2(0,1)$$

and

$$\|R(\lambda, A)\| \le C|\lambda|^{-1}, \qquad |\arg\lambda| \le \pi/2 + \alpha, \ |\lambda| \to \infty. \tag{1.15}$$

Hence, the operator A satisfies condition 1 of Theorem 5.4.1.

By virtue of [72, p.350/14]

$$s_j(J, W_2^{2m}(0,1), L_2(0,1)) \sim j^{-2m}.$$

Hence, for $p > 1/2m$

$$J \in \sigma_p(W_2^{2m}(0,1), L_2(0,1)). \tag{1.16}$$

Since $W_2^{2m}((0,1), L_p u = 0, p = 1, \ldots, 2m)$ is a subspace of $W_2^{2m}(0,1)$, then, by virtue of Lemma 2.3.3, from (1.16) follows

$$J \in \sigma_p(W_2^{2m}((0,1), L_p u = 0, p = 1, \ldots, 2m), L_2(0,1)),$$

i.e., for $p > 1/2m$ condition 2 of Theorem 5.4.1 is fulfilled. Since $p > 1/2m$ then condition 3 of Theorem 5.4.1 follows from the estimate (1.15). So, applying Theorem 5.4.1 to problem (1.14) we complete the proof. ∎

Let G be a bounded domain in the Euclidean space $\mathbb{R}^r$ with an $(r-1)$-dimensional boundary Γ. Consider in $[0, T] \times G$ a principally initial boundary value problem for the parabolic equation

$$D_t u(t, x) + \sum_{|\alpha|=2m} a_\alpha(x) D_x^\alpha u(t, x) + Bu(t, \cdot)|_x = 0, \quad (t, x) \in [0, T] \times G, \tag{1.17}$$

$$\sum_{|\alpha|=m_p} b_{p\alpha}(x') D^\alpha u(t, x') + T_p u(t, \cdot)|_{x'} = 0, \quad (t, x') \in [0, T] \times \Gamma, \ p = 1, \ldots, m, \tag{1.18}$$

$$u(0, x) = \varphi_0(x), \qquad x \in G \tag{1.19}$$

where $m_p \leq 2m - 1$, $D_t = \frac{\partial}{\partial t}$, $D_x^\alpha = D_1^{\alpha_1} \cdots D_r^{\alpha_r}$, $D_k = -i\frac{\partial}{\partial x_k}$. The corresponding spectral problem is

$$\lambda u(x) + \sum_{|\alpha|=2m} a_\alpha(x) D_x^\alpha u(x) + Bu|_x = 0, \quad x \in G, \tag{1.20}$$

$$\sum_{|\alpha|=m_p} b_{p\alpha}(x') D^\alpha u(x') + T_p u|_{x'} = 0, \quad x' \in \Gamma, \ p = 1, \ldots, m. \tag{1.21}$$

By virtue of Lemma 2.0.1 the function of form (1.13) becomes an elementary solution to the problem (1.17)–(1.18) if and only if a system of functions $u_{i0}(x), u_{i1}(x), \ldots, u_{ik_i}(x)$ is a chain of root functions of the problem (1.20)–(1.21), corresponding to the eigenvalue λ_i.

Theorem 1.4. *Let the following conditions be satisfied:*

(1) $a_\alpha \in C(\overline{G})$, $b_{p\alpha} \in C^{2m-m_p}(\overline{G})$, $\Gamma \in C^{2m}$;

(2) *there exist rays ℓ_k with the angles between the neighboring rays less than $2m\pi/r$, such that conditions 2 and 4 of Theorem 1.2 are fulfilled on the set*

$$\ell_k \cup \{\lambda| \ \lambda \in \mathbb{C}, \ |arg\lambda| \leq \pi/2\};$$

(3) *system (1.18) is normal, i.e., condition 3 of Theorem 1.2 is fulfilled;*

(4) *the operator B from $W_2^{2m}(G)$ into $L_2(G)$ is compact;*

(5) *operators T_p from $W_2^{m_p+1/2}(G)$ into $L_2(\Gamma)$ and from $W_2^{2m}(G)$ into $W_2^{2m-m_p-1/2}(\Gamma)$ are compact, $T_p = 0$, if $m_p = 0$;*

(6) $\varphi_0 \in W_2^{2m}(G, L_p u = 0, p = 1, \ldots, m)$.

Then the problem (1.17)–(1.19) has a unique solution

$$u \in C^1([0, T], W_2^{2m}(G), L_2(G))$$

and there exist numbers C_{in} such that

$$\lim_{n \to \infty} \max_{t \in [0, T]} (\|D_t u(t, \cdot) - \sum_{i=1}^{n} C_{in} D_t u_i(t, \cdot)\|_{0,2,G}$$

$$+ \|u(t, \cdot) - \sum_{i=1}^{n} C_{in} u_i(t, \cdot)\|_{2m,2,G}) = 0,$$

where $u(t, x)$ is a solution to the problem (1.17)–(1.19) and $u_i(t, x)$ are elementary solutions (1.13) to the problem (1.17)–(1.18).

Proof. Consider in $H = L_2(G)$ an operator A given by the equalities

$$D(A) = W_2^{2m}(G, L_p u = 0, p = 1, \ldots, m),$$
$$Au = \sum_{|\alpha|=2m} a_\alpha(x) D^\alpha u(x) + Bu|_x.$$

Then the problem (1.17)–(1.19) may be rewritten in the form (1.14). The rest of the proof is analogous to the proof of Theorem 1.3, but instead of Theorems 3.2.1 and 3.1.7 we have to use Theorems 4.1.5 and 4.1.1. In addition, by virtue of [72, p.350/14]

$$s_j(J, W_2^{2m}(G), L_2(G)) \sim j^{-2m/r}.$$

Hence, for $p > r/2m$

$$J \in \sigma_p(W_2^{2m}(G), L_2(G)). \tag{1.22}$$

Since $W_2^{2m}(G, L_p u = 0, p = 1, \ldots, m)$ is a subspace of $W_2^{2m}(G)$, then, by virtue of Lemma 2.3.3, from (1.22) follows

$$J \in \sigma_p(W_2^{2m}(G, L_p u = 0, p = 1, \ldots, m), L_2(G)). \blacksquare$$

6.2. Boundary value problems for elliptic equations of the 4-th order

6.2.1. Fredholm property of boundary value problems for elliptic equations.
A lot of monographs and articles have been devoted to the questions of solvability of regular elliptic boundary value problems. Nevertheless the results on the solvability of boundary value problems in the cylindrical domains for elliptic equations of the 4-th order presented here are new. They are new even if the problem is differential, since boundary conditions are non-local. Algebraic conditions determining solvability are found.

Let us consider in the domain $[0,1] \times [0,1]$ a principally boundary value problem for an elliptic equation of the 4-th order

$$L(t,x,D_x,D_t)u = D_t^4 u(t,x) + a(x)D_x^2 D_t^2 u(t,x) + b(x)D_x^4 u(t,x)$$
$$+ \sum_{k \leq 3} M_k(t)D_t^k u(t,\cdot)|_x$$
$$= f(t,x), \quad (t,x) \in [0,1] \times [0,1], \tag{2.1}$$

$$L_k u = a_k D_x^{k+1} u(t,0) + \sum_{j \leq k} a_{kj} D_x^j u(t,0) = 0, \quad t \in [0,1], \quad k = 1,2,$$
$$L_k u = b_k D_x^{k-1} u(t,1) + \sum_{j \leq k-2} b_{kj} D_x^j u(t,1) = 0, \quad t \in [0,1], \quad k = 3,4, \tag{2.2}$$

$$P_k u = \gamma_k D_t^{p_k} u(0,x) + \delta_k D_t^{p_k} u(1,x) + \sum_{j \leq p_k - 1} \sum_{s \leq N_{kj}} T_{kj} D_t^j u(t_{kjs},\cdot)|_x$$
$$= \varphi_k(x), \quad x \in [0,1], \quad k = 1,...,4, \tag{2.3}$$

where $0 \leq p_1, p_2 \leq 1$, $p_3 = p_1 + 2$, $p_4 = p_2 + 2$; a_k, a_{kj}, b_k, b_{kj}, γ_k, δ_k are complex numbers, $t_{kjs} \in (0,1)$, $D_t = \frac{\partial}{\partial t}$, $D_x = \frac{\partial}{\partial x}$.

Theorem 2.1. *Let the following conditions be satisfied:*

(1) $a \in C[0,1]$, $b \in C[0,1]$, $b(x) \neq 0$;
(2) *if* $x \in [0,1]$, $\sigma = (\sigma_1, \sigma_2) \in \mathbb{R}^2$, $\sigma \neq 0$, *then*

$$\sigma_1^4 + a(x)\sigma_1^2 \sigma_2^2 + b(x)\sigma_2^4 \neq 0;$$

(3) $a_1 \neq 0$, $a_2 \neq 0$, $b_3 \neq 0$, $b_4 \neq 0$;
(4)

$$\begin{vmatrix} \gamma_1(-1)^{p_1} & 0 & \delta_1 & 0 \\ 0 & \gamma_3(-1)^{p_1} & 0 & \delta_3 \\ \gamma_2(-1)^{p_2} & 0 & \delta_2 & 0 \\ 0 & \gamma_4(-1)^{p_2} & 0 & \delta_4 \end{vmatrix} \neq 0;$$

(5) *operators* $M_k(t)$ *from* $W_2^{4-k}((0,1), L_s u = 0, m_s \leq 3 - k)$ *into* $L_2(0,1)$ *are compact for almost all* $t \in [0,1]$ *; for any* $\varepsilon > 0$ *and for almost all* $t \in [0,1]$

$$\|M_k(t)u\|_{L_2(0,1)} \leq \varepsilon \|u\|_{W_2^{4-k}(0,1)} + C(\varepsilon)\|u\|_{L_2(0,1)};$$

for $u \in W_2^{4-k}((0,1), L_s u = 0, m_s \leq 3 - k)$ functions $t \to M_k(t)u$ from $[0,1]$ into $L_2(0,1)$ are measurable;

(6) operators T_{kj} from $B_{2,p}^{4-j-1/p}((0,1), L_k u = 0, m_k < 4-j-1/p-1/2)$, where $p \neq 2$, or $p = 2$ and $m_k \neq 3 - j,$[20] into $B_{2,p}^{4-p_k-1/p}((0,1), L_s u = 0, m_s < 4 - p_k - 1/p - 1/2)$ are compact.

Then the operator $\mathbb{L} : u \to (L(t, x, D_x, D_t)u, P_1 u, P_2 u, P_3 u, P_4 u)$ from

$$W_p^4((0,1), W_2^4((0,1), L_k u = 0, k = 1, ..., 4), L_2(0,1))$$

into

$$L_{p,2}((0,1) \times (0,1)) \underset{k=1}{\overset{4}{\dotplus}} B_{2,p}^{4-p_k-1/p}((0,1), L_s u = 0, m_s < 4 - p_k - 1/p - 1/2)$$

is fredholm.

Proof. Let us denote

$$H = L_2(0,1), \qquad H_4 = W_2^4((0,1), L_k u = 0, \ k = 1, \ldots, 4).$$

Consider operators A_k, $B_k(t)$ which are defined by the equalities

$$A_2 u = a(x)D_x^2 u(x), \quad A_4 u = b(x)D_x^4 u(x),$$

$$B_k(t)u = M_k(t)u, \quad k = 0, \ldots, 3.$$

Then the problem (2.1)–(2.3) can be rewritten in the form

$$u''''(t) + A_2 u''(t) + A_4 u(t) + \sum_{k \leq 3} B_k(t)u^{(k)}(t) = f(t), \tag{2.4}$$

$$\gamma_k u^{(p_k)}(0) + \delta_k u^{(p_k)}(1) + \sum_{j \leq p_k - 1} \sum_{s \leq N_{kj}} T_{kj} u^{(j)}(t_{kjs})$$

$$= \varphi_k, \qquad k = 1, ..., 4, \tag{2.5}$$

where $u(t) = u(t, \cdot), f(t) = f(t, \cdot), \varphi_k = \varphi_k(\cdot)$ are functions with values in the Hilbert space $H = L_2(0,1)$.

---

[20] In the case where $p = 2$ and $m_k = 3 - j$ instead of $B_{2,2}^{4-j-1/2}((0,1), L_k u = 0, m_k < 3 - j)$ it should be written $B_{2,2}^{4-j-1/p}((0,1), L_k u = 0, m_k < 3 - j; \ L_k u \in \widetilde{B}_{2,2}^{1/2}(0,1), m_k = 3 - j)$ [72, p.321]. $\widetilde{B}_{p,q}^s(G) = \{u|u \in B_{p,q}^s(\mathbb{R}^r), \text{supp}(u) \subset \overline{G}\}$. It is known that [72, p.180] $W_2^{s+\varepsilon}(G) \subset B_{2,p}^s(G) \subset W_2^{s-\varepsilon}(G)$ for $\varepsilon > 0$.

Let us apply Theorem 5.3.1 to the problem (2.4)–(2.5). Obviously condition 1 of Theorem 5.3.1 is fulfilled. Since, by virtue of P. Grisvard [22]

$$H_2 = (L_2(0,1), W_2^4((0,1), L_k u = 0, k = 1, \ldots, 4))_{1/2,2}$$
$$= B_{2,2}^2((0,1), L_k u = 0, m_k \leq 1)$$
$$= W_2^2((0,1), L_k u = 0, m_k \leq 1),$$

then condition 2 of Theorem 5.3.1 is also fulfilled.

By virtue of condition 2 the equation

$$1 + a(x)\omega^2 + b(x)\omega^4 = 0 \qquad (2.6)$$

does not have real roots. Since $\omega_1(x)$, $\omega_2(x)$ are the roots of equation (2.6) situated in the upper half-plane and $\omega_3(x)$, $\omega_4(x)$ are the corresponding roots in the lower half-plane, then

$$\underline{\omega} = \inf_{x \in [0,1]} \min\{\arg \omega_1(x), \arg \omega_2(x), \arg \omega_3(x) + \pi, \arg \omega_4(x) + \pi\} > 0,$$

$$\overline{\omega} = \sup_{x \in [0,1]} \max\{\arg \omega_1(x), \arg \omega_2(x), \arg \omega_3(x) + \pi, \arg \omega_4(x) + \pi\} < \pi.$$

From Theorem 3.1.7 it follows that condition 3 of Theorem 5.3.1 is satisfied.[21] By virtue of P.Grisvard [22] (see also [72, p.321]) we have

$$(H, H_4)_{\theta,p} = (L_2(0,1), W_2^4((0,1), L_k u = 0, k = 1, \ldots, 4))_{\theta,p}$$
$$= B_{2,p}^{4\theta}((0,1), L_k u = 0, m_k < 4\theta - 1/2).$$

Consequently,

$$(H, H_4)_{1-k/4,2} = B_{2,2}^{4-k}((0,1), L_s u = 0, m_s \leq 3 - k)$$
$$= W_2^{4-k}((0,1), L_s u = 0, m_s \leq 3 - k),$$

and

$$(H, H_4)_{(4p-jp-1)/4p,p} = B_{2,p}^{4-j-1/p}((0,1), L_k u = 0, m_k < 4 - j - 1/p - 1/2).$$

Hence from this and from conditions 5 and 6 of our Theorem 2.1 it follows that conditions 5 and 6 of Theorem 5.3.1 are fulfilled. So, for the problem (2.4)–(2.5)

---

[21] In the case when $\omega_1(0) = \omega_2(0)$ or $\omega_1(1) = \omega_2(1)$ it should be proved a theorem similar to Theorem 3.1.7. In this case 2-regularity takes place since the determinant $\theta = \begin{vmatrix} (\omega_1(0)\lambda)^2 & 2(\omega_1(0)\lambda) \\ (\omega_1(0)\lambda)^3 & 3(\omega_1(0)\lambda)^2 \end{vmatrix} \neq 0.$

all conditions of Theorem 5.3.1 are fulfilled, from which follows the statement of Theorem 2.1. ∎

Remark 2.2. *Obviously if $T_{kj}v = \delta_{kj}v(x)$, where δ_{kj} are complex numbers, then condition 6 is fulfilled.*

Let us consider in the cylindrical domain $[0,1] \times G$, where $G \subset \mathbb{R}^r$ is a bounded domain, a principally boundary value problem for an elliptic equation of the 4-th order

$$L(t,x,D_x,D_t)u = D_t^4 u(t,x) + \sum_{i,j=1}^{r} D_i(a_{ij}(x)D_j D_t^2 u(t,x))$$

$$+ \sum_{i,j,k,m=1}^{r} D_i D_j(b_{ijkm}(x)D_k D_m u(t,x))$$

$$+ \sum_{k \leq 3} M_k(t)D_t^k u(t,\cdot)|_x = f(t,x), \quad (t,x) \in [0,1] \times G, \tag{2.7}$$

$$L_k u = \sum_{|\alpha| \leq m_k} b_{k\alpha}(x')D_x^\alpha u(t,x') = 0, \quad (t,x') \in [0,1] \times \Gamma, \; k = 1,2, \tag{2.8}$$

$$P_k u = \gamma_k D_t^{p_k} u(0,x) + \delta_k D_t^{p_k} u(1,x) + \sum_{j \leq p_k - 1} \sum_{s \leq N_{kj}} T_{kj} D_t^j u(t_{kjs},\cdot)|_x$$

$$= \varphi_k(x), \quad x \in [0,1], \; k = 1,...,4, \tag{2.9}$$

where $m_k \leq 3, 0 \leq p_1, p_2 \leq 1, p_3 = p_1 + 2, p_4 = p_2 + 2$; γ_k, δ_k are complex numbers, $t_{kjs} \in (0,1)$, $\Gamma = \partial G$ is a boundary of G, $D_t = \frac{\partial}{\partial t}$, $D_x^\alpha = D_1^{\alpha_1} \cdots D_r^{\alpha_r}$, $D_j = -i\frac{\partial}{\partial x_j}$.

Let us denote $H = L_2(G)$ and consider operators A_2 and A_4 which are defined by the equalities

$$D(A_2) = W_2^2(G, L_k u = 0, m_k \leq 1), \quad A_2 u = \sum_{i,j=1}^{r} D_i(a_{ij}(x)D_j u(x)),$$

$$D(A_4) = W_2^4(G, L_k u = 0, k = 1,2),$$

$$A_4 u = \sum_{i,j,k,m=1}^{r} D_i D_j(b_{ijkm}(x)D_k D_m u(x)) + s_0 u(x),$$

where $s_0 > 0$.

Theorem 2.3. *Let the following conditions be satisfied:*

(1) $a_{ij} \in C(\overline{G})$, $b_{ijkm} \in C(\overline{G})$, $b_{k\alpha} \in C^{4-m_k}(\overline{G})$, $\Gamma \in C^4$;

(2) $a_{ij}(x) = \overline{a_{ji}(x)}$; $b_{sjkm}(x) = \overline{b_{kmsj}(x)}$; A_2 and A_4 are formally symmetric operators in $L_2(G)$;

(3) *if* $x \in \overline{G}$, $\sigma \in \mathbb{R}^r$, $|\sigma| + |\lambda| \neq 0$ *then*

$$\lambda + \sum_{i,j,k,m=1}^{r} b_{ijkm}(x)\sigma_i\sigma_j\sigma_k\sigma_m \neq 0, \qquad \lambda \geq 0;$$

(4) *system* (2.8) *is normal;*

(5) *let* x' *be any point on* Γ, *the vector* σ' *be tangent and the vector* σ *be normal to* Γ *at the point* $x' \in \Gamma$. *Consider the following ordinary differential problem*

$$[\lambda + \sum_{s,j,k,m=1}^{r} b_{sjkm}(x')(\sigma'_s - i\sigma_s\frac{\mathrm{d}}{\mathrm{d}y})(\sigma'_j - i\sigma_j\frac{\mathrm{d}}{\mathrm{d}y})$$

$$\times (\sigma'_k - i\sigma_k\frac{\mathrm{d}}{\mathrm{d}y})(\sigma'_m - i\sigma_m\frac{\mathrm{d}}{\mathrm{d}y})]u(y) = 0, \qquad y \geq 0,\ \lambda \geq 0, \qquad (2.10)$$

$$\sum_{|\alpha|=m_k} b_{k\alpha}(x')(\sigma' - i\sigma\frac{\mathrm{d}}{\mathrm{d}y})^\alpha u(y)|_{y=0} = h_k, \qquad k = 1,2; \qquad (2.11)$$

the problem (2.10)–(2.11) *should have one and only one solution, including all its derivatives, tending to zero as* $y \to \infty$ *for any numbers* $h_k \in \mathbb{C}$;

(6) *for* $x \in \overline{G}$, $\sigma \in \mathbb{C}^r$

$$\sum_{i,j=1}^{r} a_{ij}(x)\sigma_i\bar{\sigma}_j \leq 0;$$

(7) *condition 4 of Theorem 2.1 is fulfilled;*

(8) *operators* $M_k(t)$ *from* $W_2^{4-k}(G, L_s u = 0, m_s \leq 3-k)$ *into* $L_2(G)$ *are compact for almost all* $t \in [0,1]$; *for any* $\varepsilon > 0$ *and for almost all* $t \in [0,1]$

$$\|M_k(t)u\|_{L_2(G)} \leq \varepsilon\|u\|_{W_2^{4-k}(G)} + C(\varepsilon)\|u\|_{L_2(G)};$$

for $u \in W_2^{4-k}(G, L_s u = 0, m_s \leq 3-k)$ *functions* $t \to M_k(t)u$ *from* $[0,1]$ *into* $L_2(G)$ *are measurable;*

(9) *operators* T_{kj} *from* $B_{2,p}^{4-j-1/p}(G, L_k u = 0, m_k < 4-j-1/p-1/2)$, *where* $p \neq 2$, *or* $p = 2$ *and* $m_k \neq 3-j$,[22] *into* $B_{2,p}^{4-p_k-1/p}(G, L_s u = 0, m_s < 4-p_k-1/p-1/2)$ *are compact.*

---

[22]In the case where $p = 2$ and $m_k = 3-j$ instead of $B_{2,2}^{4-j-1/2}(G, L_k u = 0, m_k < 3-j)$ it should be written $B_{2,2}^{4-j-1/2}(G, L_k u = 0, m_k < 3-j$; $L_k u \in \widetilde{B}_{2,2}^{1/2}(G), m_k = 3-j)$ [72, p.321]. $\widetilde{B}_{p,q}^s(G) = \{u|u \in B_{p,q}^s(\mathbb{R}^r),\ \mathrm{supp}(u) \subset \overline{G}\}$.

Then the operator $L : u \rightarrow (L(x, D_x, D_t)u, P_1u, P_2u, P_3u, P_4u)$ from

$$W_p^4((0,1), W_2^4(G, L_k u = 0, k = 1, 2), L_2(G))$$

into

$$L_{p,2}((0,1) \times G) \dotplus \sum_{k=1}^{4} B_{2,p}^{4-p_k-1/p}(G, L_s u = 0, m_s < 4 - p_k - 1/p - 1/2)$$

is fredholm.

Proof. Let us denote $H = L_2(G)$ and also consider operators $B_k(t)$ which are defined by the equalities

$$B_0(t)u = M_0(t)u - s_0 u, \qquad B_k(t)u = M_k(t)u, \quad k = 1, 2, 3.$$

Then the problem (2.7)–(2.9) can be rewritten in the form

$$u''''(t) + A_2 u''(t) + A_4 u(t) + \sum_{k \leq 3} B_k(t) u^{(k)}(t) = f(t), \qquad (2.12)$$

$$\gamma_k u^{(p_k)}(0) + \delta_k u^{(p_k)}(1) + \sum_{j \leq p_k - 1} \sum_{s \leq N_{kj}} T_{kj} u^{(j)}(t_{kjs})$$

$$= \varphi_k, \qquad k = 1, \ldots, 4, \qquad (2.13)$$

where $u(t) = u(t, \cdot)$, $f(t) = f(t, \cdot)$, $\varphi_k = \varphi_k(\cdot)$ are functions with values in the Hilbert space $H = L_2(G)$. Let us apply Theorem 5.3.2 to the problem (2.12)–(2.13). By virtue of Theorems 4.1.5 and 4.1.1 it follows that for some $s_0 > 0$ condition 1 of Theorem 5.3.2 is satisfied. Consequently,

$$\|(A_4 + \lambda I)^{-1}\| \leq (1 + \lambda)^{-1}, \qquad \lambda \geq 0,$$

and

$$\|A_4^{it}\| \leq \sup_{\lambda \in (0,\infty)} |\lambda^{it}| = 1, \qquad t \in \mathbb{R}.$$

Then, by virtue of [72, p.103] we have $[H, H(A_4)]_q = D(A_4^q)$ and by virtue of R. Seeley [61]

$$D(A_4^q) = [L_2(G), W_2^4(G, L_k u = 0, k = 1, 2)]_q$$

$$= W_2^{4q}(G, L_k u = 0, m_k < 4q - 1/2).$$

Then from conditions 2, 6 and 8 it follows that conditions 2 and 4 of Theorem 5.3.2 are fulfilled. So, for the problem (2.12)–(2.13) all conditions of Theorem 5.3.2 are fulfilled, from which follows the statement of Theorem 2.3. ∎

Remark 2.4. *In the case when the equation (2.7) is biharmonic, i.e., $\Delta^2 u = f$, conditions 1,2,3,5 and 6 of Theorem 2.3 should be replaced by the following conditions:*

(1)′ $\Gamma \in C^4$;

(2)′ *operators* $A_2 : D(A_2) = W_2^2(G, L_k u = 0, m_k \leq 1)$, $A_2 u = \Delta u$ *and* $A_4 : D(A_4) = W_2^4(G, L_k u = 0, k = 1, 2)$, $A_4 u = \Delta^2 u$ *are formally symmetric in* $L_2(G)$;

(5)′ *let* x' *be any point on* Γ, *the vector* σ' *be tangent and the vector* σ *be normal to* Γ *at the point* $x' \in \Gamma$. *Consider the following ordinary differential problem*

$$[\lambda + (\sum_{j=1}^{r}(\sigma'_j - i\sigma_j\frac{d}{dy})^2)^2]u(y) = 0, \qquad y \geq 0, \; \lambda \geq 0, \tag{2.10}'$$

$$\sum_{|\alpha|=m_k} b_{k\alpha}(x')(\sigma' - i\sigma\frac{d}{dy})^\alpha u(y)|_{y=0} = h_k, \qquad k = 1, 2; \tag{2.11}'$$

the problem (2.10)′–(2.11)′ *should have one and only one solution, including all its derivatives, tending to zero as* $y \to \infty$ *for any numbers* $h_k \in \mathbb{C}$.

6.2.2. An isomorphism of boundary value problems for elliptic equations.
Consider in the semi-infinite strip $[0, \infty) \times [0, 1]$, a principally boundary value problem for an elliptic equation of the 4-th order,

$$\begin{aligned}L(x, D_x, D_t)u &= D_t^4 u(t, x) + a(x)D_x^2 D_t^2 u(t, x) + b(x)D_x^4 u(t, x) \\ &\quad + M_2 D_t^2 u(t, \cdot)|_x + M_0 u(t, \cdot)|_x \\ &= f(t, x), \qquad (t, x) \in [0, \infty) \times [0, 1],\end{aligned} \tag{2.14}$$

$$\begin{aligned}L_k u &= a_k D_x^{k+1} u(t, 0) + \sum_{j \leq k} a_{kj} D_x^j u(t, 0) = 0, \; t \in [0, \infty), \; k = 1, 2, \\ L_k u &= b_k D_x^{k-1} u(t, 1) + \sum_{j \leq k-2} b_{kj} D_x^j u(t, 1) = 0, \; t \in [0, \infty), \; k = 3, 4,\end{aligned} \tag{2.15}$$

$$\begin{aligned}P_1 u &= \gamma u(0, x) + \delta D_t u(0, x) = \varphi_1(x), \qquad x \in [0, 1], \\ P_2 u &= \gamma D_t^2 u(0, x) + \delta D_t^3 u(0, x) = \varphi_2(x), \qquad x \in [0, 1],\end{aligned} \tag{2.16}$$

where a_k, a_{kj}, b_k, b_{kj}, γ, δ are complex numbers, $D_t = \frac{\partial}{\partial t}$, $D_x = \frac{\partial}{\partial x}$ and the corresponding spectral problem is

$$\begin{aligned}&\lambda^4 u(x) + \lambda^2(a(x)D_x^2 u(x) + M_2 u|_x) + b(x)D_x^4 u(x) \\ &+ M_0 u|_x = 0, \qquad x \in [0, 1],\end{aligned} \tag{2.17}$$

$$L_k u = a_k D_x^{k+1} u(0) + \sum_{j \le k} a_{kj} D_x^j u(0) = 0, \qquad k = 1, 2,$$

$$L_k u = b_k D_x^{k-1} u(1) + \sum_{j \le k-2} b_{kj} D_x^j u(1) = 0, \qquad k = 3, 4,$$

(2.18)

Theorem 2.5. *Let the following conditions be satisfied:*

(1) *conditions 1–3 of Theorem 2.1 are fulfilled;*

(2) *the spectral problem (2.17)–(2.18) does not have eigenvalues on the straight line* $\mathrm{Re}\lambda = 0$;

(3) *at least one of the two numbers* γ *and* δ *is not equal to zero;* $\mathrm{Re}\gamma\delta^{-1} \le 0$ *when* $\delta \ne 0$;

(4) *operators* M_k *from* $W_2^{4-k}((0,1), L_s u = 0, m_s \le 3 - k)$ *into* $L_2(0,1)$ *are compact.*

Then the operator $\mathbb{L} : u \to (L(x.D_x, D_t)u, P_1 u, P_2 u)$ *from*

$$W_p^4((0, \infty), W_2^4((0,1), L_k u = 0, k = 1, \ldots, 4), L_2(0,1))$$

into

$$L_{p,2}((0,\infty) \times (0,1)) \overset{2}{\underset{k=1}{\dotplus}} B_{2,p}^{4-p_k-1/p}((0,1), L_s u = 0, m_s < 4 - p_k - 1/p - 1/2),$$

where $p_2 = p_1 + 2$, $p_1 = 0$ *if* $\delta = 0$, $p_1 = 1$ *if* $\delta \ne 0$, *and* $p \ne 2$, *or if* $p = 2$ *then* $m_s \ne 3 - p_k,$[23] *is an isomorphism.*

Proof. Let us denote

$$H = L_2(0,1), \qquad H_4 = W_2^4((0,1), L_k u = 0, \ k = 1, \ldots, 4).$$

Consider operators A_k which are defined by the equalities

$$A_2 u = a(x)D^2 u(x) + M_2 u|_x,$$

$$A_4 u = b(x)D^4 u(x) + M_0 u|_x.$$

(2.19)

Then the problem (2.14)–(2.16) can be rewritten in the form

$$u''''(t) + A_2 u''(t) + A_4 u(t) = f(t),$$

(2.20)

---

[23] In the case where $p = 2$ and $m_s = 3 - p_k$ instead of $B_{2,2}^{4-p_k-1/2}((0,1), L_s u = 0, m_s < 3 - p_k)$ it should be written $B_{2,2}^{4-p_k-1/2}((0,1), L_s u = 0, m_s < 3 - p_k; L_s u \in \tilde{B}_{2,2}^{1/2}(0,1), m_s = 3 - p_k)$ [72, p.321]. $\tilde{B}_{p,q}^s(G) = \{u | u \in B_{p,q}^s(\mathbb{R}^r), \mathrm{supp}(u) \subset \bar{G}\}$.

$$\gamma u(0) + \delta u'(0) = \varphi_1,$$
$$\gamma u''(0) + \delta u'''(0) = \varphi_2, \tag{2.21}$$

where $u(t) = u(t,\cdot), f(t) = f(t,\cdot), \varphi_k = \varphi_k(\cdot)$ are functions with values in the Hilbert space $H = L_2(0,1)$. The rest of the proof coincides with a part of the proof of Theorem 2.1, namely, we apply Theorem 5.3.3 instead of Theorem 5.3.1 to problem (2.20)–(2.21). ∎

Consider in the semi-infinite domain $\Omega = [0,\infty) \times G$ a principally boundary value problem for an elliptic equation of the 4-th order

$$L(x, D_x, D_t)u = D_t^4 u(t,x) + \sum_{i,j=1}^{r} D_i(a_{ij}(x)D_j D_t^2 u(t,x))$$

$$+ \sum_{i,j,k,m=1}^{r} D_i D_j(b_{ijkm}(x)D_k D_m u(t,x))$$

$$+ M_2 D_t^2 u(t,\cdot)|_x + M_0 u(t,\cdot)|_x$$

$$= f(t,x), \qquad (t,x) \in [0,\infty) \times G, \tag{2.22}$$

$$L_k u = \sum_{|\alpha| \le m_k} b_{k\alpha}(x')D_x^\alpha u(t,x') = 0, \quad (t,x') \in [0,\infty) \times \Gamma, \ k=1,2, \tag{2.23}$$

$$P_1 u = \gamma u(0,x) + \delta D_t u(0,x) = \varphi_1(x), \qquad x \in G,$$
$$P_2 u = \gamma D_t^2 u(0,x) + \delta D_t^3 u(0,x) = \varphi_2(x), \qquad x \in G, \tag{2.24}$$

where $m_k \le 3$.

Let us denote $H = L_2(G)$ and consider operators A_2 and A_4 which are defined by the equalities

$$D(A_2) = W_2^2(G, L_k u = 0, m_k \le 1),$$

$$A_2 u = \sum_{i,j=1}^{r} D_i(a_{ij}(x)D_j u(x)) + M_2 u|_x, \tag{2.25}$$

$$D(A_4) = W_2^4(G, L_k u = 0, k=1,2),$$

$$A_4 u = \sum_{i,j,k,m=1}^{r} D_i D_j(b_{ijkm}(x)D_k D_m u(x)) + M_0 u|_x. \tag{2.26}$$

Theorem 2.6. *Let the following conditions be satisfied:*

(1) *conditions 1–6 of Theorem 2.3 are fulfilled;*

(2) $(A_4 u, u) \ge c^2(u,u), \quad c \neq 0, \ u \in W_2^4(G, L_s u = 0, s=1,2);$

(3) at least one of the two numbers γ and δ is not equal to zero; $\operatorname{Re}\gamma\delta^{-1} \leq 0$ when $\delta \neq 0$;

(4) operators M_k from $W_2^{4-k}(G, L_s u = 0, m_s \leq 3-k)$ into $L_2(G)$ are compact;
$$M_0 u = M_0^* u, \quad u \in W_2^4(G, L_s u = 0, s = 1, 2);$$

$$\operatorname{Re}(M_2 u, u) \leq 0 \text{ and } \operatorname{Re}(M_2^* u, u) \leq 0, \qquad u \in W_2^2(G, L_s u = 0, m_s \leq 1).$$

Then the operator $\mathbb{L} : u \to (L(x.D_x, D_t)u, P_1 u, P_2 u)$ from

$$W_p^4((0, \infty)), W_2^4(G, L_k u = 0, k = 1, 2), L_2(G))$$

into

$$L_{p,2}((0, \infty) \times G) \dotplus_{k=1}^{2} B_{2,p}^{4-p_k-1/p}(G, L_s u = 0, m_s < 4 - p_k - 1/p - 1/2),$$

where $p_2 = p_1 + 2$, $p_1 = 0$ if $\delta = 0$, $p_1 = 1$ if $\delta \neq 0$, and $p \neq 2$, or if $p = 2$ then $m_s \neq 3 - p_k$,[24] is an isomorphism.

Proof. The problem (2.22)–(2.24) can be rewritten in the form

$$u''''(t) + A_2 u''(t) + A_4 u(t) = f(t), \qquad t > 0, \tag{2.27}$$

$$\gamma u(0) + \delta u'(0) = \varphi_1,$$
$$\gamma u''(0) + \delta u'''(0) = \varphi_2, \tag{2.28}$$

where $u(t) = u(t, \cdot), f(t) = f(t, \cdot), \varphi_k = \varphi_k(\cdot)$ are functions with values in the Hilbert space $H = L_2(G)$.

All conditions of Theorem 5.3.5 for the problem (2.27)–(2.28) can be checked as in Theorem 2.3. Then, from Theorem 5.3.5 follows the statement of Theorem 2.6. ∎

6.2.3. Completeness of elementary solutions of boundary value problems for elliptic equations. Let us consider in the semi-infinite strip $[0, \infty) \times [0, 1]$, a principally boundary value problem for an elliptic equation of the 4-th order,

$$L(x, D_x, D_t)u = D_t^4 u(t, x) + a(x)D_x^2 D_t^2 u(t, x) + b(x)D_x^4 u(t, x)$$
$$+ M_2 D_t^2 u(t, \cdot)|_x + M_0 u(t, \cdot)|_x$$
$$= 0, \qquad (t, x) \in [0, \infty) \times [0, 1], \tag{2.29}$$

---

[24] In the case where $p = 2$ and $m_s = 3 - p_k$ instead of $B_{2,2}^{4-p_k-1/2}(G, L_s u = 0, m_s < 3 - p_k)$ it should be written $B_{2,2}^{4-p_k-1/2}(G, L_s u = 0, m_s < 3 - p_k; L_s u \in \widetilde{B}_{2,2}^{1/2}(G)), m_s = 3 - p_k)$ [72, p.321]. $\widetilde{B}_{p,q}^s(G) = \{u | u \in B_{p,q}^s(\mathbb{R}^r), \operatorname{supp}(u) \subset \overline{G}\}$.

$$L_k u = a_k D_x^{k+1} u(t,0) + \sum_{j \leq k} a_{kj} D_x^j u(t,0) = 0, \ t \in [0,\infty], \ k = 1,2,$$

$$L_k u = b_k D_x^{k-1} u(t,1) + \sum_{j \leq k-2} b_{kj} D_x^j u(t,1) = 0, \ t \in [0,\infty), \ k = 3,4,$$

(2.30)

$$u(0,x) = \varphi_1(x), \qquad x \in [0,1],$$

$$D_t^2 u(0,x) = \varphi_2(x), \qquad x \in [0,1],$$

(2.31)

and the corresponding spectral problem (2.17)–(2.18), where a_k, a_{kj}, b_k, b_{kj} are complex numbers; $D_t = \frac{\partial}{\partial t}$, $D_x = \frac{\partial}{\partial x}$.

By virtue of Lemma 2.0.1 the function of the form

$$u_i(t,x) = e^{\lambda_i t}\left(\frac{t^{k_i}}{k_i!} u_{i0}(x) + \frac{t^{k_i-1}}{(k_i-1)!} u_{i1}(x) + \cdots + u_{ik_i}(x)\right) \tag{2.32}$$

becomes an elementary solution to the problem (2.29)–(2.30) if and only if a system of functions $u_{i0}(x), u_{i1}(x), \ldots, u_{ik_i}(x)$ is a chain of root functions of the problem (2.17)–(2.18), corresponding to the eigenvalue λ_i.

Theorem 2.7. *Let the following conditions be satisfied:*

(1) *conditions 1–3 of Theorem 2.1 are fulfilled;*

(2) *the spectral problem (2.17)–(2.18) does not have eigenvalues on the straight line* $\mathrm{Re}\lambda = 0$;

(3) *operators* M_k *from* $W_2^{4-k}((0,1), L_s u = 0, m_s \leq 3 - k)$ *into* $L_2(0,1)$ *are compact.*

(4) $\varphi_k \in B_{2,p}^{4-p_k-1/p}((0,1), L_s u = 0, m_s < 4 - p_k - 1/p - 1/2), \ k = 1,2,$ *where* $p_1 = 0, p_2 = 2,$ *and* $p \neq 2,$ *or if* $p = 2$ *then* $m_s \neq 3 - p_k$.[25]

Then the problem (2.29)–(2.31) has a unique solution

$$u \in W_p^4((0,\infty), W_2^4((0,1), L_k u = 0, k = 1, \ldots, 4), L_2(0,1))$$

and there exist numbers C_{in} *such that*

$$\lim_{n \to \infty} \int_0^\infty \left(\left\| D_t^4 u(t,\cdot) - \sum_{i=1}^n C_{in} D_t^4 u_i(t,\cdot) \right\|_{0,2}^p \right.$$

$$\left. + \left\| u(t,\cdot) - \sum_{i=1}^n C_{in} u_i(t,\cdot) \right\|_{4,2}^p \right) dt = 0,$$

---

[25] See the footnote on p.221.

where $u(t, x)$ is a solution to the problem (2.29)–(2.31) and $u_i(t, x)$ are the elementary solutions (2.32) to the problem (2.29)–(2.30).

Proof. Apply Theorem 5.4.4 to the problem (2.29)–(2.31). Consider in $H = L_2(0, 1)$ operators A_2 and A_4, defined by the equalities (2.19). Then, the problem (2.29)–(2.31) can be rewritten in the form

$$u''''(t) + A_2 u''(t) + A_4 u(t) = 0, \qquad t > 0, \tag{2.33}$$

$$u(0) = \varphi_1, \qquad u''(0) = \varphi_2, \tag{2.34}$$

where $u(t) = u(t, \cdot)$, $f(t) = f(t, \cdot)$, $\varphi_k = \varphi_k(\cdot)$ are functions with values in the Hilbert space $H = L_2(0, 1)$. Conditions 1,2,4 and 5 of Theorem 5.4.4 have been checked in Theorem 2.1 for

$$H_4 = W_2^4((0, 1), L_k u = 0, k = 1, \dots, 4).$$

By virtue of [72, p.350]

$$s_j(J; W_2^4(0, 1), L_2(0, 1)) \sim j^{-4},$$

i.e., for $q > 1/4$

$$J \in \sigma_q(W_2^4(0, 1), L_2(0, 1)). \tag{2.35}$$

Since $W_2^4((0, 1), L_k u = 0, k = 1, \dots, 4)$ is a subspace of $W_2^4(0, 1)$, then, by virtue of Lemma 2.3.3., from (2.35) follows

$$J \in \sigma_q(W_2^4((0, 1), L_k u = 0, k = 1, \dots, 4), L_2(0, 1)), \quad q > 1/4,$$

i.e., condition 3 of Theorem 5.4.4 is fulfilled. So, for the problem (2.33)–(2.34) all conditions of Theorem 5.4.4 have been checked and the theorem statement follows. ∎

Consider in the semi-infinite domain $\Omega = [0, \infty) \times G$ a principally boundary value problem for an elliptic equation of the 4-th order

$$L(x, D_x, D_t)u = D_t^4 u(t, x) + \sum_{i,j=1}^{r} D_i(a_{ij}(x)D_j D_t^2 u(t, x))$$

$$+ \sum_{i,j,k,m=1}^{r} D_i D_j(b_{ijkm}(x)D_k D_m u(t, x))$$

$$+ M_2 D_t^2 u(t, \cdot)|_x + M_0 u(t, \cdot)|_x$$

$$= 0, \qquad (t, x) \in [0, \infty) \times G, \tag{2.36}$$

$$L_k u = \sum_{|\alpha| \le m_k} b_{k\alpha}(x') D_x^\alpha u(t, x') = 0, \quad (t, x') \in [0, \infty) \times \Gamma, \ k = 1, 2, \tag{2.37}$$

$$
\begin{aligned}
u(0, x) &= \varphi_1(x), & x \in G, \\
D_t^2 u(0, x) &= \varphi_2(x), & x \in G,
\end{aligned}
\tag{2.38}
$$

and the corresponding spectral problem

$$\lambda^4 u(x) + \lambda^2 \Big(\sum_{i,j=1}^r D_i(a_{ij}(x) D_j u(x)) + M_2 u|_x \Big)$$

$$+ \sum_{i,j,k,m=1}^r D_i D_j(b_{ijkm}(x) D_k D_m u(x)) + M_0 u|_x$$

$$= 0, \quad x \in G, \tag{2.39}$$

$$L_k u = \sum_{|\alpha| \le m_k} b_{k\alpha}(x') D_x^\alpha u(x') = 0, \quad x' \in \Gamma, \ k = 1, 2. \tag{2.40}$$

By virtue of Lemma 2.0.1 the function of the form (2.32) becomes an elementary solution to the problem (2.36)–(2.37) if and only if a system of functions $u_{i0}(x), u_{i1}(x), \dots, u_{ik_i}(x)$ is a chain of root functions of the spectral problem (2.39)–(2.40), corresponding to the eigenvalue λ_i.

Consider in $H = L_2(G)$ operators A_2 and A_4 defined by the equalities (2.25)–(2.26).

Theorem 2.8. *Let the following conditions be satisfied:*

(1) $a_{ij} \in C(\overline{G})$, $b_{ijkm} \in C(\overline{G})$, $b_{k\alpha} \in C^{4-m_k}(\overline{G})$, $\Gamma \in C^4$;

(2) $a_{ij}(x) = \overline{a_{ji}(x)}$; $b_{sjkm}(x) = \overline{b_{kmsj}(x)}$; A_2 and A_4 are formally symmetric operators in $L_2(G)$;

(3) *if* $x \in \overline{G}$, $\sigma \in \mathbb{R}^r$, $|\sigma| + |\lambda| \ne 0$ *then*

$$\lambda + \sum_{i,j,k,m=1}^r b_{ijkm}(x) \sigma_i \sigma_j \sigma_k \sigma_m \ne 0, \qquad \lambda \ge 0,$$

and there exist rays ℓ_k *with angles between neighbouring rays less than* π/r *such that for* $x \in \overline{G}$, $\sigma \in \mathbb{R}^r$, $|\sigma| + |\lambda| \ne 0$, $\lambda \in \ell_k$ *the following is true*

$$\lambda^4 + \lambda^2 \sum_{j,k=1}^r a_{jk}(x) \sigma_j \sigma_k + \sum_{i,j,k,m=1}^r b_{ijkm}(x) \sigma_i \sigma_j \sigma_k \sigma_m \ne 0;$$

(4) *system* (2.37) *is normal;*

(5) *let x' be any point on Γ, the vector σ' be tangent and the vector σ be normal to Γ at the point $x' \in \Gamma$. Consider the following ordinary differential problem*

$$[\lambda + \sum_{s,j,k,m=1}^{r} b_{sjkm}(x')(\sigma'_s - i\sigma_s \frac{d}{dy})(\sigma'_j - i\sigma_j \frac{d}{dy})$$

$$\times (\sigma'_k - i\sigma_k \frac{d}{dy})(\sigma'_m - i\sigma_m \frac{d}{dy})]u(y) = 0, \quad y \geq 0, \ \lambda \geq 0, \tag{2.41}$$

$$\sum_{|\alpha|=m_k} b_{k\alpha}(x')(\sigma' - i\sigma \frac{d}{dy})^\alpha u(y)|_{y=0} = h_k, \quad k = 1,2; \tag{2.42}$$

and the ordinary differential equation

$$[\lambda^4 + \lambda^2 \sum_{j,k=1}^{r} a_{jk}(x')(\sigma'_j - i\sigma_j \frac{d}{dy})(\sigma'_k - i\sigma_k \frac{d}{dy})$$

$$+ \sum_{s,j,k,m=1}^{r} b_{sjkm}(x')(\sigma'_s - i\sigma_s \frac{d}{dy})(\sigma'_j - i\sigma_j \frac{d}{dy})$$

$$\times (\sigma'_k - i\sigma_k \frac{d}{dy})(\sigma'_m - i\sigma_m \frac{d}{dy})]u(y) = 0, \quad y \geq 0, \ \lambda \in \ell_k; \tag{2.43}$$

the problems (2.41)–(2.42) and (2.43)–(2.42) should have one and only one solution, including all its derivatives, tending to zero as $y \to \infty$ for any numbers $h_k \in \mathbb{C}$;

(6) *for $x \in \overline{G}$, $\sigma \in \mathbb{C}^r$*

$$\sum_{i,j=1}^{r} a_{ij}(x)\sigma_i \bar{\sigma}_j \leq 0;$$

(7) $(A_4 u, u) \geq c^2(u,u)$, $c \neq 0$, $u \in W_2^4(G, L_s u = 0, s = 1;2)$;

(8) *operators M_k from $W_2^{4-k}(G, L_s u = 0, m_s \leq 3-k)$ into $L_2(G)$ are compact;*
$$M_0 u = M_0^* u, \quad u \in W_2^4(G, L_s u = 0, s = 1,2);$$

$$\text{Re}(M_2 u, u) \leq 0 \text{ and } \text{Re}(M_2^* u, u) \leq 0, \quad u \in W_2^2(G, L_s u = 0, m_s \leq 1);$$

(9) $\varphi_k \in B_{2,p}^{4-p_k-1/p}(G, L_s u = 0, m_s < 4 - p_k - 1/p - 1/2)$, $k = 1,2$, *where* $p_1 = 0, p_2 = 2$, *and $p \neq 2$, or if $p = 2$ then $m_s \neq 3 - p_k$.*[26]

---

[26] See the footnote on p.223.

Then the problem (2.36)–(2.38) has a unique solution

$$u \in W_p^4((0,\infty), W_2^4(G, L_k u = 0, k = 1,2), L_2(G))$$

and there exist numbers C_{in} such that

$$\lim_{n \to \infty} \int_0^\infty (\|D_t^4 u(t, \cdot) - \sum_{i=1}^n C_{in} D_t^4 u_i(t, \cdot)\|_{0,2,G}^p$$

$$+ \|u(t, \cdot) - \sum_{i=1}^n C_{in} u_i(t, \cdot)\|_{4,2,G}^p) \, dt = 0,$$

where $u(t, x)$ is a solution to the problem (2.36)–(2.38) and $u_i(t, x)$ are the elementary solutions (2.32) to the problem (2.36)–(2.37).

Proof. Apply Theorem 5.4.5 to the problem (2.36)–(2.38). The problem (2.36)–(2.38) can be rewritten in the form

$$u''''(t) + A_2 u''(t) + A_4 u(t) = 0, \qquad t > 0, \tag{2.44}$$

$$u(0) = \varphi_1, \qquad u''(0) = \varphi_2, \tag{2.45}$$

where $u(t) = u(t, \cdot)$, $f(t) = f(t, \cdot)$, $\varphi_k = \varphi_k(\cdot)$ are functions with values in the Hilbert space $H = L_2(G)$. Conditions 1 and 2 of Theorem 5.4.5 have been checked in Theorem 2.3. By virtue of [72, p.350]

$$s_j(J; W_2^4(G), L_2(G)) \sim j^{-\frac{4}{r}},$$

i.e., when $q > \frac{r}{4}$ we have

$$J \in \sigma_q(W_2^4(G), L_2(G)). \tag{2.46}$$

Since $W_2^4(G, L_k u = 0, k = 1, 2)$ is a subspace of $W_2^4(G)$, then, by virtue of Lemma 2.3.3., from (2.46) follows

$$J \in \sigma_q(W_2^4(G, L_k u = 0, k = 1, 2), L_2(G)), \qquad q > \frac{r}{4},$$

i.e., condition 3 of Theorem 5.4.5 is fulfilled.

By virtue of Theorem 4.1.1 from condition 5 follows condition 4 of Theorem 5.4.5.

So, for the problem (2.44)–(2.45) all conditions of Theorem 5.4.5 have been checked and the theorem statement follows. ∎

Reference Notes

To chapter 2. The completeness of root vectors of non-selfadjoint operators was first studied by M. V. Keldysh [29, 30]. From these works follows Theorem 2.11, which is known as the Keldysh theorem on completeness of root vectors of weakly perturbed selfadjoint operators.

The main result of 2.1 – theorem 2.1.29 – is given in the book by N. Dunford and Ja. T. Schwartz [13, ch.XI, §9.29]. As can be seen in 2.1, a nontrivial method is worked out for the proof of theorem 2.1.29. This method applies such important results of functional analysis as the Schmidt expansion of compact operators (2.1.8), the Weyl inequality (2.1.15) – the relation between the eigenvalues and s-numbers (2.1.16), the resolvent growth estimate of a Volterra operator – Theorem 2.1.23.

The main result of 2.2 – Theorem 2.2.3 – is also given in the book by N. Dunford and Ja. T. Schwartz [13, ch.XI, §9.31]. Theorem 2.2.11 follows from works of M. V. Keldysh [29, 30]. Theorems 2.2.12 and 2.2.14 and Lemma 2.2.13 were proved by S. Ya. Yakubov and K. S. Mamedov [86]. The other results in this paragraph 2.2 were mainly obtained by the author during the work for this book.

Keldysh also proved the fundamental theorem on the n-fold completeness of a system of root vectors of a polynomial operator pencil with the principal part generated by one selfadjoint operator (see [29, 30]). The ideas and methods of these works were supplemented and developed by Dzh. E. Allakhverdiev [5], A. S. Markus [46,47], S. Ya. Yakubov and K. S. Mamedov [86], A. A. Shkalikov [64] and others. Theorem 2.3 [64, p.156] was mainly proved by S. Ya. Yakubov [89]. But the results obtained by the author [90, 91] are much stronger since these works are concerned with the completeness of root vectors of a system of unbounded polynomial operator pencils, rather than of a single operator pencil and even for one operator pencil our results are stronger since (unlike [64, p.141]) we do not suppose that

$$\|A_k u\| \leq C \|A^k u\|, \quad u \in D(A^k),$$

where $A \geq \gamma^2 I$ in H.

The detailed bibliography on this question, including works published untill 1982, can be found in the review by G. V. Radzievskii [56]. In the works of A. G. Kostiouchenko and G. V. Radzievskii [34], M. G. Gasymov [16] and others, the problem of multiple completeness is given in a more general formulation through consideration not only of the Cauchy problem, but also of boundary value problems in both the segment and the semi-axis. A detailed review of these works as well is given in [56].

These abstract results cover mainly only those differential equations, the principal part of which consists of the binomial $A(x, D) + \lambda^n$, where $A(x, D)$ is an elliptic operator. Theorems on n-fold completeness of root vectors, proved in 2.3 for unbounded polynomial operator pencils, remove the above mentioned gap. These results have been partially published in works by S. Ya. Yakubov [88, 89, 91].

In chapters 3 and 4, with the help of abstract results of 2.3, the results on completeness of root functions of boundary value problems both for ordinary and partial differential equations with full principal parts are obtained.

Theorem 2.3.16 is the Keldysh theorem for a pencil [29, 30]. The results of Keldysh are also given in the book by V. A. Sadovnichii [58].

A lot of works are devoted to quadratic operator pencils not of the Keldysh type. Important results were obtained by M. G. Krein and H. K. Langer [38] (see also [19 ,ch.V, §12]), S. Ya. Yakubov [79, 82] , A. I. Miloslavskii [50], A. A. Shkalikov [62], S. S. Mirzoev [51] and others. In these works various results on the theory of quadratic operator pencils are given.

To chapter 3. In this chapter the investigative method of local regular boundary value problems for elliptic equations, given in papers by M. S. Agranovich and M. I. Vishik [3], and S. Agmon and L. Nirenberg [2], is modified for the investigation of principally regular ordinary differential equations (with a polynomial complex spectral parameter) with functional conditions that are polynomially dependent on the complex spectral parameter. The region of coerciveness is found for the complex spectral parameter. This question is closely connected with the question of the construction of the Green function, but does not coincide with it. The problem of constructing the Green function for regular boundary value problems originates from the works of Birkhoff-Tamarkin (see [7], [69]). These investigations were continued in the books by M. A. Naimark [53], M. L. Rasulov [57], A. A. Dezin [10] and others. The results of 3.1 mainly belong to the author and have been partially published in the paper [84] and in the book [85]. For non-local boundary value conditions, regular in terms of Birkhoff-Tamarkin, Theorem 3.1.7 in the case when $n = 1$, B_k are differential operators and $T_\nu = 0$ was proved by I. D. Evzerov and P. Ye. Sobolevskii [14]. Theorem 3.2.5 when $n = 1$ and $T_\nu = 0$ was proved by S. Ya. Yakubov and K. S. Mamedov [87]. Some of the results of V. N. Vizitei and A. S. Markus [77], A. I. Vagabov [75, 76] and A. A. Shkalikov [63] follow from Theorem 3.2.5. In [75] a part of the boundary value conditions contains the largest derivative only in one boundary point of an interval and in [63] the leading coefficients are constant. The known results of N. Dunford and Ja. T. Schwartz [12, ch.XIX] follow from our theorems 3.2.5–3.2.8 since, first, the problems with functional conditions are covered and, second, the equation is with a variable complex-valued function of the highest derivative. For ordinary differential equations with boundary value conditions of Sturm type in the work of S. Agmon [1] the completeness of root functions with a variable complex-valued function of the highest derivative is proved. In [63] it is proved that a system of root functions is a basis in the case of strong regular conditions, thus it naturally continues work by V. P. Mikhailov [49], L. A. Muravej [52] and others. Theorem 3.2.8 for a periodic boundary value condition has a nonempty intersection with the result of R. Seeley [59]. So, the completeness theorems of root functions of regular boundary value problems for ordinary differential equations, proved in chapter 3, have the following advantages in comparison with the numerous previous results on this theme:

1) problems (2.33)–(2.34) and (2.43)–(2.44) are differential only in their principal parts;

2) the roots of the characteristic equations (2.35) and (2.45) do not satisfy the conditions $\arg \omega_j(x) =$const, $\arg(\omega_j(x) - \omega_k(x)) =$const;

3) the conditions of p-regularity are algebraic;

4) the orders of the boundary value conditions may exceed the order of the equation, and instead of the traditional condition $m_\nu \leq m - 1$ it is supposed that $\max\{m_\nu\} - \min\{m_\nu - dn_\nu\} \leq m - 1$.

There are many papers on expansion in root functions of irregular boundary value problems for ordinary differential equations (see, for example, the book by M. A. Naimark [53, ch.II, §5.4]). But to find an order of non-coerciveness of irregular boundary value problems is a more difficult problem. Papers by Ya. S. Yakubov [94, 95, 96] have been devoted, in particular, to this question and also to completeness and expansion in root functions of some second order irregular ordinary differential boundary value problems.

To chapter 4. M. V. Keldysh [30] and F. E. Browder [9] proved the completeness of root functions of elliptic boundary value problems if the principal part is selfadjoint. S. Agmon [1] proved the completeness of root functions of elliptic boundary value problems if the principal part is non-selfadjoint (see also G. Geymonat and P. Grisvard [17]). In this chapter the completeness of root functions is proved if the principal part of differential-operator problems is an elliptic boundary value problem for differential equations with a polynomial spectral parameter. Completeness of root functions of elliptic (in terms of Douglis-Nirenberg) pseudo-differential systems on a compact manifold without boundary, was proved by A. N. Kozhevnikov [36]. Completeness of root functions of elliptic boundary value problems, when the spectral parameter is entered linearly not only in the equation, but in boundary conditions too, was proved by A. N. Kozhevnikov [36, 37], and L. A. Kotko and S. G. Krein [35]. The main result of this chapter – Theorem 4.1.8, was also published in a very partial case, namely when $m_\nu \leq 2m - 1$, in the work by A. V. Shkred [65], that was submitted for publication a year after this result was published by the author [90]. Theorem 4.1.8 also strengthens results in the author's previous work [90, 91]. Without the traditional condition $m_\nu \leq 2m - 1$ the theorem of completeness of root functions was first proved by the author [91]. The paper by M. S. Agranovich [4] contains results about the completeness of root functions of elliptic (in terms of Agmon-Douglis-Nirenberg) non-selfadjoint problems with a parameter.

Theorems 4.2.2 and 4.2.3 were published by the author [81]. For the Laplace equation the boundary value problems with non-classical spectral asymptotics were also described, but with boundary value conditions, containing pseudo-differential or abstract operators, in the books by A. A. Dezin [10], V. I. Gorbachuk and M. L. Gorbachuk [20] etc..

To chapter 5. For $p = 2$ Theorem 5.1.3 was proved by B. A. Plamenevskii [55], and Theorem 5.1.6 was proved by Yu. A. Dubinskii [11]. Theorem 5.2.2 comes from E. Hill and R. S. Phillips [23], and K. Yosida [97]. Theorem 5.2.4 comes from Yu. I. Ljubich [45].

Completeness of elementary solutions for first order differential-operator equations was proved by S. Agmon and L. Nirenberg [2], and for equations of higher order by A. A. Shkalikov [64 , p.165, Theorem 3.3]. These papers prove that any solution of a homogeneous differential-operator equation of the m-th order, belonging to $W_2^m(0, \infty)$, may be approximated by the linear combinations of elementary solutions. In the work of V. V. Vlasov [78] the minimality of elementary solutions to differential-operator equations of the Keldysh type is proved.

The results of 5.3 and 5.4 were derived by the author. We have also found the algebraic conditions on boundary value problems for differential-operator equations of the 4-th order in an interval to be fredholm. A solution to the Cauchy problem for a differential-operator equation of the parabolic type is approximated by the linear combinations of elementary solutions. In chapter 5 this problem is solved in the case when application of the Fourier method becomes very difficult. This is the case when the corresponding spectral problem is principally non-selfadjoint.

To chapter 6. A lot of monographs and articles have been devoted to the solvability questions of regular elliptic boundary value problems. Nevertheless the results on the solvability of boundary value problems in the cylindrical domains for elliptic equations of the 4-th order, which are presented in chapter 6, are new.

They are new even if the problem is differential, since the boundary conditions are non-local. Algebraic conditions of the solvability are found.

We have considered elliptic boundary value problems in tube domains (hence in nonsmooth domains). The book by P. Grisvard [21] is devoted to this question, however our results are new in comparison.

In the works by V. A. Kondratiev [32], V. A. Kondratiev and O. A. Oleinik [33], and A. A. Shkalikov [64] a complementary condition arises in types of theorems as theorems 6.2.1, 6.2.4 and 6.2.6. More precisely, determinants of auxiliary "Kondratiev problems" do not have zeros on the line $\mathrm{Re}\lambda = 3$.

Equations and problems that are considered here belong to known types only in their principal parts. Algebraic conditions of solvability and completeness of elementary solutions are found. The theorems of completeness of elementary solutions are new even for simple boundary conditions. Similar theorems have been proved in the papers by D. D. Joseph [25], Y. A. Ustinov and V. I. Yudovich [74], G. Geymonat and P. Grisvard [18].

The results in chapter 6 belong to the author.

References

[1] Agmon, S., *On the eigenfunctions and on the eigenvalues of general elliptic boundary value problems*, Comm. Pure Appl. Math., **15** (1962), 119-147.

[2] Agmon, S., Nirenberg, L., *Properties of solutions of ordinary differential equations in Banach spaces*, Comm. Pure Appl. Math., **16** (1963), 121-239.

[3] Agranovich, M. S., Vishik, M. I., *Elliptic problems with a parameter and parabolic problems of general type*, Uspekhi Mat. Nauk, **19**, No.3 (1964), 53-161; English translation in Russian Math. Surveys, **19**, No.3 (1964), 53-159.

[4] Agranovich, M. S., *On non-selfadjoint problems with a parameter, elliptic in terms of Agmon-Douglis-Nirenberg*, Funk. Anal. i Yego Priloz., **24** (1990), 59-61.

[5] Allakhverdiev, D. E., *On the completeness of a system of eigenelements and associated elements of non-selfadjoint operators close to normal operators*, Dokl. Akad. Nauk SSSR, **115** (1957), 207-210.

[6] Besov, O. V., Il'in, V. P., Nikolskii, S. M., " Integral Representations of Functions and Embedding Theorems", Halsted Press, New York, v.I, 1978.

[7] Birkhoff, G. D., *Boundary value and expansion problems of ordinary linear differential equations*, Trans. Amer. Math. Soc., **9** (1908), 373-395.

[8] Bitsadze, A. V., *On an uniqueness of Dirichlet problem's solution for partial elliptic equations*, Uspekhi Matem., Nauk, **3**, No.6 (1948), 211-212.

[9] Browder, F. E., *On the eigenfunctions and eigenvalues of the general elliptic differential operators*, Proc. Nat. Acad. Sci. USA, **39** (1953), 433-439.

[10] Dezin, A. A., "General Questions of the Boundary Value Problem Theory", Nauka, Moscow, 1980.

[11] Dubinskii, Ju. A., *On some differential-operator equations of an arbitrary order*, Mat. Sborn. **90**, No1 (1973), 3-22.

[12] Dunford, N., Schwartz, J. T., "Linear Operators. Part III: Spectral operators", Interscience, New York, 1971.

[13] Dunford, N., Schwartz, J. T., "Linear Operators. Part II. Spectral Theory", Interscience, New York, 1963.

[14] Evzerov, I. D., Sobolevskii, P. Je., *Fractional powers of ordinary differential operators*, Differ. Uravn., **9**, No.2 (1973), 228-240.

[15] Fattorini, H. O., "Second Order Linear Differential Equations in Banach Spaces", North-Holland, Amsterdam, 1985.

[16] Gasymov, M.G., *On multiple completeness of a part of eigenvectors and associated vectors of polynomial operator pencils*, Izv. Akad. Nauk Armyan. SSR.

Ser. Matem., **6**, No.2-3 (1971), 131-147.

[17] Geymonat, G., Grisvard, P., *Alcuni risultati di teoria spettrale per i problemi ai limiti lineari ellittici*, Rend. Semin. Mat. Univ. Padova, **38** (1967), 121-173.

[18] Geymonat, G., Grisvard, P., *Eigenfunction expansions associated to some non-selfadjoint operators*, Lecture notes in Mathematics, n.1121, Oberwolfach-Springer-Verlag (1983), 123-136.

[19] Gohberg, I. C., Krein M. G., "Introduction to the Theory of Linear Non-Selfadjoint Operators", Amer. Math. Soc., Providence, 1969.

[20] Gorbachuk, V. I., Gorbachuk M. L., "Boundary Value Problems for Differential-Operator Equations", Naukova Dumka, Kiev, 1984.

[21] Grisvard, P. "Elliptic Problems in Nonsmooth Domains", Pitman, Boston, 1985.

[22] Grisvard, P., *Caracterization de quelques espaces d'interpolation*, Arch. Rat. Mech. Anal., **25** (1967), 40-63.

[23] Hille, E., Phillips, R. S., "Functional Analysis and Semigroups", Amer. Math. Soc., Providence, 1957.

[24] Hörmander, L., " Linear Partial Differential Operators", Springer-Verlag, Berlin, 1963.

[25] Joseph, D. D., *The convergence of biorthogonal series for biharmonic and Stokes flow edge problems*, Part I. Siam.J. of Appl. Math., **33**, No.2 (1977), 337-347.

[26] Joseph, D. D., "A new separation of variables theory for problems of Stokes flow and elasticity", Proceedings of "Trends in applications of pure mathematics to mechanics", Pitman, London, 1979.

[27] Kantorovich, L. V., Akilov, G. P., "Functional Analysis in Normed Spaces", Pergamon Press, Oxford, 1964.

[28] Kato T., "Perturbation Theory for Linear Operators", Springer- Verlag, New York Inc., 1966.

[29] Keldysh, M. V., *On eigenfunction completeness of some classes of non-selfadjoint linear operators*, Uspekhi. Matem. Nauk, **27**, No.4 (1971), 15-47.

[30] Keldysh, M. V., *On eigenvalues and eigenfunctions of some classes of non-selfadjoint equations*, Dokl. Akad. Nauk SSSR, **77**, No.1 (1951) , 11-14.

[31] Kolmogorov, A. N., Fomin, S. V., "Elements of the Theory of Functions and Functional Analysis", Nauka, Moscow, 1972.

[32] Kondratiev, V. A., *Boundary value problems for elliptic equations in domains with conic or corner points*, Trans. Moscow. Math. Soc., (1967), 227-313.

[33] Kondratiev, V. A., Oleinik O. A., *Boundary value problems for partial differential equations in non-smooth domains*, Russian. Math. Surveys, **38**, No.2 (1983), 1-86.

[34] Kostiouchenko, A. G., Radzievskii, G. V., *On summation of n-fold expansions by Abel method*, Sibir. Matem. Zhurnal, **15**, No.4 (1974), 855-870.

[35] Kotko, Z. A., Krein, S. G., *On completeness of a system of eigenfunctions and associated functions of boundary value problems with a parameter in boundary conditions*, Dokl. Akad. Nauk SSSR, **227**, No.2 (1976), 288-290.

[36] Kozhevnikov, A. N., *Spectral problems for pseudo-differential systems elliptic by Douglis-Nirenberg, and their applications*, Matem. Sborn., **92**, No.1 (1973), 60-88.

[37] Kozhevnikov, A. N., *On relation between resolvents of three elliptic boundary value problems*, Prikl. Matem., Leningrad. Inzhen.-Stroitel. Institut (1977), 52-55.

[38] Krein, M. G., Langer. H. K., *To the theory of quadratic pencils of selfadjoint operators*, Dokl. Akad. Nauk SSSR, **154**, No.6, (1964), 1258-1261.

[39] Krein, S. G. "Linear Differential Equations in Banach Space", Providence, 1971.

[40] Krein, S. G., Petunin, Y. I., Semenov, E. M., "Interpolation of Linear Operators", Amer. Math. Soc., Providence, 1982.

[41] Lidskii, V. B., *Non-selfadjoint operators with traces*, Dokl. Akad. Nauk SSSR, **125**, No.3 (1959), 485-488.

[42] Lidskii, V. B., *On summation of series by principal vectors of non-selfadjoint operators*, Trudy Moskov. Mat. Obsh, **11** (1962), 3-35.

[43] Lions, J. L., "Equations Differentielles Operationnelles et Problems aux Limites", Springer-Verlag, Berlin, 1961.

[44] Lions, J. L, Magenes E., "Non-Homogeneous Boundary Value Problems and Applications", Springer-Verlag, Berlin, 1972.

[45] Ljubich, Yu. I., *On conditions of uniqueness of the solution to an abstract Cauchy problem*, Dokl. Akad. Nauk SSSR, **130**, No.5 (1960), 966-972.

[46] Markus, A. S., " Introduction to the Spectral Theory of Polynomial Operator Pencils", Amer. Math. Soc., Providence, 1988

[47] Markus, A. S., *To the spectral theory of polynomial pencils in Banach space*, Sibir. Matem. Zhurn., **8**, No.6 (1967), 1346-1369.

[48] Matsaev, V. I., *On a method of estimation of the resolvents of non-selfadjoint operators*, Dokl. Akad. Nauk SSSR, **154**, No.5, (1964), 1034-1037.

[49] Mikhailov, V. P., *On Riesz bases in $L^2(0,1)$*, Dokl. Akad. Nauk SSSR, **144**, No.5 (1962), 981-984.

[50] Miloslavskii, A. I., *Spectral properties of one class of quadratic operator pencils*, Func. Analis i Yego Pril., **15**, No.2 (1981), 81-82.

[51] Mirzoev, S. S., *On multiple completeness of root vectors of polynomial operator*

pencils, corresponding to the boundary value problems on the semi-axis, Funk. Analis i Yego Pril., **17**, No.2 (1983), 84-85.

[52] Muravei, L. A., *Riesz bases in $L_2[-1,1]$*, Trudy MIAN im. V. A. Steklova, **91** (1967), 113-131.

[53] Naimark, M. A., "Linear Differential Operators", Ungar, New York, 1967.

[54] Pazy, A., "Semigroups of Linear Operators and Applications to Partial Differential Equations', Springer, Berlin, 1983.

[55] Plamenevskii, B. A., *On the existence and asymptotics of solutions of differential equations with unbounded operator coefficients in Banach space*, Izv. Akad. Nauk SSSR, **36**, No.6 (1972), 1348-1401.

[56] Radzievskii, G. V., *A problem on the completeness of root vectors in the spectral theory of operator-valued functions*, Uspekhi. Matem. Nauk, **37**, No.2 (1982), 81-145.

[57] Rasulov, M. L., "Applications of the Contour Integral Method", Nauka, Moscow, 1975.

[58] Sadovnichii, V. A., "The Theory of Operators', Izd. Mosk. Univ. Moscow, 1986.

[59] Seeley, R., *A simple example of spectral pathology for differential operators*, Commun. Part. Differ. Equat. **11**, No.6 (1986), 595-598.

[60] Seeley, R., *Fractional powers of boundary problems*, Actes, Congres. Intern. Math. 1970. **2** (1971), 795-801.

[61] Seeley, R., *Norms and domains of complex powers A_B^z*, Amer. Journ. Math., **93** (1971), 299-309.

[62] Shkalikov, A. A., *On eigenvectors of quadratic operator pencils to be a basis*, Matem. Zamet., **30**, No.3 (1981), 371-385.

[63] Shkalikov, A. A., *Boundary value problems for ordinary differential equations with a parameter in boundary conditions*, Trudy Sem. imeny I. G. Petrovskogo, **9** (1983), 190-229.

[64] Shkalikov, A. A., *Elliptic equations in Hilbert spaces and spectral problems connected with them*, Trudy Seminara imeny I.G. Petrovskogo, **14** (1989), 140-224.

[65] Shkred, A. V., *On a linearization of spectral problems with a parameter in a boundary condition and properties of M. V. Keldysh's generated chains*, Matem. Zamet. **46**, No.4 (1989), 99-109.

[66] Smirnov, V. I., "A Course of Higher Mathematics, Part V", Pergamon Press, Oxford, 1964.

[67] Sova, M., *Linear Differential Equations in Banach Spaces*, Rozp. CSAV mpv.85, No.6 (1975), 82p.

[68] Stewart, H. B., *Generation of analytic semigroups by strongly elliptic operators*

under general boundary conditions, Trans. Amer. Math. Soc, **259** (1980), 299-310.

[69] Tamarkin, J. D., "About Certain General Problems of Theory of Ordinary Linear Differential Equations and about Expansions of Derivative Functions into Series", Petrograd, 1917.

[70] Tanabe, H., "Equations of Evolution", Pitman, London, 1979.

[71] Titchmarsh, E. C., "Introduction to the Theory of Fourier Integrals", The Clarendon Press, Oxford, 1937.

[72] Triebel, H., "Interpolation Theory. Function Spaces. Differential Operators", North-Holland, Amsterdam, 1978.

[73] Triebel, H., "Theory of Function Spaces", Birkhauser-Verlag, Basel, 1983.

[74] Ustinov, Y. A. and Yudovich, V. I., *On completeness of a system of elementary solutions of the biharmonic equation in a semi-strip*, Prikladnaya Matematica i Mekhanika, **37** (1973), 706-714.

[75] Vagabov, A. I., *On eigenelements of irregular differential pencils with non-decomposed boundary conditions*, Dokl. Akad. Nauk SSSR, **261**, No.2 (1981), 268-271.

[76] Vagabov, A. I., *On completeness of eigenfunctions of an ordinary differential pencil of a non-Keldysh type*, Diff. Uravn., **20**, No.3 (1984), 375-382.

[77] Vizitei, V. N., Markus, A. S., *On the convergence of multiple expansions by a system of eigenvectors and associated vectors of an operator pencil*, Matem. Sborn, **66**, No.2 (1965), 287-320.

[78] Vlasov, V. V., *Multiple minimality of a part of root vector system of a M.V.Keldysh pencil*, Dokl. Akad. Nauk SSSR, **263**, No.6 (1982), 1289-1293.

[79] Yakubov, S. Ya., *On two-fold completeness of eigenelements and associated elements of a quadratic operator pencil*, Funk. Analiz i Yego Pril, **7**, No.1 (1973), 92-94.

[80] Yakubov, S. Ya., *A non-local boundary value problem for one class of equations correct in the sense of Petrovskii*, Matem. Sborn, **118**, No.2, 1982, 252-261.

[81] Yakubov, S. Ya., *A boundary value problem for the Laplace equation with non-classical spectral asymptotics*, Dokl. Akad. Nauk SSSR, **265**, No.6 (1982), 1330-1333.

[82] Yakubov, S. Ya., *Quadratic operator pencils*, Dokl. Akad. Nauk SSSR, **269**, No.1 (1983), 39-42.

[83] Yakubov, S. Ya., Aliev, B. A., *Fredholmness of a boundary value problem with an operator in the boundary conditions for an elliptic differential-operator equation of the second order*, Dokl. Akad. Nauk SSSR,**257**, No.5 (1981), 1071-1074.

[84] Yakubov, S. Ya., *Operator pencils and differential-operator equations*, Dep. VINITI, No. 6295-84 (1984).

[85] Yakubov, S. Ya. "Linear Differential-Operator Equations and their Applications", ELm, Baku, 1985.

[86] Yakubov, S. Ya., Mamedov, K. S., *On the multiple completeness of a system of eigenelements and associated elements of a polynomial operator pencil and on multiple expansions in that system*, Funk. Anal. i Yego Pril., **9**, No.1 (1975), 91-93.

[87] Yakubov, S. Ya., Mamedov, K. S., *Completeness of eigenfunctions and associated functions of some irregular boundary value problems for ordinary differential equations*, Funk. Anal. i Yego Pril., **14**, No.4 (1980), 93-94.

[88] Yakubov, S. Ya., *Multiple completeness of root functions of elliptic boundary value problems with a polynomial spectral parameter in the sense of M.V.Keldysh*, Izv. Akad. Nauk SSSR ,Ser. Mat., **50**, No.2 (1986), 425-431.

[89] Yakubov, S. Ya., *Fold completeness of root vectors of unbounded polynomial pencils*, Revue Roumaine de Math. Pures et. Appl., XXXI, No.5 (1986), 423-438.

[90] Yakubov, S. Ya., *Multiple completeness of root vectors of a system of operator pencils and boundary value problems with a parameter*, Izv. Akad. Nauk Azerb. SSR, Ser. Fiz-Techn. i Mat. Nauk, **2**, No.2 (1987), 42-46.

[91] Yakubov, S. Ya., *Multiple completeness of a system of operator pencils and elliptic boundary value problems*, Matem. Sborn., **181**, No.1 (1990), 95-113.

[92] Yakubov, S. Ya., Balaev, M. K.,*Correctness of a mixed problem for some classes of equations correct in the sense of Petrovskii*, Matem. Zamet., **42**, No.4 (1987), 527-536.

[93] Yakubov, Ya. S., *The investigation of the resolvent of an operator generated by irregular boundary value conditions and ordinary differential expressions*, Dokl. Akad. Nauk Azerb. SSR, **40**, No.8 (1984), 3-6.

[94] Yakubov, Ya. S., *The investigation of a differential operator generated by irregular boundary value conditions*, VINITI, 11.02.83 (1983), No.769-83, 43p.

[95] Yakubov, Ya. S., *The investigation of an irregular boundary value problem for a quadratic differential pencil*, VINITI, 20.06.84 (1984), No.4127-84, 45p.

[96] Yakubov, Ya. S.,*Coercive boundary value problems with a defect*, Ph.D. Thesis, Baku, 1986, 115p.

[97] Yosida, K., "Functional Analysis", Springer-Verlag, Berlin, 1965.

List of notations

1. Operators

$\overline{A}$, 10

A^*, 11

A^{-1}, 9

A_{-1}, 28

$A_{-1}(\lambda_0)$, 30

$D(A)$, 9

$R(A)$, 9

$\|A\|_{B(E,F)}$, 9

$\rho(A)$, 25

$\sigma(A)$, 25

$\gamma(A)$, 36

$\mathrm{sp}A$, 45

$(\mathrm{sp}A)_{-0}$, 36

$s_j(A; H, H_1)$, 33

$\sigma_p(H, H_1)$, 36

$s_j(A)$, 35

$\sigma_p(H)$, 37

J, 15

$\ker A$, 96

$\mathrm{coker}A$, 96

$R(\lambda, A)$, 25

$r(A)$, 33

F, 15

F^{-1}, 15

$K(t, u)$, 17

2. Elements

$\|u\|$, 8

$< u, u' >$, 10

$\|u'\|_{E'}$, 10

(u, v), 11

$u \perp v$, 11

$u_n \xrightarrow{E} u$, 8

3. Sets

$\ell(a, \varphi)$, 31

$G(a, \varphi_1, \varphi_2)$, 31

$S(a, r)$, 31

$\ell(a)$, 31

$G(a, \theta)$, 31

$M_1 + M_2$, 7

$M \perp N$, 11

$H_1^{\perp}$, 27

$\mathrm{sp}\{u_k\}$, 44

$\overline{M}|_E$, 8

4. Spaces

$N^k = N_\lambda^k$, 26

$N = N_\lambda$, 26

$H(C)$, 60

$E(A^n)$, 9

$B(E, F)$, 9

$(E_0, E_1)_{\theta,p}$, 17

$[E_0, E_1]_\theta$, 19

$H_1 \oplus H_2$, 11

$\mathbb{C}^n$, 10

E', 10

E'', 10

$E_1 \subset E_2$, 15

$E_2' \subset E_1'$, 17

$E_0 + E_1$, 17

$E_1 \dotplus E_2$, 7, 8

$\{E_0, E_1\}$, 17

5. Functional spaces

$B_{p,q}^s(\Omega)$, 20

$W_p^s(\Omega)$, 19, 20

$W_{p,\mathrm{lok}}^n((0,1), E)$, 14

$C^n([0,1], E)$, 14

$W_p^n((0,1), E)$, 14

Subject index

Author index

Printed and bound by CPI Group (UK) Ltd, Croydon, CR0 4YY

01/11/2024

01782616-0015